Characterization of Solid Polymers

Characterization of Solid Polymers

New techniques and developments

Edited by

S.J. Spells

Division of Applied Physics, School of Science,
Sheffield Hallam University, Sheffield S1 1WB, UK

CHAPMAN & HALL

London · Glasgow · Weinheim · New York · Tokyo · Melbourne · Madras

Published by Chapman & Hall, 2–6 Boundary Row, London SE1 8HN, UK

Chapman & Hall, 2–6 Boundary Row, London SE1 8HN, UK

Blackie Academic & Professional, Wester Cleddens Road, Bishopbriggs, Glasgow G64 2NZ, UK

Chapman & Hall GmbH, Pappelallee 3, 69469 Weinheim, Germany

Chapman & Hall Inc., One Penn Plaza, 41st Floor, New York NY 10119, USA

Chapman & Hall Japan, Thomson Publishing Japan, Hirakawacho Nemoto Building, 6F, 1–7–11 Hirakawa-cho, Chiyoda-ku, Tokyo 102, Japan

Chapman & Hall Australia, Thomas Nelson Australia, 102 Dodds Street, South Melbourne, Victoria 3205, Australia

Chapman & Hall India, R. Seshadri, 32 Second Main Road, CIT East, Madras 600 035, India

---

First edition 1994

Typeset in 10/12pt Palatino by Type Study, Scarborough
Printed in Great Britain at the University Press, Cambridge

ISBN 0 412 58490 5 (HB)

A catalogue record for this book is available from the British Library

∞ Printed on permanent acid-free text paper, manufactured in accordance with ANSI/NISO Z39.48–1992 and ANSI/NISO Z39.48–1984 (Permanence of Paper).

Contents

Colour plates appear between pages 308 and 309

Contributors

A.J. Brandolini
Edison Research Laboratory
Mobil Chemical Company,
PO Box 3029,
Edison, New Jersey 08818–3029,
USA

D. Briggs
ICI PLC, Wilton Research Centre,
PO Box 90,
Wilton, Middlesbrough,
Cleveland TS90 8JE,
UK

D.L. Dorset
Medical Foundation of Buffalo, Inc.
Research Institute,
73 High Street
Buffalo, New York 14203–1196,
USA

C.R. Dybowski
Department of Chemistry and
Biochemistry,
University of Delaware,
Newark, Delaware 19716,
USA

M.J. Miles
H.H. Wills Physics Laboratory,
Tyndall Avenue,
University of Bristol,
Bristol BS8 1TL,
UK

A.R. Rennie
Cavendish Laboratory,
University of Cambridge,
Madingley Road,
Cambridge CB3 0HE,
UK

S.J. Spells
Division of Applied Physics,
Sheffield Hallam University,
Pond Street,
Sheffield S1 1WB,
UK

G. Ungar
Department of Engineering Materials,
University of Sheffield,
PO Box 600,
Mappin Street,
Sheffield S1 4DU,
UK

R.J. Young
Manchester Materials Science Centre,
University of Manchester and UMIST,
Grosvenor Street,
Manchester M1 7HS,
UK

Preface

The last decade or so has seen a dramatic increase in the amount of detailed structural information available from a range of experimental techniques. Exciting new techniques such as atomic force microscopy have become widely available, while the potential of established methods like X-ray diffraction and electron microscopy has been greatly enhanced by powerful new sources and analytical methods. Progress in computing has also had a widespread impact: in areas such as neutron scattering, large data sets can now be manipulated more readily. The software supplied with commercial instruments generally provides more sophisticated analytical facilities, while time-resolved X-ray studies rely on rapid data handling capabilities.

The polymer scientist is faced with an expanding array of experimental tools for addressing both fundamental science and industrial problems. This work reviews some recent developments in structural techniques, with the aim of presenting the current 'state of the art' in a selection of areas.

Chapter one describes the advances in electron microscopy, with examples of the high-resolution images now possible, together with the accompanying structural analysis. Recent progress in the novel techniques of scanning tunnelling and atomic force microscopy is reviewed in Chapter two. The use of synchrotron beams for X-ray studies has become a major area of polymer research, encompassing a wide range of X-ray optical systems and detectors. Instrumentation currently available is reviewed in Chapter three, which also considers experimental results so far obtained. The application of neutron scattering to a wide variety of physical problems is discussed in Chapter four, which includes a review of the expanding field of reflection studies from surfaces. Chapters five and six focus on newer developments in vibrational spectroscopy: Chapter five surveys the use of isotopic labelling in the infra-red spectroscopy of polyethylene, while Chapter six considers the progress made in studying the mechanical behaviour of fibres, films and composites, using Raman spectroscopy. Developments in solid-state NMR are reviewed in Chapter seven and in Chapter eight the increasingly prominent field of surface characterization is reviewed, with coverage of X-ray photoelectron spectroscopy and secondary ion mass spectrometry.

The book is intended for polymer scientists both in industry and academia, who are considering using some of these newer techniques, or who simply wish to broaden their awareness of current developments.

Stephen J. Spells
Sheffield
May 1993

1

Electron crystallography of linear polymers

D.L. Dorset

1.1 INTRODUCTION

As a tool for the qualitative description of morphology, the electron microscope has enjoyed a long use in polymer science, as can be appreciated by examination of a recent monograph [1]. Similarly, the selected area electron diffraction mode of the electron microscope has been frequently employed for the determination of unit cell constants and symmetry for thin lamellar crystals [2]. This exploits the enhanced scattering cross-section of matter for electrons which enables these thin lamellar plates to be examined as single crystals [3]. What is not generally appreciated is that the electron microscope can also be employed as an instrument for the quantitative determination of linear polymer crystal structures [4].

The emerging field of electron crystallography, as practiced by a few research groups around the world [5] is therefore beginning to make a significant impact in the polymer community for the understanding of linear chain packing and the types of disorder that can be incorporated into a crystal lattice. Unlike X-ray crystallography, where one is restricted to measured data from reciprocal space, the electron crystallographic technique combines information from the back focal (diffraction) and image planes of the electromagnetic objective lens of the electron microscope. This chapter will discuss the recent breakthroughs in the recording and analysis of direct electron microscope images for the visualization of polymer chain packing and also the employment of crystallographic phasing techniques for the analysis of electron diffraction intensities. As will be shown, the latter technique increases the resolution of the polymer chain image, enabling one to discern molecular architecture on an atomic level.

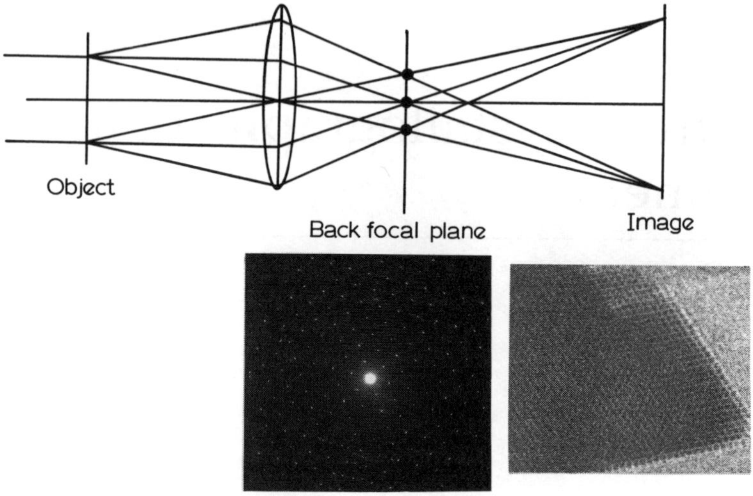

Fig. 1.1 Geometry of electron diffraction and electron microscopy. An election transparent 'phase object', a crystal on a grid, scatters incident electrons which are eventually observed as an image at the **image plane** of an electron microscope (electromagnetic) objective lens. Before this plane is reached one finds the back focal plane where the electron diffraction pattern is formed. The sample here is copper perchlorophthalocyanine.

1.2 ELECTRON CRYSTALLOGRAPHY

As shown in Fig. 1.1, the two places on the optic axis of an electron microscope objective lens where structural information can be obtained are: the back focal plane, where the electron diffraction pattern is formed, and the image plane, the locus of the phase contrast image [6]. Ideally, the information contained in these two planes can be interrelated by the Fourier transform (*FT*) operators. Hence, if the function $f(\mathbf{r})$ describes the image, then the diffraction pattern $F(\mathbf{s})$ is found by

$$F(\mathbf{s}) = \int_{-\infty}^{\infty} f(\mathbf{r}) \exp 2\pi i \mathbf{r} \cdot \mathbf{s} \, d\mathbf{r} = FT f(\mathbf{r}) \tag{1.1}$$

and inversely

$$f(\mathbf{r}) = \int_{-\infty}^{\infty} F(\mathbf{s}) \exp(-2\pi i \mathbf{r} \cdot \mathbf{s}) \, d\mathbf{s} = FT^{-1} F(\mathbf{s}) \tag{1.2}$$

A recording medium (such as a photographic film) when it is inserted at either plane, measures the power of the signal. At the image plane, it can be shown that the intensity of the image of a phase object is linearly proportional to the object density $\phi(x,y)$. That is [6]

$$I(x,y) = (\exp - i\sigma\phi(x,y)\Delta z)^2 \simeq 1 - 2\sigma\phi(x,y)\Delta z \tag{1.3}$$

Thus the density and location of mass can be used unequivocally to calculate the structure factors $F(\mathbf{s})$ of the diffraction pattern, where $F(\mathbf{s}) = |F(\mathbf{s})| \exp i\alpha$ combines a wave amplitude and crystallographic phase for each reflection. This uses the Fourier transform operation in equation (1.1). The analysis is not so straightforward when the diffraction pattern is recorded as a power spectrum with intensities $I(\mathbf{s}) = |F(\mathbf{s})|^2$. Although the wave amplitudes are directly measured, the phase terms are lost, frustrating the direct calculation of an image by the reverse Fourier transform in (1.2). This is the source of the famous crystallographic phase problem which will be considered further later.

The direct interpretation of intensity data recorded in an electron microscope depends on several conditions being satisfied. First of all, the measured signal from the irradiated object must adequately conform to the kinematical assumption, i.e. that each electron is only scattered once. In terms of coherent scattering, this is favored by the object being made up of light atoms and being adequately thin [6]. The former requirement is met by the composition of most organic compounds and the latter by the thickness of most polymer lamellae, which is generally less than 100 Å. Selection of a suitably large electron accelerating voltage can also contribute to the near kinematical scattering requirement [7]. An incoherent expression of multiple scattering, due to strongly scattered diffracted beams from upper layers acting as primary beams for successive crystalline layers [8], can also be controlled by crystallization conditions.

If the intensity data from a diffraction pattern are being analyzed, another possible perturbation, due to elastic crystal curvature [9], must be kept in mind. Since an electron beam is a source with high spatial coherence under the illumination conditions used for selected area diffraction, slight bends of the crystal within this coherence width of the electron source may result in an effective diffraction **incoherence**, i.e. the intensity data may not simply correspond to the total unit cell contents if the projected unit cell length exceeds a certain limiting value. Fortunately, it has been shown that the fiber repeats of many polymer samples lie within this value so that electron diffraction data from lamellar crystals grown by self-seeding can be utilized for crystal structure determination [10]. For other materials, such as alkanes [11,12], this assumption cannot be made, and other crystallization techniques, such as epitaxial orientation on organic and inorganic substrates [13], must be employed to project shorter unit cell lengths parallel to the incident beam. As will be demonstrated by examples below, such epitaxial crystallization, which gives an orthogonal view on to the polymer chain axes, can be of advantage for the collection of diffraction data from polymer samples. Hence data, which fall outside the ±60° tilt range of most electron microscope goniometer stages, can be recorded from epitaxially oriented specimens when they cannot be obtained from the more usual chain-folded lamellar crystals.

When images of polymer crystals are being analyzed, the aberrations of the electron microscope objective lens must be taken into account [6]. Spherical aberration, C_s, can be partially compensated by an underfocus term Δf to optimize the contrast envelope for the diffracted waves in the back focal plane. The contrast envelope is essentially an additional phase term for the structure factors, $F(\mathbf{s})$, defined

$$\chi(\mathbf{s}) = \frac{2\pi}{\lambda}\left(C_s\frac{|\mathbf{s}|^4\lambda^4}{4} - \Delta f\frac{|\mathbf{s}|^2\lambda^2}{2}\right) \tag{1.4}$$

where λ is the electron wavelength and $|\mathbf{s}|$ the reciprocal spacing of the reflection ($\mathbf{s}$). The sine of $\chi(\mathbf{s})$ is often plotted to depict this transfer function. Hence, at Scherzer focus, the optimal resolution of the electron microscope is found, i.e. where the value of this function changes contrast sign. Typical 100 kV electron microscopes generally have a useful resolution at about 3.0 Å but newer instruments with exceptionally good C_s values (e.g. 0.44 mm) have been built to give a useful resolution of about 1.8 Å. Since this only changes the phases of the scattered beams, it has no effect on the electron diffraction intensities which can be analyzed to the resolution of the pattern, provided an independent means can be found for determining crystallographic phases.

Finally, an often-touted limitation to the electron crystallography of organic materials, radiation damage, caused by inelastic interactions of the electrons with the specimen [14], has received considerable attention in the history of electron microscopy, leading to early conclusions that the recording of useful high-resolution images from such specimens would be impossible [15]. Fortunately, this opinion has been reversed by experiments on very beam-sensitive crystals – such as *n*-paraffins [16] and polyethylene [17] – which have demonstrated image resolution at 2.5 Å. Although this was initially achieved on a liquid-helium-cooled electron microscope [16], equivalent results are obtained at room temperature [17], particularly when innovative procedures are used to record an image [18]. The design of low-dose illumination techniques to 'make every electron count' led to this successful employment of the electron microscope to observe lattice images of organic crystals [19]. As will be discussed later, various types of image averaging will then obtain the best representation of the unit cell contents from a noisy electron micrograph. Radiation damage is scarcely a problem for electron diffraction experiments, provided that proper illumination conditions and suitably sensitive recording media are used.

1.3 CRYSTAL STRUCTURE ANALYSIS

The objective of any structure analysis is to obtain the clearest possible representation of the unit cell contents which then can be interpreted in terms of molecular geometry and packing. In practice, the procedure for image analysis ends up at the same place as an electron diffraction analysis at equivalent resolution, i.e. an average representation of the crystal structure is produced. Although, ideally, the visualization of local differences in molecular conformation and packing would be desirable, the constraints of radiation damage, mentioned earlier, leave one with a noisy image in which the details of individual unit cells often cannot be discerned.

1.3.1 Image analysis

There are two image-averaging techniques commonly employed for low-dose electron micrographs: Fourier-peak-filtration and correlation averaging.

Fourier-peak-filtration [20] assumes that the desirable signal for constructing an average representation of the unit cell is found at the Bragg peaks of the Fourier transform and that any other continuous scattering is due to noise caused, e.g. by radiation damage or other undesirable alterations of the specimen. Hence, on an optical bench (Fig. 1.2) it is possible to carry out such a filtration by imposing a mask at the back focal plane of the lens used for producing a magnified image of the photographic negative. This mask will have holes where the diffraction spots of

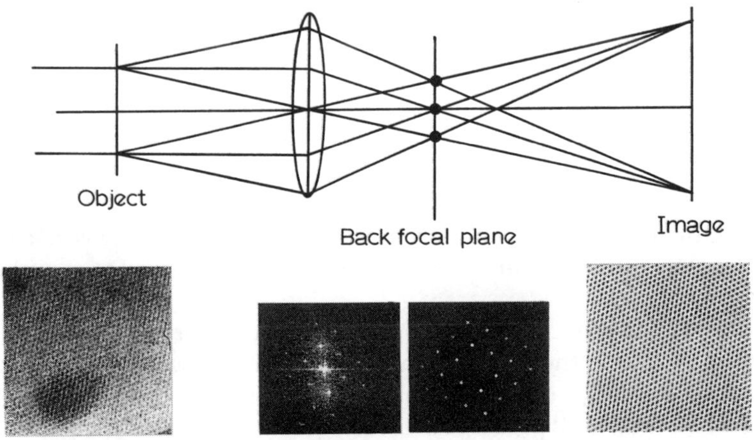

Fig. 1.2 Spatial filtering of an electron microscope image. The electron microscope image formed in Fig. 1.1 is illuminated by a laser on an optical bench. At the back focal plane of the (glass) objective lens, a diffraction pattern is formed and a mask can be superimposed to pass only the Bragg reflections. The resultant filtered image formed at the image plane of this lens is free of noise sources due to non-periodic distortion to the micrograph.

the two-dimensional lattice will occur so that only these points in reciprocal space are permitted to form the image.

Such an operation can be carried out computationally. For example, the area of the negative with sufficient crystalline order is scanned on a microdensitometer to produce a two-dimensional pixel array which can be converted to the diffraction pattern, e.g. by a fast Fourier transform. Image-analysis software exists [21] which then permits one to construct an effective mask for the diffraction pattern. The reverse Fourier transform to the average image will thus be made without the continuous signal.

A second, commonly employed technique is correlation averaging [22]. Here one can scan an experimental electron microscope image (e.g. by calculating diffraction patterns of image sub-areas) to find regions which have the same crystal packing and/or average resolution. A cross-correlation of a suitable reference image area with the total image can also identify regions of the experimental micrograph which have the same crystalline order. As indicated by the computed diffraction patterns, it may be necessary to realign some domains. A superposition of suitably aligned sub-areas, therefore, can result in an average image superior to the one obtained by peak filtration, particularly for samples co-crystallizing in two polymorphic forms.

1.3.2 Electron diffraction structure analysis

If the electron diffraction intensities are all that one has to work with, then one must find some device to construct trial images of the structure and test them against the observed diffraction magnitudes $|F_o|$ by use of the crystallographic R-factor

$$R = \frac{\Sigma||F_o| - k|F_c||}{\Sigma|F_o|} \tag{1.5}$$

where F_c is obtained from the trial structure via equation (1.1) and k is a scale factor. It is obvious that an infinity of models can be constructed, making such a 'trial and error' approach somewhat futile for complicated structures.

A second approach to this problem is to interpret the Patterson function, which is the Fourier transform of the power spectrum in the back focal plane [6], i.e. using intensity values in place of $|F(\mathbf{s})|$ in equation (1.2) and assuming all phase values are equal to zero. This is equivalent to the auto-correlation function of the image – hence a superposition of all inter-atomic vectors (assuming atomic resolution data!) to a common origin. Although this was a standard technique for 'solving' X-ray crystal structures for many years [23], and has also been used in electron crystallography, it can also be difficult to find a unique structure if it is very complicated. This is particularly true for electron diffraction data, since the often-attempted location of heavy atom labels is frustrated by the narrow range of electron diffraction atomic scattering factors relative to their X-ray values.

In polymer science, the most frequent way of constructing a trial molecular model is to determine first the complete X-ray crystal structure(s) of monomer or oligomer fragments to determine what part of the repeating unit can be safely assumed to be conformationally rigid. These units are then concatenated through 'linkage bonds' and the rigid units are then rotated around these links to evaluate numerous chain conformations [24,25]. While the crystallographic *R*-factor is being minimized, a simultaneous minimization of internal energy (calculated by atom–atom non-bonded potential functions) is also sought. It is fortunate that two figures of merit can be used since the statistical significance of the *R*-factor is problematic when one is forced to work with small data sets [26].

Recently, an alternative approach to the determination of linear polymer crystal structures from electron diffraction intensities has made use of the direct phasing techniques commonly employed in X-ray crystallography [28]. Here one estimates the probability that certain linear combinations of phases

$$\psi = \phi_{\mathbf{h}_1} + \phi_{\mathbf{h}_2} + \phi_{\mathbf{h}_3} + \ldots \tag{1.6}$$

subject to the constraints to the Miller indices,

$$\sum_i \mathbf{h}_i \equiv 0, \tag{1.7}$$

will have a certain value. This probability is found to be directly related to the normalized structure factor magnitudes of the reflections $|E_{\mathbf{h}}|$, which can be calculated from the observed intensities. One solves the resulting system of simultaneous equations in phase after assigning the values of a small phase subset (less than or equal to 3) to define the unit cell origin. Although this approach may be thought to be inappropriate for small data sets, it is found to be extremely effective for electron diffraction intensities from organic microcrystals, even when these intensities are perturbed somewhat by dynamical scattering. An initial structure found in the first potential map can then be refined via Fourier techniques.

1.4 EXAMPLES OF STRUCTURE ANALYSES

1.4.1 High-resolution images

A representative, but not exhaustive, list of polymer crystal lattices imaged in the electron microscope is given in Table 1.1. As mentioned above, the most impressive aspect of this work is that materials,

Table 1.1 High-resolution electron images of linear polymer crystals

Polymer	*Resolution (Å)*
polyethylene[17]	3.7
polytetrafluoroethylene[42]	4.9
poly(4-methyl-pentene-1)[43]	4.3
poly(β-hydroxybutyrate)[44]	3.5
α-chitin[45]	5.1
β-chitin[46]	4.8
cellulose III[47]	4.3
poly(*p*-xylylene) α and β forms [48]	2.8
isotactic polystyrene[49]	5.5
poly(tetramethyl-*p*-silphenylene siloxane)[50]	3.2
poly(*p*-phenylene sulfide)[51]	4.3

previously thought to be inaccessible to high-resolution electron microscopy due to radiation sensitivity, can be visualized to e.g. 4.0 Å resolution. A review of the procedures used for obtaining these polymer lattice images has been published by Tsuji [29].

One may wonder what use can be made of these images, since the stereochemistry of the molecule cannot be directly visualized. In his impressive study of collapsed polyethylene lamellae, Revol [17] was able to directly visualize crystalline regions corresponding to alternate bands of tilted and untilted molecular chain stems. A similar study [30] was subsequently carried out for very long *n*-alkanes which also form sectorized crystals, even though the molecular chains are unfolded (Fig. 1.3). The first direct representation of edge dislocations was also found in images of an aromatic polymer [31]. The presence of polarized molecular domains could also be visualized in fibers of a conductive polymer [32]. In another polymer, the lattice image was able to account for the existence of a hexagonal unit cell with 16 molecules, i.e. the required positional disorder of one molecule at the origin was clearly apparent [33]. Finally, it has been shown also that even liquid crystalline polymers, e.g. in the smectic phase [34], can be imaged in the electron microscope.

1.4.2 Electron diffraction structure analysis

Quantitative structure analyses of complex polymers from electron diffraction data began with the work of Vainshtein and Tatarinova on polypeptides [35]. Until recently, most of the structure determinations were based on measured zonal diffraction data and most analyses were carried out with an assumed monomer structure, minimizing both the crystallographic *R*-factor and internal packing energy, as stated above.

Fig. 1.3 Higher resolution images of *n*-dooctacontane $n - C_{82}H_{166}$ which forms sectorized lozenge crystals when grown from dilute solution [30]. As also found for collapsed polyethylene pyramids [17], the bright bands in the bright field image indicate regions where the chains are tilted and the dark bands are regions where the chain axes are perpendicular to the support film. The high-resolution images were obtained on a cryomicroscope with a magnification of 66,000 × at 4 K.

A catalog of these structures based on electron diffraction data is given in Table 1.2.

Although the monomer and oligomer structures determined by X-ray crystallographic analyses provide good starting models for electron diffraction studies based on conformational refinement, this procedure is very time-consuming. It would be better to carry out a direct structure analysis with the measured electron diffraction intensities if it were possible. This possibility has been effectively demonstrated recently for three-polymer structures in Table 1.2 using 3-D intensity data. The importance of sampling the major regions of the unit cell transform in the data collection is underscored by the problems experienced in the direct

Table 1.2 Quantitative electron diffraction structure: analyses of linear polymers

Polymer	*Data*	R-*factor*	*Phasing technique*
Two-dimensional determinations:			
poly(tetrafluoroethylene) (oligomer)[52]	*hk*0	0.11	Theoretical model
poly(diacetylene)[53]	0*kl*	0.12	Patterson map and chain rotation
trans poly(acetylene)[54]	0*kl*	0.09	Chain rotation
poly(ethylene sulfide)[55]	*hk*0	0.33 (0.19 when corrected)[56]	Conformational refinement
α-poly(3,3-bis-chloromethyl oxacyclobutane)[57]	*hk*0	0.25	Conformational refinement, previous model
cellulose triacetate[58]	*hk*0	0.26	Conformational refinement
nigeran (anhydrous)[59]	*hk*0	0.26	Conformational refinement
poly(trimethylene terephthalate)[60]	*hk*0	0.35 (0.25 when corrected)[10]	Conformational refinement
poly(hexamethylene[61] terephthalate)	*hk*0	0.17	Conformational refinement
poly(1,11-dodecadiyne)[62]			
macromonomer	*h*0*l*	0.13	Conformational refinement
cross-linked	*h*0*l*	0.13	Conformational refinement
poly(γ-methyl-L-glutamate)			
α-form[63]	*hk*0	not given	Theoretical model
β-form[64]	*h*0*l*	0.38	Theoretical model
dextran – low temp.[65]	*hk*0	0.26	Conformational refinement
–high temp.[66]	*hk* 0	0.18	Conformational refinement
Three-dimensional determinations:			
polyethylene[67,68]	*hkl*	0.21	Rotational search, direct methods
poly(ϵ-caprolactone)[36,69]	*hkl*	0.20	Fiber X-ray models, direct methods
poly(sulfur nitride)[70]	0*kl* + *hk*0	0.19	Patterson function
cis(polyacetylene)[71]	*hkl*	0.13	Rotational refinement
Valonia cellulose[72]	*hkl*	0.23	Conformational refinement
mannan I[73]	*hkl*	0.22	Conformational refinement, direct methods[27]
V_H amylose[74]	*hkl*	0.24	Conformational refinement
poly(*trans*-cyclohexanediyl dimethylene succinate)[75]	*hkl*	0.24	Conformational refinement direct methods[27]
γ-poly(pivalolactone)[76]	*hkl*	0.14	Conformational refinement

phase determination for mannan I [27]. Although the phase assignments are correct, series termination errors frustrate any attempt to locate accurate atomic positions in the potential maps. When direct phasing

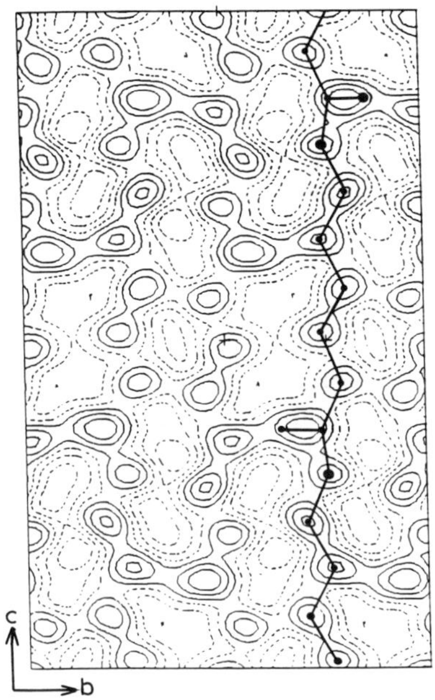

Fig. 1.4 Electrostatic potential map for poly(ε-caprolactone). The crystal structure was solved by direct phasing of 47 3-D electron diffraction data [36].

works well (as shown in Fig. 1.4), the atomic positions are clearly resolved and can be refined by Fourier techniques to derive a final structure with reasonable bond distances and angles. In the case of poly(ε-caprolactone) [36], the electron diffraction structure analysis was based on a more complete intensity data set than obtained from fiber X-ray patterns and resulted in a molecular model with uniformly consistent thermal parameters (unlike the best fiber X-ray analysis [37]).

Electron diffraction data are also useful for describing defect structures, particularly when the continuous diffuse scattering signal is analyzed. (Hence, one must be careful when rejecting a continuous signal in the diffraction pattern during image analysis. Some sources of continuous diffuse scattering are structurally significant!). In a recent study [38] of polymethylene chain packing (including polyethylene) we have found that the continuous diffuse diffraction in the projection down the chain axis is well explained by thermal lattice vibrations, since it is temperature-sensitive and well matched by a thermal diffuse scattering model. A projection on to the chain axes, achieved by epitaxial orientation, produces a continuous scattering signal which is not very temperature-sensitive. The intensity distribution is well matched by static longitudinal displacements along the chain axes which are 'frozen-in' to the crystal packing when the material is crystallized from the melt.

1.5 FUTURE PROSPECTS

The use of electron microscope data for determining polymer crystal structures has had an auspicious beginning. It has been demonstrated that experimental conditions can be established where the measured image or diffraction data can be interpreted directly in terms of molecular structure. Refinements of structures with high-resolution diffraction data thus result in molecular geometries which are very similar to equivalent X-ray determinations [27].

It is also clear that the relationship between electron image and diffraction data can be exploited. In some structure analyses carried out in this laboratory, e.g. the packing of linear paraffin chains, the low-angle crystallographic phase information for the 'lamellar' reflections available from high-resolution, low-dose electron microscope images were quite necessary for the direct structure analyses, since the evaluation of phase invariants only enabled one to assign values to strong methylene 'sub-cell' reflections [39]. The complete structure, therefore, required the combination of the two separate phase determinations. In other cases, where phase invariants by themselves were used, multi-solution evaluation of algebraic quantities to account for phase ambiguities, resulting from poorly defined linkages between phase relationships, may have been unnecessary, had a few more low-angle phases been available from direct images [27].

This latter point is underscored by our recent structure of copper perchlorophthalocyanine [40] using 198 $h'k0$ electron diffraction data obtained at 1200 kV. Since the unit cell is centered, only one reflection could be used to define the origin. Thus, in linking sets of reflections by phase invariant relationships, $n = 3$ algebraic unknowns were necessary requiring the calculation of $2^n = 8$ potential maps. The best potential map correctly showed the positions of the heavy Cu and Cl atoms, a result consistent with the best high-resolution electron microscope images published by Uyeda *et al.* [41]. Had we obtained such images first, the direct phasing would have been greatly facilitated and the initial structural model would have been derived more easily to be refined by Fourier techniques. Given the existence of 'molecular resolution' images (Table 1.1), it is important to use the phase information from them for the analysis of electron diffraction data from linear polymer structures.

Another aspect of direct phasing which is currently being explored is the case where atomic resolution electron diffraction data from a projection, nevertheless, do not allow one to determine the crystal structure, because of nearly eclipsed atomic positions. The use of maximum entropy techniques [77] to determine the best phase set (and hence the best potential map) will be of considerable benefit when atomic positions cannot be identified to allow refinement of the structure, e.g. by

Fourier techniques. A preliminary study with 1 Å resolution data from perchlorocoronene [78] demonstrates how the analysis can be made with combined electron diffraction and image data.

REFERENCES

1. Woodward, A.E. (1988) *Atlas of Polymer Morphology*, Hanser, Munich.
2. Geil, P.H. (1973) *Polymer Single Crystals*, Krieger, Huntington, NY.
3. Vainshtein, B.K. (1964) *Structure Analysis by Electron Diffraction*, Pergamon, Oxford, pp. 3–4.
4. Dorset, D.L. (1989) Electron diffraction from crystalline polymers, in *Comprehensive Polymer Science, Vol. 1 Polymer Characterization*, (eds C. Booth and C. Price), Pergamon, Oxford, pp. 651–68.
5. Fryer, J.R. and Dorset, D.L. (eds) (1991) *Electron Crystallography of Organic Molecules*, Kluwer, Dordrecht.
6. Cowley, J.M. (1981) *Diffraction Physics*, 2nd edn, North-Holland, Amsterdam.
7. Jap, B.K. and Glaeser, R.M. (1980) The scattering of high energy electrons. II Quantitative validity domains of the single-scattering approximations for organic crystals. *Acta Crystallogr.* **A36** 57–67.
8. Cowley, J.M., Rees, A.L.G. and Spink, J.A. (1951) Secondary electron scattering in electron diffraction. *Proc. Phys. Soc. (London)*, **A64**, 609–19.
9. Cowley, J.M. (1961) Diffraction intensities from bent crystals. *Acta Crystallogr.*, **14**, 920–7.
10. Moss, B. and Dorset, D.L. (1982) Refinement of linear polymer crystal structures determined from electron diffraction data. *J. Polym. Sci. Polym. Phys. Ed.* **20**, 1789–804.
11. Dorset, D.L. (1980) Electron diffraction intensities from bent molecular organic crystals. *Acta Crystallogr.* **A36**, 592–600.
12. Moss, B. and Dorset, D.L. (1983) Dynamical electron diffraction from elastically bent organic crystals. *Acta Crystallogr.* **A39**, 609–15.
13. Wittmann, J.C., Hodge, A.M. and Lotz, B. (1983) Epitaxial crystallization of polymers onto benzoic acid: polyethylene and paraffins, aliphatic polyesters, and polyamides. *J. Polym. Sci.–Polym. Phys. Ed.* **21**, 2495–509.
14. Zeitler, E. (1991) Radiation damage in electron crystallography, in *Electron Crystallography of Organic Molecules*, (eds J.R. Fryer and D.L. Dorset), Kluwer, Dordrecht, pp. 355–60.
15. Glaeser, R.M. (1975) Radiation damage and biological electron microscopy, in *Physical Aspects of Electron Microscopy and Microbeam Analysis*, (eds B.M. Siegel and D.R. Beaman), Wiley, New York, pp. 205–29.
16. Zemlin, F., Reuber, E., Beckmann, E. *et al.* (1985) Molecular resolution electron micrographs of monolamellar paraffin crystals. *Science*, **229**, 461–2.
17. Revol, J.F. and Manley, R. St. J. (1986) Lattice imaging in polyethylene single crystals. *J. Mater. Sci. Lett.* **5**, 249–51.
18. Downing, K.H. and Glaeser, R.M. (1986) Improvement in high resolution image quality of radiation-sensitive specimens with reduced spot size of the electron beam. *Ultramicroscopy*, **20**, 269–78.
19. Fujiyoshi, Y., Kobayashi, T., Ishizuka, K. *et al.* (1980) A new method for optimal-resolution electron microscopy of radiation-sensitive specimens. *Ultramicroscopy*, **5**, 459–68.
20. Misell, D.L. (1978) *Image Analysis Enhancement and Interpretation*, North-Holland, Amsterdam, pp. 142–7.

21. Heel, M.V. and Keegshra, W. (1981) IMAGIC: a fast, flexible and friendly image analysis software system. *Ultramicroscopy*, **7**, 113–30.
22. Frank, J. (1980) The role of correlation techniques in computer image processing, in *Computer Processing of Electron Microscope Images*, (ed. P.W. Hawkes), Springer, Berlin, pp. 187–222.
23. Buerger, M.J. (1959) *Vector Space and Its Application in Crystal-Structure Investigation*, Wiley, New York.
24. Brisse, F. (1989) Electron diffraction of synthetic polymers; The model compound approach to polymer structure. *J. Electron Microsc. Techn.* **11**, 272–9.
25. Perez, S. and Chanzy, H. (1989) Electron crystallography of polysaccharides. *J. Electron Microsc. Techn.* **11**, 280–5.
26. Hamilton, W.C. (1964) *Statistics in Physical Science. Estimation, Hypothesis Testing and Least-Squares*, Ronald, New York, pp. 157–60.
27. Dorset, D.L. Is electron crystallography possible? The direct determination of organic crystal structures. *Ultramicroscopy*, **38**, 23–40.
28. Hauptman, H.A. (1972) *Crystal Structure Determination. The Role of the Cosine Seminvariants*, Plenum, New York.
29. Tsuji, M. (1989) Electron microscopy, in *Comprehensive Polymer Science. Vol. 1 Polymer Characterization*, (eds C. Booth and C. Price), Pergamon, Oxford, pp. 785–840.
30. Dorset, D.L., Hanlon, J., McConnell, C.H. *et al.* (1990) Why do polyethylene crystals have sectors? *Proc. Natl. Acad. Sci. U.S.A.* **87**, 1696–700.
31. Isoda, S., Tsuji, M., Ohara, M. *et al.* (1983) Direct observation of dislocation in polymer single crystals. *Makromol. Chem. Rapid Commun.* **4**, 141–4.
32. Kawaguchi, A., Isoda, S., Petermann, J. and Katayama, K. (1984) High resolution electron microscopy of $(SN)_x$. *Colloid Polym. Sci.* **262**, 429–34.
33. Isoda, S., Tsuji, M., Ohara, M. *et al.* (1983) Structural analysis of β-form poly(*p*-xylylene) starting from a high-resolution image. *Polymer* **24**, 1155.
34. Voigt-Martin, I.G. and Durst, H. (1987) Direct observation of smectic layers in side chain liquid crystal polymer films. *Liquid Crystals*, **2**, 601–10.
35. Vainshtein, B.K. and Tatarinova, L.I. (1967) Electron-diffraction studies of synthetic polypeptides, in *Conformation of Biopolymers Vol. 2*, (ed. G.N. Ramachandran), Academic, New York, pp. 569–82.
36. Dorset, D.L. (1991) Electron crystallography of linear polymers: direct structure analysis of poly(ε-caprolactone). *Proc. Natl. Acad. Sci. U.S.A.* **88**, 5499–502.
37. Chatani, Y., Okita, Y., Tadokoro, H. and Yamashita, Y. (1970) Structural studies of polyesters III. Crystal structure of poly-ε-caprolactone. *Polym. J.*, **1**, 555–62.
38. Dorset, D.L., Hu, H. and Jäger, J. Continuous diffuse scattering from polymethylene chains – an electron diffraction study of crystalline disorder. *Acta Crystallogr.* **A47**, 543–49.
39. Dorset, D.L. and Zemlin, F. (1990) Direct phase determination in electron crystallography. The crystal structure of an *n*-paraffin. *Ultramicroscopy*, **33**, 227–36.
40. Dorset, D.L., Tivol, W.F. and Turner, J.N. Electron crystallography at atomic resolution – *ab initio* structure analysis of copper perchlorophthalocyanine. *Ultramicroscopy*, **38**, 41–5.
41. Uyeda, N., Kobayashi, T., Ishizuka, K. and Fujiyoshi, Y. (1978) High voltage electron microscopy for image discrimination of constituent atoms in crystals and molecules. *Chem. Scripta*, **14**, 47–61.
42. Chanzy, H., Folda, T., Smith, P., Gardner, K.H. and Revol, J.F. (1986) Lattice

imaging in polytetrafluoroethylene single crystals. *J. Mater. Sci. Lett.* **5**, 1045–7.
43. Pradere, P., Revol, J.F., Nguyea, L. and Manley, R. St. J. (1988) Lattice imaging of poly-4-methyl-pentene-1 single crystals: use and misuse of Fourier averaging techniques. *Ultramicroscopy* **25** 69–80.
44. Revol, J.F., Chanzy, H.D., Deslaudes, Y. and Marchessault, R.H. (1989) High resolution electron microscopy of poly(β-hydroxybutyrate). *Polymer,* **30**, 1973–6.
45. Revol. J.-F. (1989) Lattice resolution in α-chitin. *Int. J. Biol. Macromol.* **11**, 233–5.
46. Revol, J.-F. and Chanzy, H. (1986) High-resolution electron microscopy of β-chitin microfirils. *Biopolymers,* **25**, 1599–1607.
47. Sugiyama, J., Harada, H. and Saihi, H. (1987) Crystalline morphology of *Valonia macrophysa* cellulose I revealed by direct lattice imaging. *Int. J. Biol. Macromol.* **9**, 122–30.
48. Tsuji, M., Isoda, S., Ohara, M., Kawaguchi, A. *et al.* (1982) Direct imaging of molecular chains in a poly(ρ-xylylene) single crystal. *Polymer,* **23**, 1568–74.
49. Tsuji, M., Roy, S.K. and Manley, R. St. J. (1984) Direct lattice imaging in single crystals of isotactic polystyrene. *Polymer* **25**, 1573–6.
50. Tsuji, M., Ohara, M., Isoda, S., Kawaguchi, A. *et al.* (1989) Direct imaging of the molecular arrangement in a radiation-sensitive polymer single crystal; poly(tetramethyl-ρ-silphenylene siloxane). *Phil. Mag.* **B54**, 393–403.
51. Uemura, A., Tsuji, M., Kawaguchi, A. and Katayama, K. (1988) High-resolution electron microscopy of solution-grown crystals of poly(*p*-phenylene sulphide). *J. Mater. Sci.* **23**, 1506–9.
52. Dorset, D.L. (1977) Perfluoroalkane. A model for the hexagonal methylene subcell? *Chem. Phys. Lipids,* **20**, 13–19.
53. Day, E. and Lando, J.B. (1980) Structure determination of a poly(diacetylene) monolayer. *Macromolecules,* **13**, 1483–7.
54. Shimamura, K., Karasz, F.E., Hirsch, J.A. and Chien, J.C.W. (1981) Crystal structure of *trans*-polyacetylene. *Makromol. Sci. Rapid Commun.* **2**, 473–80.
55. Hasegawa, H., Claffey, W. and Geil, P.H. (1977) Analysis of the crystal structure of poly(ethylene sulfide) by electron diffraction. *J. Macromol. Sci.–Phys.* **B13**, 89–100.
56. Moss, B. and Dorset, D.C. (1983) Poly(ethylene sulfide): re-evaluation of the electron diffraction structure analysis. *J. Macromol. Sci.–Phys.* **B22**, 69–77.
57. Claffey, W., Gardner, K., Blackwell, J., Lando, J. *et al.* (1974) Structure analysis of polymer single crystals by electron diffraction. *Phil. Mag.* **30**, 1223–32.
58. Roche, E., Chanzy, H., Boudeulle, M., Marchessault, R.H. *et al.* (1978) Three-dimensional crystalline structure of cellulose triacetate II. *Macromolecules,* **11**, 86–94.
59. Perez, S., Roux, M., Revol, J.F. and Marchessault, R.H. (1979) Dehydration of nigeran crystal: crystal structure and morphological aspects. *J. Mol. Biol.* **129**, 113–33.
60. Poulin-Dandurand, S., Perez, S., Revol, J.F. and Brisse, F. (1974) The crystal structure of poly(trimethylene terephthalate) by X-ray and electron diffraction. *Polymer,* **20**, 419–26.
61. Brisse, F., Palmer, A., Moss, B., Dorset, D. *et al.* (1984) Poly(hexamethylene terephthalate) Part II: the crystal structure of Form I and Form II from electron and X-ray diffraction and packing analyses. *Eur. Polym. J.* **20**, 791–7.
62. Thakur, M. and Lando, J.B. (1983) Structure determination of the macromonomer poly(1,11-dodecadiyne) and its cross-linked product. *Macromolecules,* **16**, 143–6.

63. Tatarinova, L.I. and Vainshtein, B.K. (1962) Issledovanie poli-γ-metil-L-glutamata v α-forme metodom difraktsii elektronov. *Visokomolek. Soed,* **4**, 261–9.
64. Vainshtein, B.K. and Tatarinova, L.I. (1967) The β-form of poly-γ-methyl-L-glutamate. *Sov. Phys. Crystallogr.* **11**, 494–8.
65. Guizard, C., Chanzy, H. and Sarko, A. (1985) The molecular and crystal structure of dextrans: a combined electron and X-ray diffraction study II. Low temperature, hydrated polymorph. *J. Mol. Biol.* **183**, 397–408.
66. Guizard, C., Chanzy, H. and Sarko, A. (1984) Molecular and crystal structure dextrans: a combined electron and X-ray diffraction study. 1. The anhydrous high-temperature polymorph. *Macromolecules,* **17**, 100–7.
67. Hu, H. and Dorset, D.L. (1989) Three-dimensional electron diffraction structure analysis of polyethylene. *Acta Crystallogr.* **B45**, 283–90.
68. Dorset, D.L. (1991) Electron diffraction structure analysis of polyethylene. A direct phase determination. *Macromolecules,* **24**, 1175–8.
69. Hu, H. and Dorset, D.L. (1990) Crystal structure of poly(ε caprolactone). *Macromolecules* **23**, 4604–7.
70. Boudeulle, M. (1975) β-poly(sulfur-nitride) $(SN)_x$. *Cryst. Struct. Commun.* **4**, 9–12.
71. Chien, J.C.W., Karasz, F.E. and Shimamura, K. (1982) Crystal structure of pristane and iodine-doped *cis*-polyacetylene. *Macromolecules,* **15**, 1012–7.
72. Claffey, W. and Blackwell, J. (1976) Electron diffraction of *Valonia* cellulose. A quantitative interpretation. *Biopolymers,* **15**, 1903–15.
73. Chanzy, H., Perez, S., Miller, D.P. *et al.* (1987) An electron diffraction study of mannan I. Crystal and molecular structure. *Macromolecules,* **20**, 2407–13.
74. Brisson, J., Chanzy, H. and Winter, W.T. (1991) The crystal and molecular structure of V_H amylase by electron diffraction analysis. *Int. J. Biol. Macromol.* **13**, 31–9.
75. Brisse, F., Rémillard, B. and Chanzy, H. (1984) Poly(1,4-*trans*-cyclohexane diyl dimethylene succinate): A structural determination using X-ray and electron diffraction. *Macromolecules* **17**, 1980–7.
76. Meille, S.V., Brückner, S. and Lando, J.B. (1989) The structure of γ-poly-pivalolactone: a combined analysis of single-crystal electron diffraction data and powder X-ray diffraction profiles with the Rietveld method. *Polymer,* **30**, 786–92.
77. Bricogne, G. (1984) Maximum entropy and the foundations of direct methods. *Acta Crystallogr.* **A40**, 410–45.
78. Dong, W., Baird, T., Fryer, J.R., Gilmore, G.J. *et al.* (1992) Extension of electron microscope resolution to 1 Å by entropy maximisation and likelihood estimation. *Nature,* **355**, 605–9.

2

New techniques in microscopy

M. J. Miles

2.1 INTRODUCTION

Microscopes based on and including the scanning tunnelling microscope (STM) form a new class of instruments, collectively known as scanning probe microscopes (SPM or SXM). For their invention of the STM [1, 2] in 1981, Gerd Binning and Heinrich Rohrer were awarded the 1986 Nobel Prize for Physics. In the same year, Binnig, Quate and Gerber published a description of the atomic force microscope, AFM [3]. This was the first, and remains the most important, offspring in a family of scanning probe microscopes derived from the STM concept and technology.

2.2 THE PRINCIPLE OF THE SCANNING TUNNELLING MICROSCOPE

Vacuum tunnelling of electrons was an important topic in the theoretical development of quantum mechanics in the 1920s and '30s. The vacuum acts as a barrier to the passage of electrons between two conductors. The wave functions of the electrons extend from the surface of the conductor into the vacuum and decay exponentially with increasing distance from the surface with a decay length of about 0.1 nm depending on the work function. If two conductors are brought sufficiently close for there to be significant overlap of the wave functions, then electrons can tunnel through the barrier from one conductor to the other. It is surprising that the first direct experimental measurement of the effect was not until 1982 with the experiment of Binnig *et al.* [4, 5]. The experimental set-up (Fig. 2.1) consisted of a flat conducting surface and a sharp conducting tip *in vacuo*. With a potential between the surface and the tip, the current tunnelling through the barrier formed by the vacuum gap was measured as a function of the gap width. The use of a tip and a flat surface avoids the problem of parallel alignment that would occur if two planar surfaces were used. Plots of current against distance showed the predicted

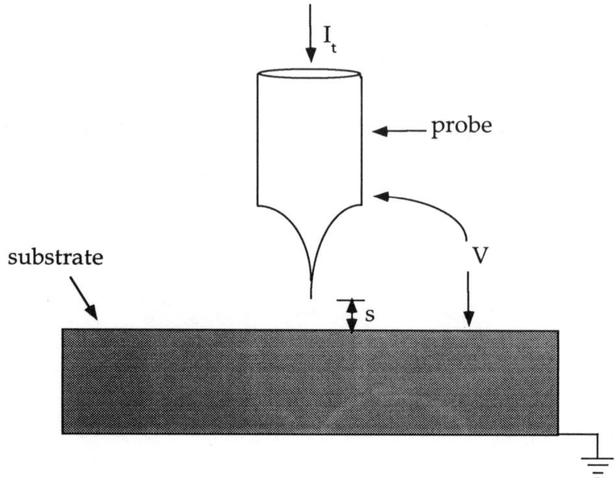

Fig. 2.1 Sharp conducting tip and flat conducting surface separated by gap, *s*, small enough that wavefunctions of electrons in the two conductors overlap sufficiently for a measurable tunnel current, i_t to flow when a bias voltage, *V*, is applied.

exponential dependence from which a work function of 4 eV for the metal surface was calculated:

$$I_t \propto V \exp\left(-\frac{4\pi}{h}(2m\varnothing)^{1/2}s\right) \tag{2.1}$$

where I_t is the tunnel current and V the bias voltage. h, m, and $\varnothing$ are Planck's constant, the electron mass, and the work function of the metal, respectively. s is the gap distance.

This arrangement of tip and flat surface is the basis of the STM. The exponential dependence of the tunnel current on the distance between the tip and the flat surface is the origin of the high resolution of the STM. Clearly, there will be great sensitivity in the *z*-direction normal to the surface. Distance changes normal to the surface of the order of 1pm can be detected by the STM, and this leads to the need for care in the design of the tip-specimen assembly in order to avoid relative movement. In particular, the STM head must be isolated from vibration. The exponential *z*-distance dependence is also responsible for the high lateral resolution since the atom at the very point of the tip contributes the most to the tunnelling current. Atoms in the layer behind the tip atom contribute about one order of magnitude less owing to the exponential decrease in current with distance, and so the tip atom dominates and the whole tip approximates to a monatomic probe.

By scanning the probe across the surface in an *x–y* raster and simultaneously measuring the current, an image of the surface can be built up [2]. There are two basic modes in which these data can be collected. The most common mode of operation is to set a predetermined value for the tunnel current and cause the tip to approach or retract in the z-direction in order to maintain this value of current as it scans over the surface. By recording the movements of the x-, y-, and z- transducers moving the probe, the topography of the surface can be imaged, if the surface is homogeneous. This is known as the constant-current mode. In practice, piezo-electric transducers are usually used in the form of either three separate orthogonal units [2] or as a single hollow tube [6]. The feedback loop controlling the motion of the probe in the z-direction has an associated time constant that restricts the speed at which a scan can be performed. For high-speed scans, the constant-height mode of operation is often used. The feedback loop is either disabled or set with a long time constant, the probe scanned over the surface, and the tunnel current recorded as a function of x and y. Typical bias voltages range from a few millivolts to a few volts, and tunnel currents from around a picoamp to tens of nanoamps. The z data as a function of x and y can be presented by a false colour or gray scale 2-D picture, or as a 3-D representation of the surface. Images can be recorded in minutes or in a fraction of a second. The fastest scan rates can be used only for small scan areas and very flat, stable specimens with the instrument operating in the constant-height mode.

STM tips can be produced by electrochemically etching or mechanically cutting wire of W, Au, Pt, Pt/Ir, etc. It is particularly important to have a sharp tip, i.e. small included angle, for imaging high, steep-sided objects. Melmed [7] has described and explained many of the electrochemical techniques. Another approach to producing a sharp tip is to grow a microtip by electron beam deposition at the apex of an existing electrochemically etched tip [8]. The sharpness of the tips can be assessed by electron microscopy, and the atomic structure at the very end of the tip can be seen in the field ion microscope [9].

One of the major advantages of the STM is that it can image with the probe *in vacuo*, in a gaseous environment, and even in a liquid. It is usually operated at room temperature but STMs capable of operating at liquid-helium temperatures [10] and at 800°C [11] have been constructed.

The tunnel current detected is the result of electrons tunnelling from filled energy states across the gap into empty states. This can be from the tip to the specimen surface or *vice versa* depending on the sign of the potential difference between the two. As the potential difference is increased, more states can contribute to the tunnel current, so that the current at a given potential, V, is given by:

$$I_t \propto \int_{E_F}^{E_{F+eV}} \rho_s(E)\rho_t(E)\mathrm{d}E \tag{2.2}$$

where E_F is the Fermi energy and ρ_s and ρ_t are the local densities of electronic states in the sample and tip, respectively.

2.2.1 STM of organic molecules

The imaging mechanisms described earlier are appropriate for metals and semiconductors. Imaging of organic molecules which are insulators in the bulk is more complicated. A thin layer of the organic specimen must be deposited on a flat conducting substrate. The maximum thickness of organic specimen that can be imaged depends on its conductivity, but is usually limited to a few nanometers. If individual molecules are to be imaged on the substrate, then they must in some way be immobilized on the surface to prevent movement during scanning. The uniform distribution and immobilization of molecules on the substrate is usually the most difficult step in the imaging of organic and biological molecules by STM. One of the most common means is the physi- or chemisorption of the molecule to the surface by suitably matching molecule and substrate. Other methods include the production of ordered arrays of molecules which are self-stabilizing. Such arrays can, for example, be two-dimensional crystals or adsorbed monolayers of molecules. To illustrate some of the practical aspects of STM imaging of polymer molecules, we begin with the imaging of small organic molecules such as benzene and alkanes which are common components of polymers.

STM images have been obtained of an array of benzene molecules co-adsorbed with carbon monoxide on the (111) surface of a rhodium crystal [12, 13]. The molecules formed a 3×3 superlattice with respect to the rhodium. The carbon monoxide molecules were barely visible (Fig. 2.2). The benzene molecules showed two structures depending on bias voltage. With bias of greater than +1.0 V on the specimen, the benzene molecules appeared nearly cylindrical with a height of about 0.2 nm. At low bias, <0.5 V, the images of the benzene molecules took on a three-fold symmetry with a central depression. If the images of the adsorbed molecules are to be understood in terms of their electronic structure, then the energy levels of the unoccupied states in benzene and carbon monoxide must be considered. Empty states a few electron volts above the Fermi level are expected for both benzene and carbon monoxide, but this would not explain the images of benzene at low bias. However, an extended Hückel-theory calculation for benzene bonded to rhodium shows that the combination of the highest filled benzene state with metal orbitals in a bonding configuration shifted the occupied states a few volts further below the Fermi level while the same combination in an antibonding configuration resulted in an empty rhodium–benzene state just above the Fermi level, and so could account for the images at low-positive specimen bias.

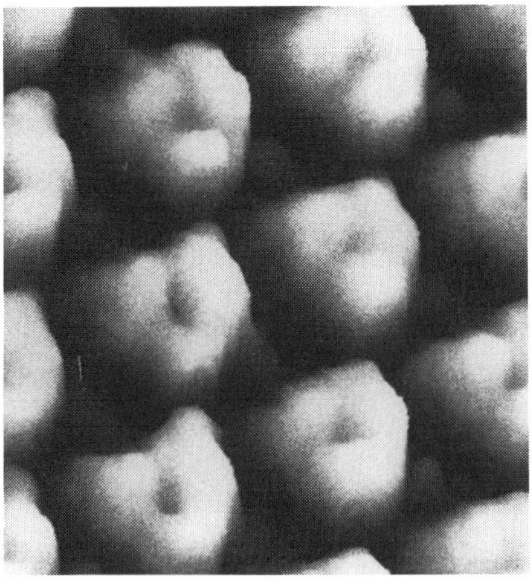

Fig. 2.2 Three-dimensional presentation of STM image of benzene molecules co-adsorbed with carbon monoxide on Rh(111) surface. (Reproduced with permission from Ohtani, H., Wilson, R.J., Chiang, S. and Mate, C.M. (1988) *Phys. Rev. Lett.* **60**, 2398.)

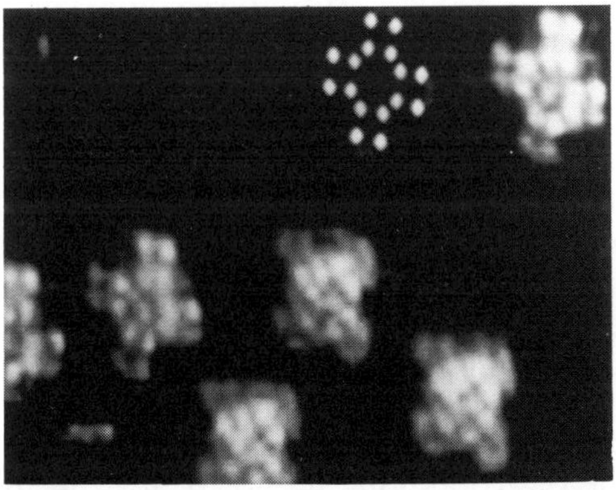

Fig. 2.3 STM image of copper phthalocyanine molecules adsorbed on the copper (100) surface and the HOMO for CuPhth calculated for 0.2 nm above the molecule. (Reproduced with permission from Lippel, P.H., Wilson, R.J., Miller, M.D. *et al.* (1989) *Phys. Rev. Lett.* **62**, 171.)

The importance of the choice of substrate for the immobilization of organic molecules is illustrated by images of copper phthalocyanine, Cu-phth, adsorbed to the (100) surface of a copper crystal [14]. Cu-phth molecules were sublimed on to the freshly-cleaned copper surface in ultra-high vacuum conditions to form a sub-monolayer coverage (Fig. 2.3). Images of Cu-phth deposited on the (111) surface of gold, silicon,

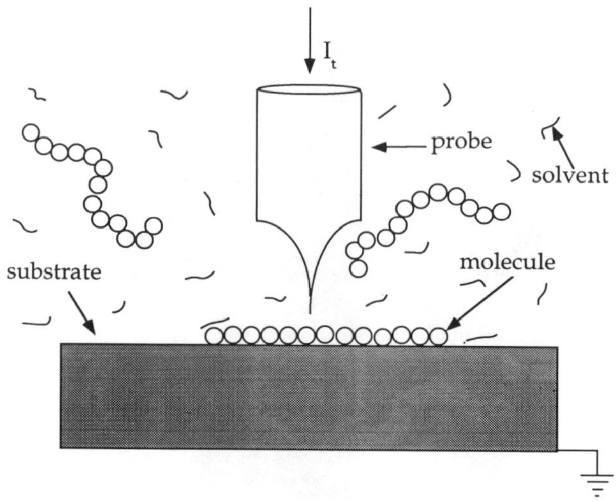

Fig. 2.4 Schematic representation of the STM tip scanning in solution an adsorbed monolayer of alkane molecules on HOPG substrate.

and silver were considerably less detailed, probably as a result of molecular motion due to poorer adsorption. Simple Hückel calculations of charge distribution around the Fermi level for the free molecule show reasonable agreement for both the highest occupied (HOMO) and the lowest unoccupied molecular orbitals (LUMO). Such calculations should include the substrate as hybrid states can be formed.

Sautet and Joachim have calculated the STM images for both benzene on rhodium [15] and copper phthalocyanine on copper [16] using their elastic scattering quantum chemistry (ESQC) technique [17]. The agreement with observed STM images is good.

The STM imaging of long alkane molecules is a major step towards imaging polymers. The adsorption isotherm measurements of Everett and Findenegg [18] for alkanes adsorbing from solution to graphite provide a basis for the conditions necessary for the formation of an adsorbed monolayer. Using concentrations of about 1 mg ml^{-1} for alkanes $C_{20}H_{42}$ to $C_{44}H_{90}$ in octane or decane, adsorbed monolayers on the surface of highly oriented pyrolitic graphite, HOPG, can be imaged by the STM [19–24] with the tip scanning in the solution (Fig. 2.4). Figure 2.5 shows an image of an adsorbed array of $C_{44}H_{90}$ molecules from a solution in decane. Alternate repeat units along the molecules are seen. The molecules are aligned parallel to each other in an all-*trans* conformation to form rows which are reminiscent of polymer lamellae. The alkane molecules have their long axes perpendicular to the interfaces between the rows. The molecular direction coincides with a crystal axis in the basal plane of the graphite substrate, and domains of aligned alkanes are

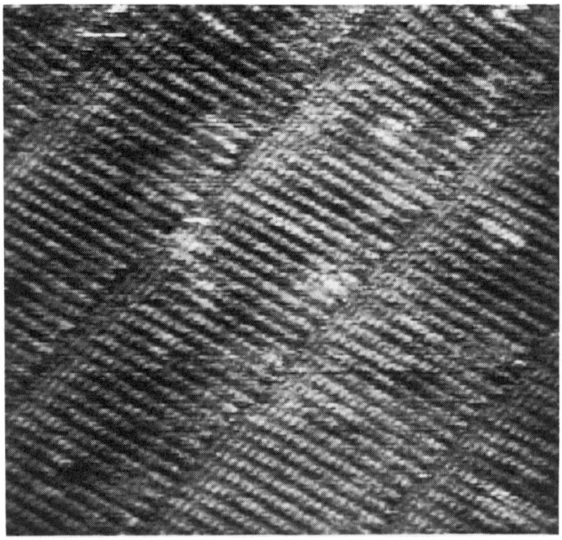

Fig. 2.5 Monolayer of tetratetracontane ($C_{44}H_{90}$) molecules adsorbed on HOPG from solution in decane. (Reproduced by permission of Sautière, L., Thomson, N.H. and Miles, M.J.)

observed with relative orientations of 120°, corresponding to the symmetry of the graphite. The domains can be several microns in lateral dimensions. Towards the end of some of the molecules, the spots representing the repeat units sometimes become more 'dash-like'. This can be understood in terms of motion of the ends of the molecules occurring during the scan, due to the ends being less constrained than the central regions of the molecules. This type of image is consistent with Monte Carlo simulations (Fig. 2.6) [23, 24]. A small modulation in intensity is often observed along the molecules. This can be ascribed to the beating of the two lattices in the direction of the long axis of the alkanes. The repeat distances for graphite and alkane backbone are 0.246 nm and 0.254 nm, respectively. No modulation along the lamellar rows is found for these alkanes, indicating that the lattices are commensurate in this direction. The repeat distance between alternate rows of carbon atoms in the graphite is 0.426 nm. Comparing this value with the repeat distances of 0.420 nm and 0.480 nm between alkane molecules in directions perpendicular and parallel to the planar zigzag in the 3-D triclinic structure, suggests that an arrangement with the planar structure perpendicular to the graphite substrate would be favoured, as the parallel arrangement would require a >10% compression [22, 23]. For the longer alkane, pentacontane ($C_{50}H_{102}$), the ordered, lamellar-like structure is sometimes replaced by a less-ordered arrangement in which directional

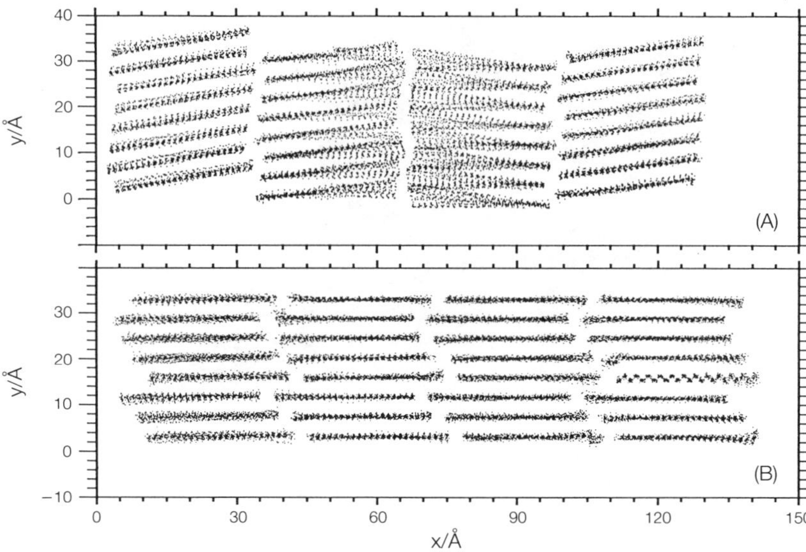

Fig. 2.6 Density plot of positions of alkane carbon backbone in time interval between 60 ps and 80 ps of Monte Carlo simulation. (Reprinted with permission from Hentschke, R., Schürmann, B.L. and Rabe, J.P. (1992) *ACS Polymer Preprints*, **33**, 609; Hentschke, R., Schürmann, B.L. and Rabe, J.P. (1992) *J. Chem. Phys.* **96**, 6213.)

order is preserved but there is no alignment of molecular ends to form the lamellar-like structure [21].

In the case of alkyl derivatives such as fatty acids, Buchholz and Rabe [22] have suggested that if the head group requires more space, then the all-*trans* planar zigzag may be parallel to the graphite substrate rather than perpendicular, as was the case for alkane molecules. STM images [22] of adsorbed monolayers of stearic acid again show lamellar-like structures, but unlike the alkanes there is a repetitive modulation along the lamellae, indicating that commensurability along the lamellae has been lost. The modulation in intensity repeats every five molecules. This is consistent with the planar zigzag being parallel to the graphite surface as the 10% mismatch, derived for the alkanes in this arrangement, would result in an accumulated shift corresponding to one carbon row on the graphite every five molecules along the lamellae. Buchholz and Rabe [21, 22, 25, 26, 27, 28] have examined several alkyl derivatives. They found that a dialkylbenzene also shows a modulation along the lamellar structures [27]. By using tunnelling conditions which allowed simultaneous imaging of the benzene groups and the underlying graphite lattice (Fig. 2.7), the incommensurability can be made manifest [28]. Alkanols did not show such modulation, but formed herringbone structures [20, 25]. Comparison of the graphite lattice before adsorption

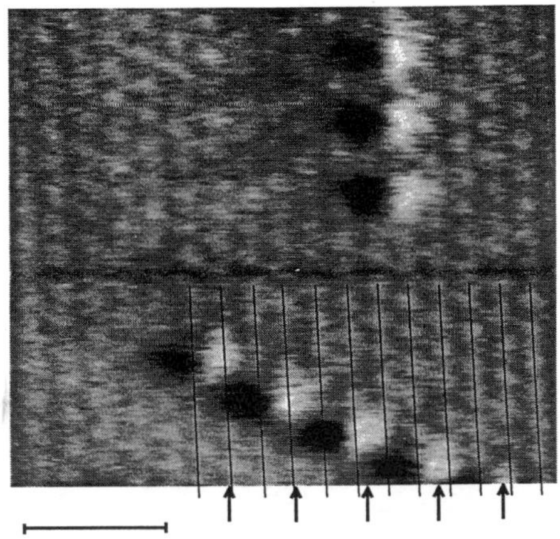

Fig. 2.7 STM image showing a domain boundary in a didodecylbenzene monolayer, taken with tunnelling conditions which allow the underlying HOPG lattice to be imaged simultaneously with the benzene rings (bright spots). (Reproduced with permission from Rabe, J.P., Buchholz, S. and Ashadskaya, L. (1992) *ACS Polymer Preprints*, **33**, 798; Rabe, J.P. (1992) *Ultramicroscopy*, **42–44**, 41.)

with the orientation of the adsorbed alkanol molecules showed that both the long axis of the molecules and the lamellar boundaries were parallel to the basis vectors of the graphite lattice. The angle of 60° between the molecular axis and the lamellar boundaries corresponds to a displacement of neighbouring alkyl chains of two methylene units and with the planar zigzag perpendicular to the surface only every second methylene unit would be in contact with the graphite. The herringbone tilt favours the hydrogen bonding between molecular ends. For the longer alkanols, tetracosanol and triacontanol, herringbone structures were not observed, the long axes of the molecules being parallel.

The monofunctional alkanols formed small domains with many defects in their structures [21]. The authors attribute this to the dense packing at the boundaries, resulting in a low line tension (interfacial energy). Using high scan rates, Buchholz and Rabe [22] have been able to follow the dynamical molecular rearrangements of these adsorbed layers. For example, with each STM frame requiring only 1.5 s, it has been possible to observe co-operative molecular reorientations that have occurred in less than 10 ms during a scan of octadecanol adsorbed on the graphite (Fig. 2.8). Similarly, co-operative rotations about the long axes of a group of didodecyl benzene molecules accompanied by lateral shifts have been observed to cause a translation of the domain boundary.

At first sight, STM imaging of three-dimensional crystallites of alkanes, well known as bulk insulators, would seem to be impossible. However, Michel *et al.* [29] have reported images of crystallites of *n*-heptadecane,

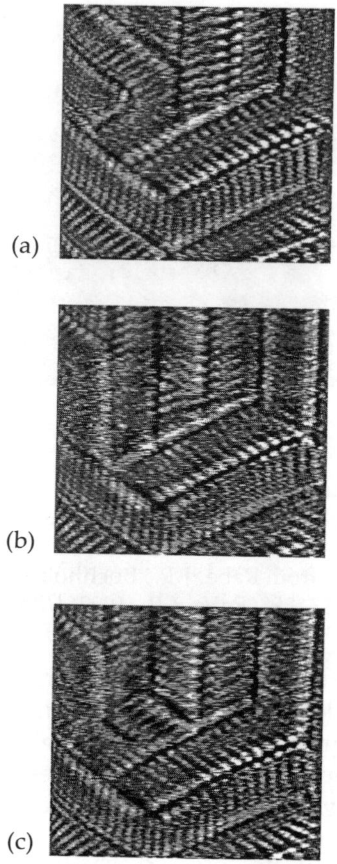

Fig. 2.8 STM images showing a monolayer of octadecanol in which molecules can be seen to flip orientation (b) during scanning from the arrangement in (a) to that in (c). (Reproduced with permission from Rabe, J.P. and Buchholz, S. (1991) *Science*, **253**, 424.)

n-octadecane, and *n*-hexatriacontane grown after deposition of saturated solutions on either HOPG or (111) gold surfaces. Crystallites of thickness corresponding to the extended length of the respective molecule were observed as well as multilayers having thicknesses which were integral multiples of chain lengths. Melting experiments were performed in which alkanes were imaged below and above their crystal melting temperatures. Heptadecane crystallites of 20 nm thickness were reproducibly imaged at temperatures below their melting point of 18°C. On heating above 18°C, the large heptadecane objects melted away, leaving smaller amorphous structures. Similarly, large crystallites of *n*-octadecane melted at about the expected melting temperature. After one hour at this melting temperature, layers with a thickness of 2.1–2.5 nm were formed and the amorphous parts became flat, uniform layers.

The measured bulk conductivity of solid alkanes is some nineteen orders of magnitude (!) lower than would be required to produce such images if it is assumed that these STM images result from conduction through the crystallites. Imaging of such large alkane objects is not routinely possible, and Michel *et al.* [29] discuss possible image-contrast mechanisms for these materials. They can identify no single specific mechanism responsible for the contrast, but propose that electron transfer through the alkane structures may be via intrinsic, extrinsic, or atmospheric conditions or defect-induced surface states.

2.2.2 STM of liquid-crystal molecules

The cyanobiphenyls are a family of thermotropic liquid-crystal-forming molecules. The molecule known as 8CB or 4-*n*-octyl-4′-cyanobiphenyl has a smectic phase at room temperature and was the first to be imaged by STM [30, 31]. Interestingly, these first studies were reported before the studies on alkanes and alkyl derivatives. Liquid-crystal-forming molecules are good candidates for STM [32] because of the possibility of immobilization that such ordered structures may offer once adsorbed to the substrate. Many STM studies of 8CB have since been reported [33–40]. As with the alkanes, the tip scans in the liquid and the monolayer adsorbed to HOPG (or MoS_2) substrate is imaged with the appropriate tunnel conditions. A smectic-like structure consisting of rows of parallel molecules is imaged. Although the observed structure is apparently that of a smectic liquid crystal, the Fourier transform shows it to be a two-dimensional crystal. The images consist of bright and dark rows. The bright rows can be attributed to the cyanobiphenyl groups and the dark rows to the alkyl chains, the molecules lying flat on the substrate surface rather than standing on end. In the original images, the repeat spacing between the bright rows was 3.8 nm. This was interpreted as a structure in which the molecules are arranged head-to-head and tail-to-tail with some interdigitation. Individual phenyl rings are visible in some images and confirm this structure. Other images (Fig. 2.9) have a smaller row repeat distance of 2.4 nm, and are seen to consist of molecules arranged head-to-tail. Another repeating structure exists along each row. For the case of 8CB, this consists of eight molecules (interdigitated structure) or four molecules (head-to-tail structure) forming a block which is followed by a jog or offset before the next block of molecules. For the case of 10CB in which the alkyl chain has 10 carbon atoms, the blocks contain 10 (or five) molecules before the jog [34]. This can be understood as a balance between the adsorption of alkyl groups to the nearly commensurate graphite lattice, as was the case for alkane molecules adsorbing to the

Fig. 2.9 STM images of adsorbed monolayer of 8CB (4-*n*-octyl-4′-cyanobiphenyl) smectic-like structure showing phenyl rings and alkyl chains. (Reproduced with permission from McMaster, T.J., Carr, H.J., Miles, M.J. *et al.* (1990) *J. Vac. Sci. Technol. A*, **8**, 672.)

graphite from solution, and intermolecular interactions of the cyanobiphenyl groups which is relieved by a dislocation of the structure in the form of a jog. The influence of the substrate is again illustrated [36, 37] when molybdenum disulphide (MoS_2) is used as a substrate instead of HOPG. The structure for 8CB on MoS_2 was found to consist of an alternating pattern of head-to-head and tail-to-tail. Laterally, the pattern alternates between head and tail along a row. 10CB shows the same arrangement as on graphite, but the intermolecular spacing within the rows is greater, consistent with the larger lattice constant of MoS_2.

Images of the shorter-chain alkyl group hexyl cyanobiphenyl again show rows of molecules with head-to-head/tail-to-tail arrangement. The interpretation of the images is a structure in which the phenyl rings in neighbouring molecules alternate between lying flat on the graphite and being tilted up from the surface. The alkyl chains are not visible in the images probably because of thermal motion.

STM studies of low-molecular-weight organic liquid crystals have been performed for a range of molecules consisting of two cyclic structures and one or two alkyl groups. In some molecules the alkyl is attached to the phenyl group via an ether linkage. Another variation is to have the two phenyl rings joined by an ether linkage. In liquid-crystal displays, it is usually a mixture of molecular species that is used. As a step towards

(a)

(b)

Fig. 2.10 (a) STM images of an adsorbed monolayer from a mixture of alkoxy phenylbenzoate and alkoxybiphenyl benzoate, showing a herringbone-type structure; (b) alternating row thickness suggesting a head-to-head and tail-to-tail arrangement of alternating molecular species in the mixture. (Reproduced with permission from Sautière, L., Day, S. and Miles, M.J. (1992) *Ultramicroscopy*, **42–44**, 1054.)

using STM to study the behaviour of liquid-crystal molecules at a surface in devices, a recent STM study [41, 42] involved a two-component liquid-crystal mixture consisting of alkoxy phenyl benzoate and alkoxy biphenyl benzoate molecules in the ratio of 2:3. The mixture was heated above the isotropic transition temperature before deposition on HOPG. After about one hour, an adsorbed monolayer consisting of smectic-like rows could be observed in the STM similar to those of the cyanobiphenyl liquid crystals. Several different structures were found which could be divided into groups having row spacings of either ~3.3 nm or 4.0 nm. Unlike the cyanobiphenyls, no jog defect was present along the rows. Several alternating row structures were seen, such as, the herringbone structure [Fig. 2.10(a)] and the alternating row thickness [Fig. 2.10(b)], which could be understood as a head-to-head and tail-to-tail arrangement of alternating molecular species.

The molecular immobilization facilitated by liquid-crystalline order has been used in STM studies of high-molecular-weight molecules. Poly(γ-benzyl-L-glutamate), PBLG, is a synthetic polypeptide specifically produced for liquid-crystal formation. It consists of an α-helical backbone decorated at the end of flexible spacings with benzyl groups, aromatic groups which are known to give high contrast in STM. McMaster *et al.* [43, 44] have examined adsorbed layers of PBLG of molecular weight 70,000 on graphite. Deposition was from solution in dimethyl formamide, a solvent which promotes lyotropic liquid-crystal structures in PBLG. STM images show lyotropic arrays of rod-like molecules [Fig. 2.11(a)]. A periodic structure along each of the molecules is also visible. The Fourier transform of such images shows that just as with the low-molecular-weight liquid crystals the adsorbed PBLG structure is crystalline rather than liquid crystalline. Fourier filtering enhances the helical character of the molecular images [Fig. 2.11(b)]. This approach to inducing liquid-crystalline order in solution in order to promote an ordered adsorbed layer for good molecular immobilization has been applied to biological rod-like molecules such as the high-molecular-weight submit protein of wheat gliadins.

These examples are main-chain liquid-crystal polymers. STM imaging of side-chain liquid-crystal polymers has not so far been reported.

2.2.3 STM of non-conducting crystalline polymers

STM imaging of uncoated crystals of non-conducting polymers is severely limited to the thickness of the specimen that can be imaged. It is, of course, possible to obtain images of specimens coated with metallic conductors, but the grain size of such a coating is usually about 5 nm, and so this is the resolution limit. Piner *et al.* [45] used STM to obtain images of solution-grown polyethylene crystals. The crystals were deposited on HOPG and coated with either gold or chromium. Typical lozenge-shaped

---

Fig. 2.11 (a) STM image of PBLG molecules deposited on HOPG in an ordered array; inset shows Fourier transform; (b) Fourier-filtered image of PBLG array. (Reproduced with permission from McMaster, T.J., Carr, H.J., Miles, M.J. *et al.* (1991) *Macromolecules*, **24**, 1428.)

(a)

(b)

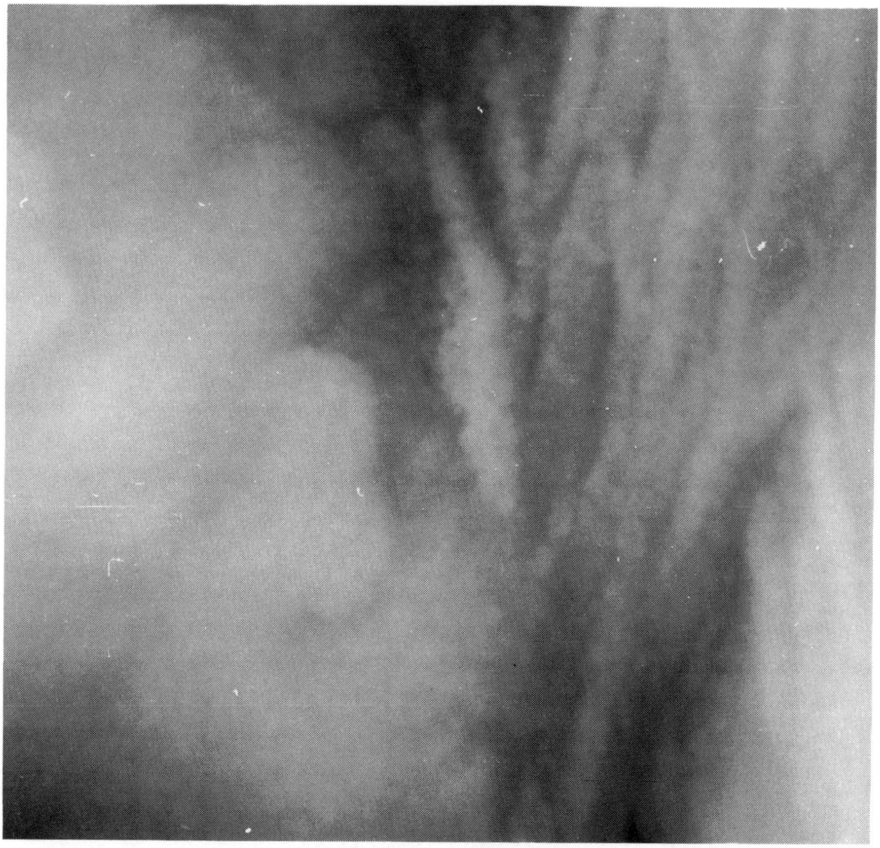

Fig. 2.12 STM image of coated surface of poly(4-methyl-1-pentene) showing sheaves of lamellae. (Reproduced with permission from Williamson, R.L. and Miles, M.J.)

crystals were seen. Small outgrowth structures were observed on the crystals and there was evidence of sectorization.

Reneker *et al.* [46] examined the structure resulting from quenching a thin layer of an entangled solution of ultra-high molecular weight polyethylene on mica. After evaporation of the solvent, the deposited polymer was coated with Pt/C. The surface structure of this preparation contained elongated patches about 100 nm long and 20 nm wide.

Structures typical of melt-crystallized polymers have been obtained by Williamson from a specimen of poly(4-methyl-pentene-1). The surface of the specimen was coated with C and Pt and then examined in the STM. Sheaves of lamellae, typical of those found at the centre of spherulites (Fig. 2.12), were frequently found. The resolution in such images is the same as that obtained by making a replica of the surface and examining it in a TEM. The replication step can thus be omitted for STM.

Films of polyimide, PI, are used as a substrate in liquid-crystal devices. After the surface of the PI film has been rubbed, nematic liquid crystals will align on it. The mechanism of this alignment is of importance in order to optimize the process, ideally to produce a single aligned domain and also to understand why surface-stabilized ferro-electric liquid crystals exhibit only poor alignment on rubbed PI. Suzuki *et al.* [47] have used STM to image the gold-coated surface of rubbed and non-rubbed polyimide films on indium-tin oxide substrate. The images showed the non-rubbed surface consisted of spherical grains 20 to 25 nm in diameter. A surface rubbed only once showed a highly-aligned structure. With increased rubbing the structure first became paste-like and then spherical structures again appeared.

In order to obtain molecular-resolution images, Fujiwara *et al.* [48] have prepared Langmuir–Blodgett films of polyimide, using polyamic acid as a starting material, deposited them on graphite, and pressure annealed them. The STM images show an ordered array of polymer molecules on the HOPG substrate. Molecular dynamics calculations of energy-optimized structures on graphite suggest a zigzag structure for the polyimide with bends of 119° at the ether linkage every 1.78 nm. A similar zigzag structure can be recognized in the STM image. The next stage would be to observe the orientation and structure of liquid-crystal molecules on this polyimide film.

A technique for producing ultra-thin, highly oriented polymer fibres [49] has been used to produce thin, oriented specimens for STM studies. Fuchs *et al.* [50] used STM to study such specimens of the non-conducting polymer poly(1-butene), PB-1, deposited on HOPG. Some images apparently show oriented molecules on the graphite lattice. However, the pitch and the lateral spacing are over one-third smaller than the values obtained from diffraction measurements on bulk samples. If these images are to be related to the polymer film, then the authors suggest two possible explanations: (i) the substrate might interact sufficiently strongly to distort the $2 \times 3_1$ conformation of the PB-1 molecules and also their packing separation, or (ii) the molecular image may be an interference or Moiré pattern resulting from an overlay of the electronic structures of the HOPG and polymer specimen. The same experimental approach has been applied to the study of ultra-thin polyethylene (PE) fibres [51]. In this case the high-resolution images obtained were more consistent with the bulk values for pitch and lateral spacing. The fibre thicknesses are again too large for the tip to scan over the surface if the specimen is acting as an insulator. It was suggested that the tip penetrates the sample and sweeps away all but the final monomolecular layer, which, especially in the case of PE (see section on alkanes), may be strongly adsorbed to the HOPG substrate with only slight distortion.

The studies of coated and uncoated non-conducting polymer specimens using STM are now being superseded by studies using the atomic force microscope (AFM) except for cases where the electronic properties of the specimen are of interest or where the sample is intrinsically extremely thin or where the specimen would be damaged by the AFM probe.

2.2.4 STM of conducting polymers

The fundamental difficulties of image contrast and specimen thickness in using STM to study insulating polymers and organic molecules are largely removed when it is applied to the study of conducting polymers. Some of the conducting polymers to have been studied are polyanaline [52–55], polydiacetylene [56], polypyrrole [57], SN_x, and polythiophene [58, 59]. Most of these utilize methods of electrochemical deposition and polymerization for the preparation of thin films on conducting substrates suitable for STM.

Lacaze *et al.* [58] have followed the electropolymerization and deposition of poly(1-naphthol) and of poly(2,6-dimethylphenol) on to HOPG and gold surfaces. They observed rather different growth mechanisms depending on the surface. Growth on the HOPG surface appears to be nucleated at steps in the graphite, but growth prefers to continue on the already deposited polymer rather than the graphite surface. Some periodic structure, indicative of crystallization, could be observed in the poly (2,6-dimethylphenol) films. However, using gold as a substrate, the surface was covered by a uniform thin film of polymer. Subsequent deposition was irregular.

The morphology of electrochemically polymerized and deposited films of polyanaline at various times of deposition on a gold surface has been studied by Jeon *et al.* [53, 55]. They also used the tunnelling spectroscopy capability of the STM to measure current versus voltage (I–V) plots for samples produced at different electrochemical potentials. The measured conductivity (dI/dV) at the Fermi level ($V = 0$) increased with increased doping, corresponding to the electrochemical potential during film production. I–V measurements over the specimen indicated an inhomogeneous material consisting of metallic islands of 20 to 30 nm in size. The STM topographic images show grains of this size. This demonstrates the advantage of combining STM topographic and spectroscopic data in relating electrical properties to micromorphology. Occasionally, I–V curves revealed negative differential resistance, indicating charge trapping in local states.

Polydiacetylene specimens for STM were prepared by spin coating a

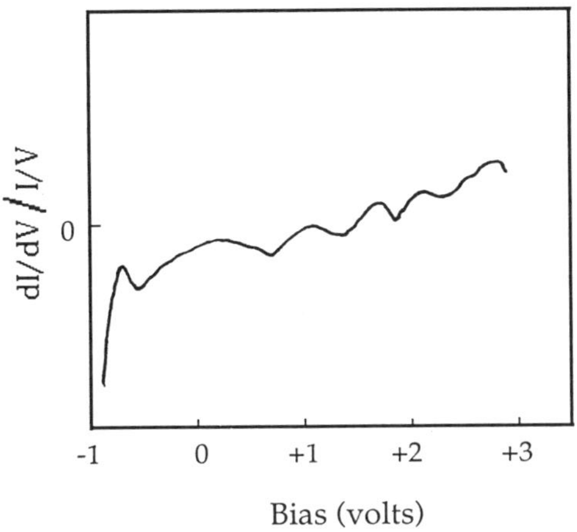

Fig. 2.13 d*I*/d*V* versus *V* plot for an SN_x crystal imaged by STM in air, indicating a multi-band-gap structure. (Reproduced with permission from Stejny, J. and Miles, M.J.)

solution of monomer in acetone on freshly cleaved HOPG, allowing the solvent to evaporate and then photopolymerizing the layer with an ultra-violet source. The STM images [56] apparently show regions containing isolated polymer molecules and other regions which resemble a two-dimensional crystal lattice. No *I–V* measurements on this specimen were reported.

The conducting polymer poly(sulphur nitride), or SN_x, can be produced in macroscopic, three-dimensional crystals which appear metallic and are fibrous in nature. Stejny and Miles have obtained STM images of the surfaces of these crystals. No molecular structure was resolved, but rather an indication of parallel domains approximately 2.5 nm wide. These may correspond to the proposed micro-twin structure. *I–V* measurements over much of the specimen were too noisy and unstable for a meaningful interpretation. In some regions, however, *I–V* measurements became very stable and differentiation of this plot gave a density-of-states versus energy plot (Fig. 2.13) which appeared to show a multi-band-gap structure.

2.2.5 STM manipulation of polymers and lithography

A spectacular demonstration of the STM's ability to create a structure on a surface was the manipulation of xenon atoms on a single-crystal nickel surface at 4 K so as to produce a pattern of atoms which formed the initials 'IBM' [60]. Several reviews have discussed surface manipulation at the nanometric scale [61]. The main area of current interest for polymer modification at this scale is in lithography. In the conventional process, thin polymer films are produced by spin coating on a silicon substrate, and patterns are written in the polymer using an electron beam, which initiates a chemical reaction resulting usually in cross-linking or chain-scission. McCord and Pease [62] have demonstrated that an STM tip can be used as the source of the electron beam to expose thin (tens of nanometers) films of poly(methyl-methacrylate), PMMA. The STM has the advantages of relatively low-voltage, high-current density, and very high spatial resolution. In order that the electrons have sufficient energy to penetrate and expose the resists, the STM must be operated *in vacuo* in the field-emission regime using bias voltages of tens of voltages. The PMMA film must be greater than 20 nm to act as an effective mask and prevent the subsequent silicon etch from passing through the polymer during this process. The authors report that PMMA can be used as a positive or negative resist depending on bias voltage and current dose. During the developing process a positive resist remains in the exposed areas whereas a negative resist is removed. Using electron energies greater than 25 eV and current doses greater than 10^{-2} C cm^{-2}, PMMA was found to undergo cross-linking and therefore act as a negative resist. Unexposed film could be removed by development in acetone. At lower energies and doses, PMMA behaved as a positive resist undergoing chain-scission and thus became soluble in the cellosolve/methanol developer. The high-resolution writing ability of the STM tip is seriously degraded by the scattering of electrons and the production of secondary electrons in the polymer film. These electrons can diverge from the initial beam before reaching the silicon substrate. If the polarity of the bias is reversed, electrons are emitted from the silicon and pass up through the resist gaining sufficient energy to expose the film if they are not inelastically scattered. Any scattered or secondary electrons will converge towards the tip rather than diverge and so the resolution is improved. Results close to the highest demonstrated in polymer resists have been achieved.

Marrian *et al.* [63] used 5 V pulses of 1 ms duration in an STM to expose a negative *e*-beam resist on a *p*-type silicon substrate, and then observe the induced surface structure before development. The resist, P4BCMU, was polydiacetylene with urethane substituents. The STM image of the surface showed that each pulse resulted in a raised circular feature about 6 nm high and 12 nm in diameter, presumably the result of polymer cross-linking.

In a similar manner, Heckl and Smith [64] have used a 10 μs pulse of 4 V to apparently polymerize glutaraldehyde monomers into clusters in an otherwise thin, highly uniform film deposited on graphite from solution. The STM, operated under conditions (700 mV; 0.1 nA) which caused no further reactions, showed that each pulse resulted in the formation of many clusters. The clusters, which had a diameter of about 2 nm, were interpreted as resulting from a polymerization reaction involving five to 15 glutaraldehyde monomers.

Piezo-electric and ferro-electric polymers offer interesting properties for study by STM. The local piezo-electric activity of metal-coated, thin films of vinylidene fluoride-trifluoroethylene (VDF-TrFE) has been studied by STM [65] to a spatial resolution of about 10 nm. The possibility of using uncoated films of VDF-TrFE for data storage has been considered. Information could be written by locally poling the polymer in the electric field between the conducting substrate beneath the polymer film and the sharp tip of a scanning probe microscope. The thickness of the films produced means that a variation on the atomic force microscope has been used rather than an STM in order to image the surface.

2.3 THE PRINCIPLE OF THE ATOMIC FORCE MICROSCOPE

Scanning force microscopy (SFM) is the collective term used to describe the scanning probe microscopies that image by measuring a force between the tip and the specimen. They are distinguished by the nature of the force measured. The three most common SFMs are: (i) the magnetic force microscope, (ii) the electrostatic force microscope, and (iii) the atomic force microscope (AFM). The term atomic force is usually meant to cover van der Waals and related forces. The principle and operation of AFMs is very similar to the STM, the quantity measured being the force between the atoms on the tip and the specimen rather than the tunnel current. This removes the restriction that the specimen be an electrical conductor, which is, of course, a major advantage for the study of polymers. For the imaging of individual molecules, removal of this restriction allows the substrate, on to which the molecule will adsorb, to be chosen from a wider range of materials.

The AFM, the first of the SFMs, was invented by Binnig, Quate and Gerber [3]. The sharp tip is fixed to the end of a cantilever so that the force between the atoms on the end of the tip and the specimen causes the cantilever to bend. The force is determined by measuring the extent of bending of the cantilever. The cantilever spring must be sufficiently soft that deflection corresponding to forces of the order of van der Waals forces can be detected. Binnig *et al.* [3] calculated that a spring constant, k, of $<10\ \mathrm{N\ m^{-1}}$ was needed in order to avoid damage to the surface. At the same time, its resonant frequency must be high, in the region of tens of kHz, in order to be decoupled from low-frequency building vibrations, as was also the case for the STM. These two requirements are achieved by using small, low-mass cantilevers. In the AFM of Binnig *et al.* [3] the displacement of the cantilever is detected with an STM tip. With this method of detection, atomic-resolution images of HOPG were achieved

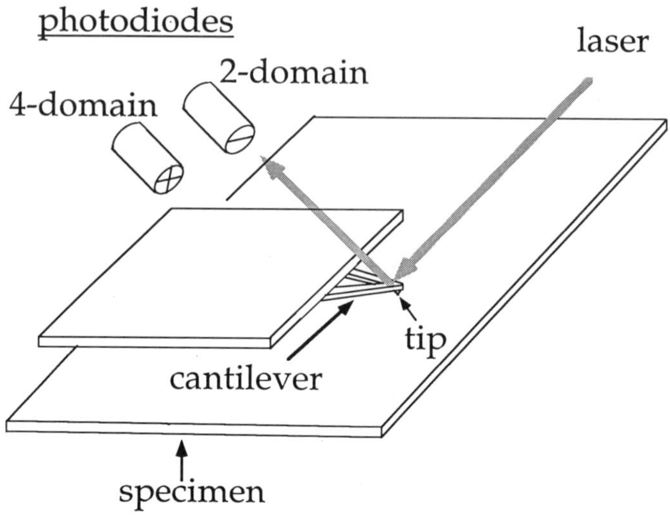

Fig. 2.14 Optical lever method for an atomic force microscope, using a two-domain photodiode for detecting the deflection of the cantilever. A four-domain photodiode is used when the twist of the cantilever resulting from lateral force is also measured.

using microfabricated, silicon dioxide cantilevers [66]. The tunnelling-tip detection, although a sensitive detector of cantilever displacement, has some disadvantages: (i) the tip exerts $\sim 10^{-7}$ N force on the cantilever, (ii) the tip or cantilever may become contaminated and the tunnel junction unstable, and (iii) the cantilever surface is relatively rough (a few nanometers root-mean-square roughness) and so any relative lateral movement of the cantilever and tip due to thermal drift will result in false measurements of the AFM tip force. Optical methods are now the most common means of detecting the cantilever displacement [67]. The force exerted by the photon flux on the cantilever is small and the detection is averaged over an area which is large compared with the surface roughness of the cantilever.

The principal optical methods used to detect the deflection of the cantilever are (i) various interferometric techniques and (ii) the optical lever. The former is preferred for UHV applications. The latter is currently employed on commercial AFMs for air operation. The optical lever (Fig. 2.14) uses a two-domain photodiode to measure the displacements of the reflected beam resulting from deflections of the cantilever. The specimen can be scanned at constant height and the force recorded through the deflections of the cantilever, or at a preset constant force with the feedback loop maintaining a constant deflection of the cantilever. These

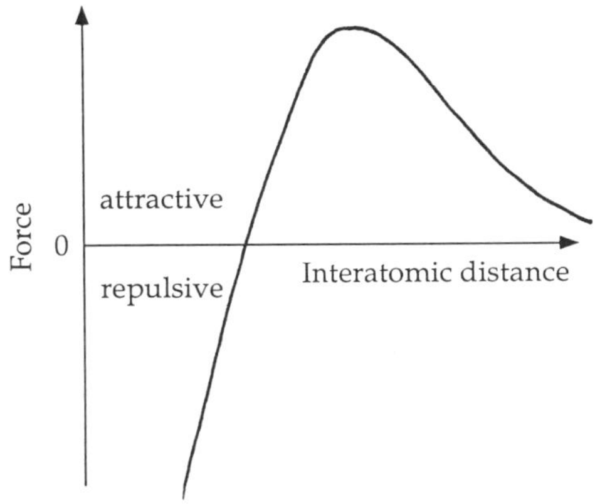

Fig. 2.15 Force versus distance plot showing the van der Waals attractive region and the repulsive region resulting from the Pauli exclusion principle.

methods are analogous to the constant-height and constant-current modes of the STM.

In order to understand the forces acting on the tip as it approaches the specimen, it is simplest to start with clean surfaces *in vacuo*, and consider an atom on the tip and an atom on the specimen experiencing a van der Waals attractive force. Figure 2.15 shows the dependence of the force on the separation. At large distances, the force can be taken as zero. As the tip and specimen approach, the attractive force between them increases and then rapidly decreases and becomes repulsive. The repulsive force has a strong distance dependence and so the last atom at the tip will dominate the force measured, as was the case for the tunnel current in the STM. Therefore, when the AFM is operated in the repulsive-force mode the highest resolution will be achieved. An important difference between the forces in AFM and currents in STM is that at any separation forces of both signs are acting on the AFM tip.

Although the attractive van der Waals force is usually small compared to the repulsive force, its rate of decrease with increasing separation is also smaller. This means that when integrating over the whole volume of the tip for the resultant force, the van der Waals attractive component may be significant. It has been estimated that for non-destructive scanning of 'hard' surfaces, the tip force should be less than 10 nN, and for 'soft' biological surfaces it should be less than 10 pN.

In air, most surfaces are covered with a water layer. At small tip-specimen separations the layers join forming a meniscus between the tip and the specimen, this results in a large capillary force ~100 nN pulling the tip and specimen together. In order to avoid this capillary force, AFM imaging should be performed in UHV or under liquid. A thorough description of the forces involved in the operation of the AFM in various liquids with both conducting and insulating tips on both conducting and insulating specimens has been given by Weisenhorn *et al.* [68]. It should be noted that capillary forces will also be present in STM, where a common manifestation of their presence is the sweeping of molecules (particularly hydrophilic biomolecules) around on the surface during scanning.

In addition to the normal forces, the tip also experiences lateral or frictional forces which tend to twist the cantilever. In order to detect this twisting, a second pair of domains is required in the photodiode. Therefore, for lateral force microscopy (LFM) a four-domain photodiode is employed (Fig 2.14). As well as being used to image regions with different 'frictional' properties, lateral-force information is useful in a diagnostic role to understand the interactions of the tip with the surface.

Another mode of operation of the AFM is to apply an oscillation to the specimen in the z-direction so that the tip-specimen separation has a superimposed sinusoidal modulation. By recording the resultant oscillations in the detected force, the relative local stiffness of the specimen can be determined at each pixel point over the surface. This is sometimes called elasticity mapping.

So far the description of the principle and operation of the AFM has been concerned with the repulsive or contact mode. The AFM may also be used in the non-contact or attractive mode which corresponds to operating in the region where the van der Waals force is dominant. This mode of operation is also important for imaging based on other forces which decay more slowly with distance than the repulsive force. In particular, it is used in the magnetic force microscope (MFM). In the AFM, the non-contact mode is important for imaging 'soft' specimens without damaging them. Because of the weakness of the force and its weaker distance dependence, the sensitivity and the signal-to-noise ratio, S/N, of the force detection system must be improved. By vibrating the cantilever and using phase-sensitive detection techniques, improvements of several orders of magnitude can be achieved. The microfabricated cantilevers have high Q values. The resonant frequency of the cantilever is shifted slightly as the spring constant associated with interaction of the tip and the specimen adds to the spring constant of the cantilever itself. By oscillating the cantilever at the frequency corresponding to the steepest slope of its resonance, the greatest sensitivity can be attained. Values as low as 3×10^{-13} N force sensitivity and 1.5×10^{-4} N m^{-1} force derivative have been reported. [69].

Cantilevers with integrated tips are made using the microfabrication techniques developed by the electronics industry. Cantilevers with integrated tips can be made in Si_3N_4 [70, 71]. A pyramidal tip is formed in an etch pit on the Si (100) surface. Etching can be used to sharpen the tip, i.e. to reduce the included angle at the end of the tip. It is important that the tip is sharp for imaging of steep-sided objects as the resulting image will be a convolution of the tip shape and the object. An alternative to etching tips sharper is to grow a sharp structure on the end of a pyramidal tip. In a method similar to that used for STM [8], the electron beam of a scanning electron microscope has been used to deposit thin carbon tips on microfabricated pyramidal tips. The included angle of these tips is only 15° compared to 114° of the pyramid, and so they image steep, high structures more accurately [72].

2.3.1 AFM of polymers

Much of the pioneering work in the development and application of AFM to organic and biological systems has taken place in the Hansma laboratory (UCSB). This was again the case for the first AFM imaging at molecular resolution of a polymer [73]. The polymer specimen was made by preparing the monomer as a polymerizable surfactant *n*-(2-aminoethyl)-10,12-tricosadiynamide and forming a monolayer at an air–water interface. This monolayer was compressed to form the crystalline state. Polymerization was achieved with ultra-violet radiation, and the polymerized film was transferred to a glass support for examination in the AFM. Operating in the constant-force mode, such monolayers were seen to consist of tightly packed crystalline domains 1 to 100 μm in diameter. Quantitative values for the lateral spacings of the molecules were obtained and some indication of structure along the molecules was found.

The first of the AFM images of the more extensively studied synthetic polymers were obtained by Patil *et al.* [74]. This study was on dendritic crystals of linear polyethylene (PE) formed on cooling a hot solution in xylene. The suspended crystalline precipitate was dried down on mica and imaged in air in the constant-force mode. Images typical of the apices of dendritic polymer crystals [75] are shown. Lozenge-shaped, chain-folded crystals with spiral terraces, familiar from TEM studies, were also found. The resolution was not sufficient to obtain information at a molecular level. Lin and Meier have obtained AFM images of solution-grown single crystals of PE, including that in Fig. 2.16 showing multiple layers of overgrowth. Snétivy and Vansco [76] used AFM to study the morphology of poly(ethylene oxide), PEO, crystals grown in toluene solutions. A suspension of the crystals was deposited on a glass slide and the solvent allowed to evaporate. After crystallization for one week, the crystals observed were mostly of square habit with edge length ~50 μm. Screw-dislocation growth spirals were found, at the centre of which it was possible to see the diagonal edge of the growing top layer. Measurement of the step height at the edge of crystals gave their thickness as

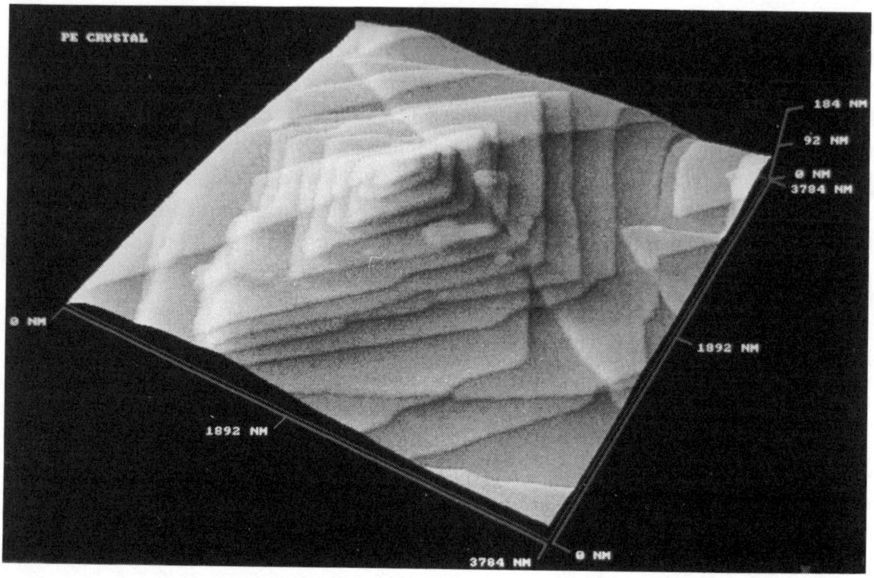

Fig. 2.16 AFM image of solution-grown polyethylene crystals. (Reproduced with permission from Lin, F. and Meier, D.)

12.5±0.5 nm. This information is difficult to obtain from TEM images but relatively simple to measure by AFM. Another advantage of AFM, and scanning probe microscopy in general, is the ability to follow changes with time without perturbations caused by radiation damage or vacuum conditions in the TEM. In the case of these PEO crystals, it was observed that after about one hour following removal from the solution the characteristic morphology of the crystals began to deteriorate. The crystals began to disintegrate as material contracted in some areas resulting in a tree-like pattern. The edges of the crystals became thicker by between 5 to 8 nm, forming a picture-frame structure. This suggests that a refolding process is underway. Discontinuities in the width of the frame at the crystal corners once refolding has begun indicated the existence of independent chain folding domains which are consistent with the folding scheme proposed by Kovacs [77].

High-resolution AFM images have been reported for linear and cyclic alkanes [78]. The *a–b* surfaces of single crystals of *n*-tritriacontane ($C_{33}H_{68}$) and *n*-hexatriacontane ($C_{36}H_{74}$) gave a two-dimensional lattice of features which had periodicities and symmetry consistent with the *a–b* plane (001) of the orthorhombic structure of alkane crystals. Electron diffraction confirmed orthorhombic symmetry. The AFM is apparently imaging the methyl group at the chain ends. A major goal in the application of AFM to polymer crystals is the imaging of the structure of the fold surface. The

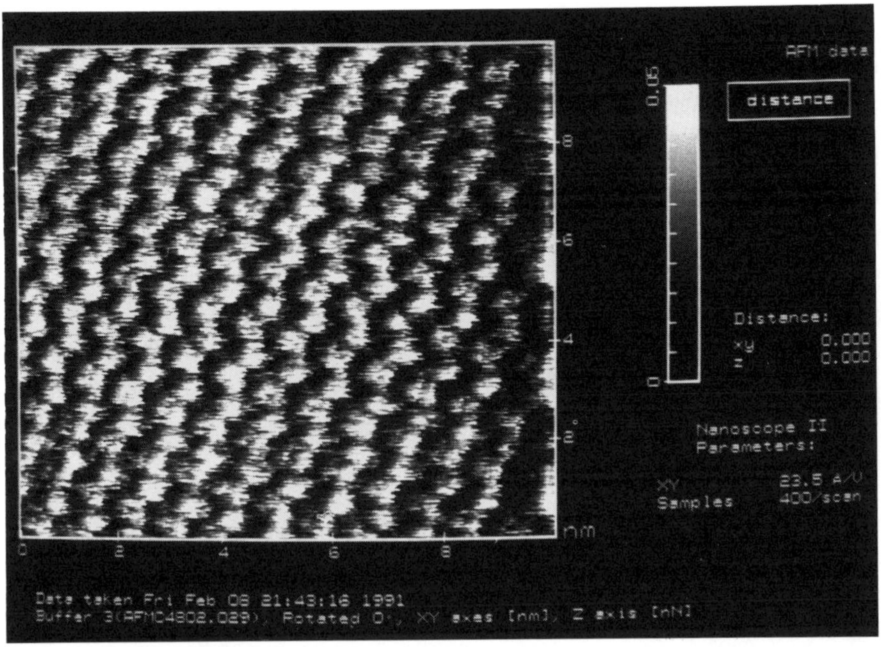

Fig. 2.17 AFM image of 001 fold surface of cyclo-octatetracontane. (Reproduced with permission from Stocker, W., Bar, G. Kunz, M. *et al.* (1991) *Polym. Bulletin*, **26**, 215.)

most favourable circumstances to achieve such images is when the folding is tight and regular. Crystals of cyclic alkanes provide such fold surfaces in a convenient form. Stocker *et al.* [78] have examined the fold surfaces of cyclo-octatetracontane, $(CH_2)_{48}$, and cyclodoheptacontane, $(CH_2)_{72}$, crystals, which are expected to have a monoclinic structure with adjacent re-entry folds (ggtgg). The structures of the 001 and $00\bar{1}$ fold planes are different. The image of the fold surface shown in Fig. 2.17 is in excellent agreement with the structure expected for the 001 surface. The cyclic alkanes appear to provide a more stable fold-surface structure for AFM imaging than the polyethylene crystals studied so far.

2.3.2 AFM of polymer fibres and oriented films

The polymer crystal surfaces parallel to the chain direction have been studied with AFM for several polymers using oriented fibres and films. Eng *et al.* [79] have imaged highly oriented, melt-spun polyethylene fibres. Their images (Fig. 2.18), which are remarkably similar to TEM pictures of the sample, show stacked lamellae grown on extended-chain

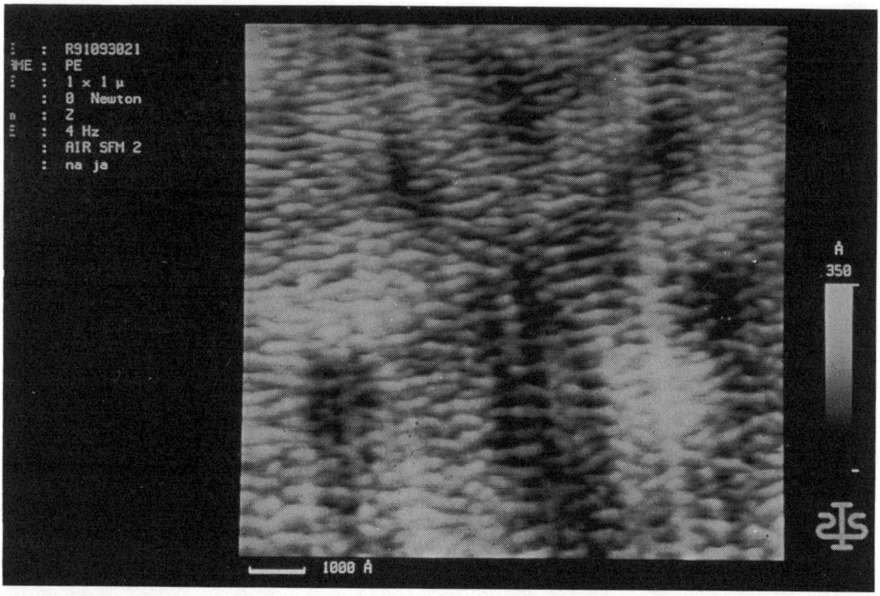

Fig. 2.18 Highly oriented, melt-spun, polyethylene fibres imaged by AFM show the stacked-lamellar structure familiar from TEM. (Reproduced with permission from Eng, L., Fuchs, H., Jandt, K. and Petermann, J. (1992) *Helvetica Physica Acta*, **65**, 871.)

needle crystals. An internal surface of a cold-extruded rod of polyethylene was prepared for AFM by Magonov *et al.* [80] by ultramicrotomy. As expected, a fibrillar structure was observed. The authors do not report any evidence of stacked lamellae. At high magnification, ridges running roughly parallel and in the extrusion direction were seen, and measurements of their diameters gave values between 0.4 and 0.5 nm (Fig. 2.19). This corresponds to the van der Waals diameter of a polyethylene molecule when the hydrogen atoms are included. In some regions, extended chains of PE in the all-*trans* conformation were seen with the expected repeat distance along the chain of 0.25 nm. For a drawn solution-cast film of ultra-high molecular weight polyethylene, Snétivy *et al.* [81] imaged regions of ordered molecules at many areas on the surface of the specimen. The values of the lateral spacing of molecules in the different regions fell into two groups: 0.73 ± 0.018 nm and 0.49 ± 0.023 nm. These values are in good agreement with molecular spacings on the *a–c* and *b–c* planes, respectively. Thus, the AFM can distinguish between two crystallographic planes present at the surface of a bulk specimen.

The ability of thin, highly oriented films of poly(tetrafluorethylene), PTFE, to induce orientation in layers of a variety of materials subsequently deposited on its surface [82] was investigated by Hansma

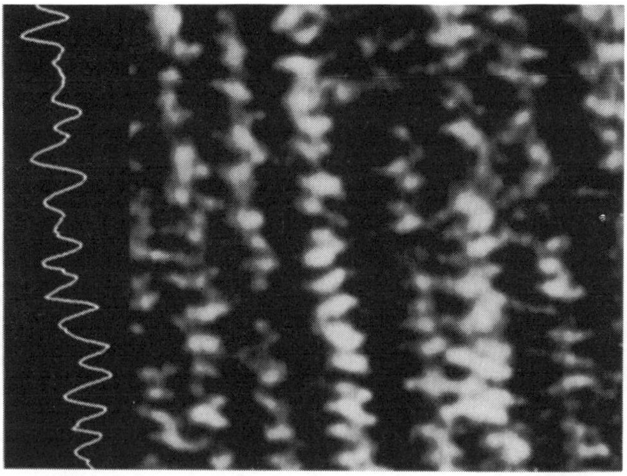

Fig. 2.19 High-magnification AFM image of surface of cold-extruded polyethylene showing aligned molecules. (Reproduced with permission from Magonov, S.N., Quarnström, K., Elings, V. and Cantow, H.-J. (1991) *Polym. Bulletin,* **25**, 689.)

et al. [83] using AFM. The method involves producing a thin, oriented film by dragging a PFTE rod over a cleaned glass slide at 300°C. The images showed parallel ridges with a wide range of heights and widths. Such structures suggest that graphoepitaxy may be involved in the alignment mechanism. At higher magnifications, regions of aligned molecules qualitatively similar to those seen for PE were observed (Fig. 2.20) with evidence of structure along their length consistent with the expected spacing for the $-CF_2-$ groups. The authors also noted that this PTFE surface, unlike the surfaces of many other materials could not be damaged by the AFM tip. Oriented sheets of poly(butene-1), PB-1 and polypropylene had earlier been discovered to promote oriented growth of certain metals that were evaporated on to their surfaces [84]. This epitaxial growth for the metals Sn and Te on PB-1 has been investigated with AFM by Eng *et al.* [85]. Thin melt-spun specimens were deposited on HOPG for study by both AFM and STM. The AFM images showed a fibrillar structure with fibrils ~20 nm in diameter. TEM observations had suggested graphoepitaxy as the orientation mechanism with the possibility also of metal–methyl group interactions; the polymer films having the greatest epitaxial effect on the deposited metal, isotactic polypropylene, (iPP) and PB-1, have side groups ending in methyl groups. However, no Te was found on the PB-1 by AFM. The Te particles may have been hidden in the trenches between the fibres and so have been inaccessible to the AFM probe, or they may have been insufficiently

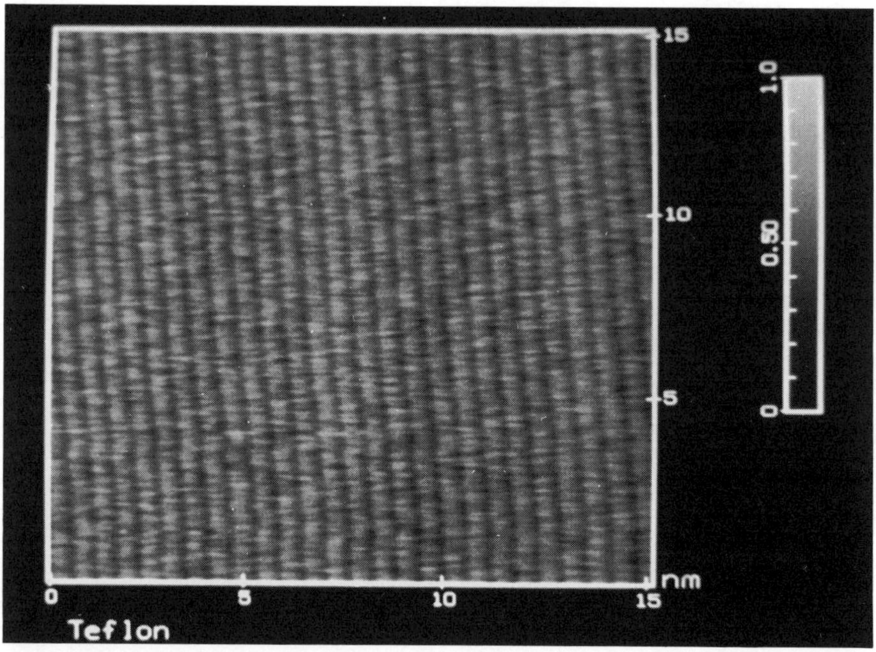

Fig. 2.20 AFM image of a thin, oriented PTFE film showing aligned molecules with periodicity along their length consistent with the expected spacing for the $-CF_2-$ groups. (Reproduced with permission from Hansma, H., Motamedi, F., Smith, P. *et al.* (1992) *Polymer*, **33**, 647.)

bound to the PB-1 and have been swept away by the scanning probe. AFM images of Sn particles on PB-1 show them to exist between the filaments.

For the polymers poly(oxymethylene), POM [86], and poly(*p*-phenylene terephthalamide), PPTA or Kevlar (Du Pont), Snétivy *et al.* [86, 87] have derived information on the respective helical structures from AFM data. The POM specimens were ultramicrotome cleaved from stretched rods, and atomic-scale AFM images [86] of these surfaces and their corresponding autocorrelation functions indicated a lateral intermolecular spacing of 0.46 ± 0.04 nm. POM crystals have hexagonal symmetry with $a = 0.446$ nm from X-ray crystallography. In the chain direction, two periodicities were found: $c_1 = 0.445 \pm 0.03$ nm and $c_2 = 2.10 \pm 0.21$ nm. In bulk isotropic POM, the molecules have a 9_5-helical conformation with a pitch of 1.739 nm determined by X-ray diffraction. However, wide-angle X-ray diffraction (WAXD) of oriented POM samples suggests that the helix can be stretched to have a pitch of 1.90 nm, which is within the error of the AFM value for c_2. In the

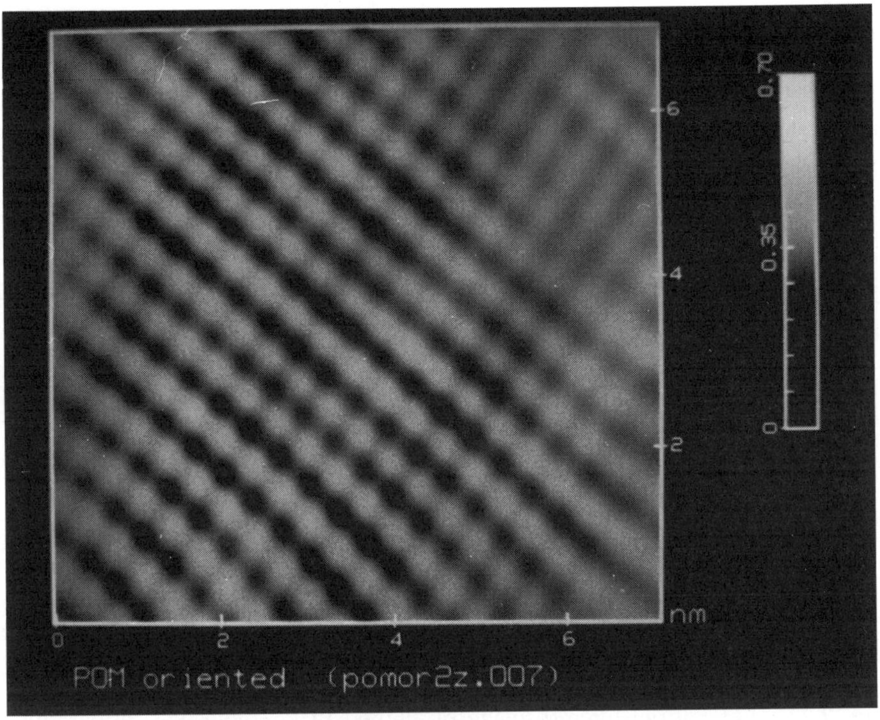

Fig. 2.21 A Fourier-filtered, molecular-resolution AFM image of POM, in which the five-turn periodicity along the molecule due to its 9_5 helical symmetry is apparent. (Reproduced with permission from Snétivy, D. and Vancso, G.J. (1992) *Macromolecules*, **25**, 3320.)

Fourier-filtered image (Fig. 2.21), the spots correspond to high points on each turn of the POM helix. The brightness of each of the spots along the molecule varies systematically. Different brightnesses correspond, to a first approximation, to different heights of the helix at each turn. The heights of the spots repeat every five turns and this is consistent with the 9_5 helical symmetry of POM. By comparing the observed pattern of spot brightnesses with the 9_5 structure of POM, the authors deduced that the AFM tip is further from the helix axis for oxygen atoms than for methyl groups.

The PPTA specimen was prepared by spinning a fibre from a solution of the polymer in sulphuric acid at 80°C and precipitating it in a water bath [87]. A surface suitable for AFM was obtained by ultramicrotoming the fibre, which had a diameter of 100 μm. The images revealed molecular domains having different orientations with respect to the fibre axis. The intermolecular spacing of 0.51 nm and the repeat along the molecules of

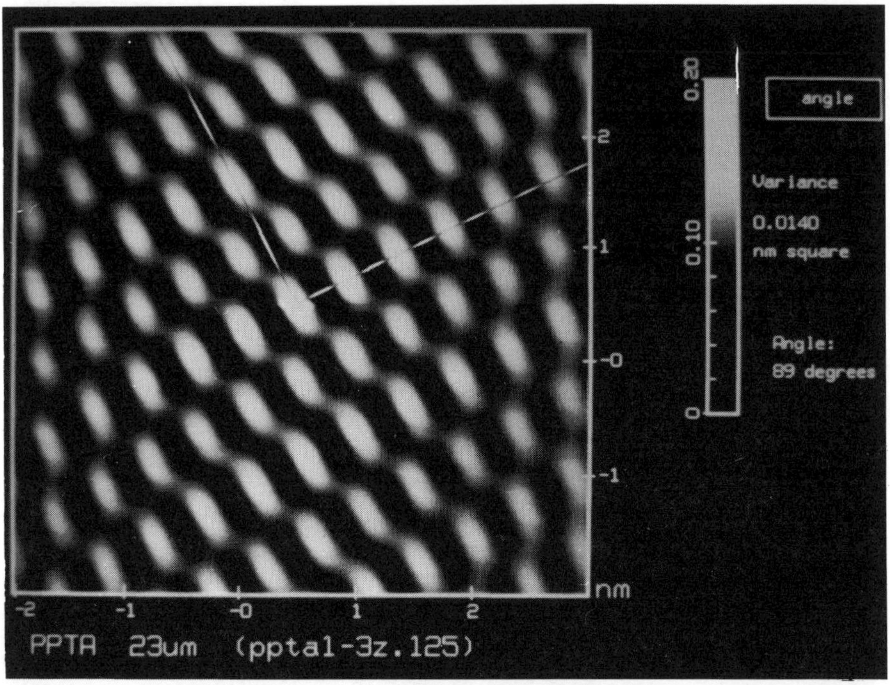

Fig. 2.22 AFM Fourier-filtered image of aligned PPTA molecules, apparently showing the expected zigzag appearance due to alternating orientation of the phenylene groups. (Reproduced with permission from Snétivy, D., Rutledge, G.C. and Vancso, G.J. (1992) *ACS Polymer Preprints*, **33**, 786.)

1.35 ± 0.08 nm are both in good agreement with WAXD data, which also indicate a 60° dihedral angle between the planes of neighbouring phenyl rings. The filtered AFM image (Fig. 2.22) appears to show this alternating direction of the phenylene groups along the molecule, giving it a zigzag appearance and indicating the ability of AFM to give near atomic-scale information for crystalline polymers.

AFM studies of isotactic polypropylene, (iPP) have been performed by Snétivy *et al.* [88] and Lotz *et al.* [89]. The former followed a specimen preparation technique similar to several already described, involving uniaxial stretching at 120°C, quenching at constant stress, and microtoming a fresh surface. As with similarly prepared specimens, the images showed a fibrillar structure and at higher magnifications aligned molecules were seen that had intermolecular spacings consistent with the monoclinic or α-phase, which X-ray diffraction showed was predominant in this specimen.

2.3.3 AFM of epitaxial polymer crystallization

The study by Lotz *et al.* [89] was concerned with the nature of the epitaxial crystallization of iPP on benzoic acid crystals. Low-molecular-weight iPP was co-melted with benzoic acid between coverslips and slowly crystallized in a temperature gradient. The iPP grows epitaxially on the benzoic acid, which can then be removed by dissolution in hot methanol. The lamellae grow edge on to the acid [90]. Electron diffraction has shown that (010) planes of the α-phase are parallel to the film surface. Neighbouring layers of molecules in the (010) plane consist of opposite chirality helices. Because of the different chiralities, alternate layers of molecules would not only exhibit molecules of opposite handedness but also show the opposite 'side' of the molecule and so present a different pattern of methyl groups to the surface studied by AFM, i.e. the surface which had grown epitaxially on the benzoic acid. High-resolution AFM images of the (010) face of iPP have revealed methyls in groups of four (rather than five) forming a lozenge shape. By comparison with models of the helices, it can be seen that this pattern means right-handed helices have grown epitaxially on benzoic acid. This cannot be determined from diffraction data.

So far, AFM has provided little evidence for reconstructions of relaxations of crystal structure at polymer surfaces. In a study [91] of the polydiacetylene 2,4-hexadiynylene bis(*p*-fluorobenzenesulphonate), there were indications that benzyl rings of the substituents attached to the main chain had rearranged at the surface compared to the bulk. An alternating structure for neighbouring molecules could be explained by a possible rearrangement of the surface positions of the benzyl rings to energetically more favourable positions.

The STM imaging of adsorbed monolayers of didodecylbenzene, DDB, on HOPG from solution and the observation of domain boundary movement was described earlier [27]. Eng *et al.* [92] have attempted a comparison of these results with those obtained with AFM. Islands of DDB, 50 nm thick and growing with time, probably as the solvent evaporated, were found with the AFM. Images of the adsorbed monolayer as seen in the STM were not obtained. The probe was presumably sweeping away this layer and subsequently imaging solidified DDB layers. AFM images of a solidified lamellar layer, similar to those obtained by STM but of lower resolution, could only be obtained using metal (Cr/Au) coated tips with a bias of +4V. Switching the bias off resulted in a gradual loss of resolution. The authors suggest that the bias may result in a Coulomb interaction with the specimen, giving an additional applied load on the cantilever which forces the tip through the organic film down to the DDB/HOPG interface. The first monolayer is expected to be more ordered and so more robust. However, the lamellar spacing was too large, not very reproducible, and repeated scanning of this layer resulted in a loss of the lamellar structure and the appearance of the HOPG substrate.

2.3.4 AFM of low-crystallinity polymers

A requirement for high-resolution imaging with AFM is that the specimen should be relatively rigid. In the case of crystalline polymers, the rigidity is provided by the crystal lattice. For amorphous or low-crystallinity polymers, the specimen rigidity is high if imaging is performed below the glass transition temperature. Crystallization from the glassy state of (bis-phenol A)polycarbonate has been studied at the specimen surface by AFM [93]. The specimen as prepared by solution casting showed little structure. After holding the specimen at 145°C for 36 hours, spherulitic-like objects could be found on the surface (Fig. 2.23). They consisted of ball-like patterns, the smallest balls having a diameter of ~20 nm. At higher annealing temperatures, 159°C and 178°C, another morphology was observed in which the image consisted of many nanocrystals of uniform shape and size. The authors interpret the formation of nanocrystals as a result of the formation of low-molecular-weight material by thermal degradation.

Fracture surfaces of polymers is another topic to which AFM can make a contribution. Images obtained so far are similar to those produced by the scanning electron microscope, SEM. AFM could replace SEM in this application provided the tip is sufficiently sharp to probe the depths of the steepest-sided structure of the surface.

2.3.5 AFM of co-polymers

For phase-separated specimens, the AFM offers methods for obtaining 'contrast' between the phases without staining. This can be simple surface morphology, as in the case of the diblock co-polymer styrene methylmethacrylate. In the absence of solvents, both components are glassy. Using a mixed solvent such as acetone/cyclohexane, the solvent for one component is a precipitant for the other component, and vice versa for the other solvent component. Bickmann *et al.* [94] have observed two surface morphologies with the AFM: (i) rod-like structures parallel to the surface and (ii) rods perpendicular to the surface in an hexagonal array.

In the AFM, it is also possible to image different phases with different contrast by imaging the local stiffness of the specimen. As described earlier, essentially this is done by applying a small oscillatory displacement, Δz, to the probe (or specimen) in the z-direction and measuring the resulting increment in force, ΔF. A relative value of stiffness or elasticity modulus is given by $\Delta F/\Delta z$ at each pixel. This mode of imaging has been

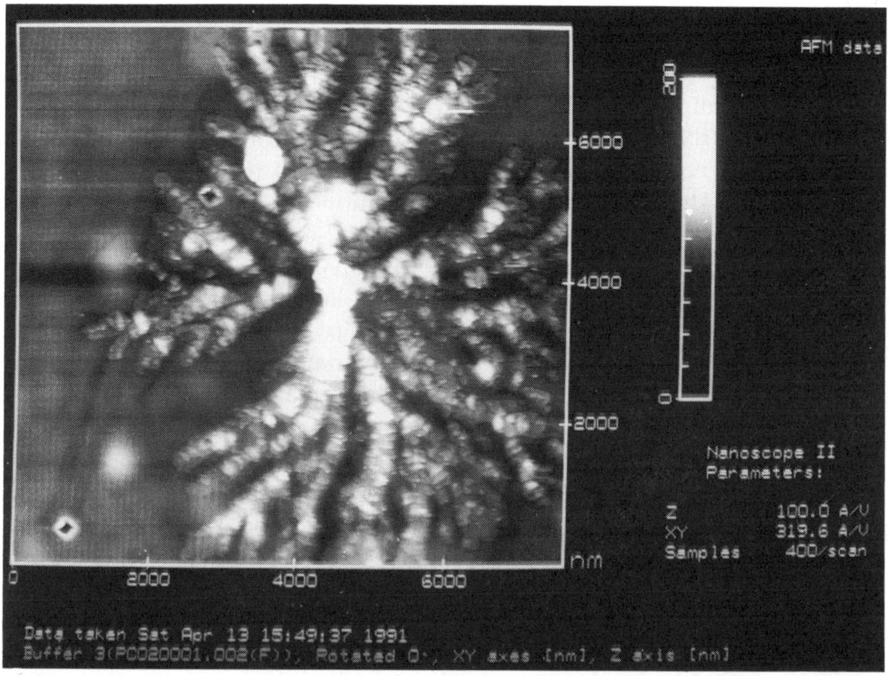

Fig. 2.23 AFM image of bis-phenol A polycarbonate after annealing showing spherulitic-type structure. (Reproduced with permission from Magonov, S.N., Kempf, S., Kimmig, M. and Cantow, H.-J. (1991) *Polym. Bulletin*, **26**, 715.)

applied to recognise the soft and hard segments in a polyurethane specimen [95]. The domain sizes observed were between 1 and 35 nm.

The AFM may also distinguish between different polymer phases on the surface of a specimen using the frictional or lateral force mode as described earlier. Such an application of the frictional or lateral force mode to phase-separated polymer specimens has not been reported so far. However, Hamada and Kaneko [96] have used frictional force microscopy to measure frictional force distributions after wear-mark tests on polycarbonate and poly(methylmethacrylate).

2.4 FUTURE WORK

The application of scanning probe microscopy to the study of polymers has only just begun. Over the next few years, there will probably be a rapid increase in the application of AFM in particular to investigate polymer surfaces, in part replacing TEM and SEM, but also in new applications as the possibilities offered by molecular and even atomic resolution imaging are realised. At present the resolution of AFM images of polymers is less than that of STM, but is likely to improve if the force sensitivity can be routinely reduced to $\sim 10^{-11}$ N. In addition to topographic images, AFM can also measure mechanical properties such as elastic modulus and friction of the polymer specimen to the same resolution as the image, so that images of the frictional or elastic nature of the surface are possible. The restriction of STM to thin or conducting specimens will limit its general application to conducting polymers, adsorbed monolayers of liquid crystals, metal-coated specimens, etc. and to studies requiring information on local density of electronic states.

The ability of both AFM and STM to follow structural changes in a specimen with time is a major advantage over electron microscopy of polymers, which usually suffer radiation damage in the beam, preventing continuous observation. By combining this advantage with the ability of AFM and STM to operate in liquids such as polymer solutions and melts, then high-resolution imaging of crystal growth is a possibility.

Many scanning probe microscopes based on the technology and in the spirit of STM have been and continue to be invented [97–105]. Of these, the photon scanning tunnelling microscope, PSTM, [104, 105] has possible applications in polymer science. It is capable of $\lambda/100$ resolution on transparent specimens. Its advantages over STM and AFM are that it does not require a conducting specimen and the fibre-optic probe does not need to apply a force. It is capable of providing information on the optical properties such as spectroscopic data, birefringence, refractive index, etc. Currently, the application of PSTM to polymers is limited to those specimens which do not strongly scatter light.

In the immediate future, AFM offers the greatest opportunities for the study of polymeric surfaces at high resolution with other scanning probes having niche applications.

REFERENCES

1. Binnig, G. and Rohrer, H. (1982) *Helvetica Physica Acta*, **55**, 726.
2. Binnig, G., Rohrer, H., Gerber, Ch. and Weibel, E. (1982) *Phys. Rev. Lett.* **49**, 57.
3. Binnig, G., Quate, C.F. and Gerber, Ch. (1986) *Phys. Rev. Lett.* **56**, 930.
4. Binnig, G., Rohrer, H., Gerber, Ch. and Weibel, E. (1982) *Appl. Phys. Lett.* **40**, 178.
5. Binnig, G., Rohrer, H., Gerber, Ch. and Weibel, E. (1982) *Physica*, **109** and **110B**, 2075.
6. Binnig, G. and Smith, D.P.E. (1986) *Rev. Sci. Instrum.* **57**, 1688.
7. Melmed, A.J. (1991) *J. Vac. Sci. Technol. B*, **9**, 601.
8. Akama, Y., Nishimura, E., Sakai, A. and Murakami, H. (1990) *J. Vac. Sci. Technol. A*, **8**, 429.
9. Schmidt, U., Rasch, H., Fries, Th. *et al.* (1992) *Surf. Sci.* **266**, 249.
10. See, for example, Renner, Ch., Niedermann, Ph., Kent, A.D. and Fischer, Ø. (1990) *J. Vac. Sci. Technol. A*, **8**, 330.

11. Kitamura, S., Sato, T. and Iwaksuki, M. (1991) *Nature*, **351**, 215; Miki, K., Morita, Y., Tokumoto, H. *et al.* (1992) *Ultramicroscopy*, **42–44**, 851.
12. Chiang, S., Wilson, R.J., Mate, C.M. and Ohtani, H. (1988) *J. Microscopy*, **152**, 567.
13. Ohtani, H., Wilson, R.J., Chang, S. and Mater, C.M. (1988) *Phys. Rev. Lett.* **60**, 2398.
14. Lippel, P.H., Wilson, R.J., Miller, M.D., Wöll, Ch. *et al.* (1989) *Phys. Rev. Lett.* **62**, 171.
15. Sautet, P. and Joachim, C. (1991) *Chem. Phys. Lett.* **185**, 23.
16. Sautet, P. and Joachim, C. (1992) *Surf. Sci.* **271**. 387.
17. Sautet, P. and Joachim, C. (1988) *Chem. Phys. Lett.* **153**, 511.
18. Everett, D.H. and Findenegg, G.H. (1969) *Nature*, **223**, 52.
19. Bernhardt, R.H., McGonigal, G.C., Schnieder, R. and Thomson, D.J. (1990) *J. Vac. Sci. Technol. A*, **8**, 667.
20. McGonigal, G.C., Bernhardt, R.H., Yeo, Y.H. and Thomson, D.J. (1991) *J. Vac. Sci. Technol. B*, **9**, 1107.
21. Rabe, J.P. and Buchholz, S. (1991) *Makromol. Chem.* (Macromol. Symposium), **50**, 261.
22. Rabe, J. P. and Buchholz, S. (1991) *Science*, **253**, 424.
23. Hentschke, R., Schürmann, B.L. and Rabe, J.P. (1992) *J. Chem. Phys.* **96**, 6213.
24. Hentschke, R., Schürmann, B.L. and Rabe, J.P. (1992) *ACS Polymer Preprints*, **33**, 609.
25. Buchholz, S. and Rabe, J.P. (1992) *Angew, Chem. Int. Ed. Engl.* **31**, 189.
26. Rabe, J.P. and Buchholz, S. (1991) *Phys. Rev. Lett.* **66**, 2096.
27. Buchholz, S. and Rabe, J.P. (1991) *J. Vac. Sci. Technol. B*, **9**, 1126.
28. Rabe, J.P., Buchholz, S. and Askadskaga, L. (1992) *ACS Polymer Preprints*, **33**, 798.
29. Michel, B., Travaglini, G., Rohrer, H. *et al.* (1989) *Phys. B-Condensed Matter*, **76**, 99.
30. Foster, J.S. and Frommer, J.E. (1988) *Nature*, **333**, 542.
31. Smith, D.P.E., Hörber, J.K.H., Gerber, Ch. and Binnig, G. (1989) *Science*, **245**, 432.
32. Frommer, J.E. (1992) *Angew. Chemie*, **31**, 1298.
33. McMaster, T.J., Carr, H.J., Miles, M.J. *et al.* (1990) *J. Vac. Sci. Technol. A*, **8**, 672.
34. Smith, D.P.E., Hörber, J.K.H., Binnig, G. and Nejoh, H. (1990) *Nature*, **344**, 641.
35. Mizutani, W., Shigeno, M., Sakakibara, Y. *et al.* (1990) *J. Vac. Sci. Technol. A*, **8**, 675.
36. Hara, M., Iwakabe, Y., Tochigi, K. *et al.* (1990) *Nature*, **344**, 228.
37. Smith, D.P.E. and Heckl, W. (1990) *Nature*, **346**, 616.
38. Poulin, J.-C. and Kagan, H.B. (1991) *C.R. Acad. Sci. Paris*, **313**, 1533.
39. Smith, D.P.E. (1991) *J. Vac. Sci. Technol. B*, **9**, 119.
40. McMaster, T.J., Carr, H.J., Miles, M.J. *et al.* (1991) *Liquid Crystals*, **9**, 11.
41. Sautière, L., Day, S. and Miles, M.J. (1992) *Ultramicroscopy*, **42–44**, 1054.
42. Sautière, L., Day. S. and Miles, M.J. (1992) *ACS Polymer Preprints*, **33**, 784.
43. McMaster, T.J., Carr, H.J., Miles, M.J. *et al.* (1990) *J. Vac. Sci. Technol. A*, **8**, 648.
44. McMaster, T.J., Carr, H.J., Miles, M.J. *et al.* (1991) *Macromolecules*, **24**, 1428.
45. Piner, R., Reiffenberger, R., Martin, D.C. *et al.* (1990) *J. Polym. Sci. C, Polym. Lett.* **28**, 399.
46. Reneker, D.H., Schneir, J., Howell, B. and Harary, H. (1990) *Polymer Communications*, **31**, 167.
47. Susuki, M., Maruno, T., Yamamoto, F. and Nagai, K. (1990) *J. Vac. Sci. Technol. A*, **8**, 631.

48. Fujiwara, I., Ishimoto, C. and Seto, J. (1991) *J. Vac. Sci. Technol.* **9**, 1148.
49. Petermann, J. and Gohil, R.M. (1979) *J. Mater. Sci.* **14**, 2260.
50. Fuchs, H., Eng, L.M., Sander, R. *et al.* (1991) *Polymer Bulletin*, **26**, 95.
51. Jandt, K.D., Bukh, M., Petermann, J. *et al.* (1991) *Polym. Bulletin*, **27**, 101.
52. Mantovani, J.G., Warmack, R.J., Annis, B.K. *et al.* (1990) *J. Appl. Polym. Sci.* **40**, 1693.
53. Jeon, D., Kim, J., Gallagher, M.C. *et al.* (1991) *J. Vac. Sci. Technol. B*, **9**, 1154.
54. Wan, M.X., Yang, J., Zhu, C. and Bai, C. (1992) *Thin Solid Films*, **208**, 153.
55. Jeon, D., Kim, J., Gallagher, M.C. and Willis, R.F. (1992) *Science*, **256**, 1662.
56. Leung, M.S., Ives, N.A. and Stupian, G.W. (1991) *J. Appl. Phys.* **69**, 2044.
57. Yang, R., Naoi, K., Evans, D.F. *et al.* (1991) *Langmuir*, **7**, 556.
58. Lacaze, E., Garbarz, J., Quillet, V. *et al.* (1992) *Ultramicroscopy*, **42–44**, 1037.
59. Li, F.B. and Albey, W.J. (1992) *Electrochimica Acta*, **37**, 393.
60. Eigler, D.M. and Schweizer, E.K. (1990) *Nature*, **344**, 524.
61. See, for example, Stroscio, J.A. and Eigler, D.M. (1991) *Science*, **254**, 1319; Staufer, U. (1992) in *Scanning Tunnelling Microscopy II* (eds H.-J. Güntherodt and R. Wiesendanger) Springer, Berlin.
62. McCord, M.A. and Pease, R.F.W. (1988) *J. Vac. Sci. Technol. B*, **6**, 93.
63. Marrian, C.R.K., Dibisz, E.A. and Colton, R.J. (1991) *J. Vac. Sci. Technol. B*, **9**, 1367.
64. Heckl, W.M. and Smith, D.P.E. (1991) *J. Vac. Sci. Technol. B*, **9**, 1159.
65. Güthner, P., Glatz-Reichenbach, J. and Dransfeld, K. (1991) *J. Appl. Phys.* **69**, 7895.
66. Binnig, G., Gerber, Ch., Stoll, E. *et al.* (1987) *Europhys. Lett.* **3**, 1281.
67. Sarid, D. (1991) *Scanning Force Microscopy*, Oxford University Press, New York and Oxford.
68. Weisenhorn, A.L., Maivald, P., Butt, H.-J. and Hansma, P.K. (1992) *Phys. Rev. B*, **45**, 11226.
69. Martin, Y., Williams, C.C. and Wickramasinghe, H.K. (1987) *J. Appl. Phys.* **61**, 4723.
70. Albrecht, T.R., Akamine, S., Carver, T.E. and Quate, C.F. (1990) *J. Vac. Sci. Technol. A*, **8**, 3386.
71. Wolter, O., Bayer, Th. and Greschner, J. (1991) *J. Vac. Sci. Technol. B*, **9**, 1353.
72. Keller, D.J. and Chih-Chung, C. (1992) *Surf. Sci.* **268**, 333.
73. Marti, O., Ribi, H.O., Drake, B. *et al.* (1988) *Science*, **231**, 50.
74. Patil, R., Kim, S.J., Smith, E. *et al.* (1990) *Polymer Communications*, **31**, 455.
75. Geil, P.H. and Reneker, D.H. (1961) *J. Polym. Sci.* **51**, 569.
76. Snétivy, D. and Vansco, G.J. (1992) *Polymer*, **32**, 432.
77. Kovacs, A.J., Straupe, C. and Gonthier, A.J. (1977) *J. Polym. Sci.* (Polym. Symposium) **59**, 31.
78. Stocker, W., Bar, G., Kunz, M. *et al.* (1991) *Polymer Bulletin*, **26**, 215.
79. Eng, L.M., Fuchs, H., Jandt, K. and Petermann, J. (1992) *Helvetica Physica*, **65**, 71.
80. Magonov, S.N., Ovarnström, K., Elings, V. and Cantow, H.-J. (1991) *Polymer Bulletin*, 1991, **25**, 689.
81. Snétivy, D., Yang, H. and Vancso, G.J. (1992) *J. Materials Chemistry*, **2**, 891.
82. Wittmann, J.C. and Smith, P. (1991) *Nature*, **352**, 414.
83. Hansma, H., Motamedi, F., Smith, P. *et al.* (1992) *Polymer* **33**, 647.
84. Petermann, J. and Broza, G. (1987) *J. Mater. Sci.* **22**, 1108.
85. Eng, L.M., Fuchs, H., Jandt, K.D. and Petermann, J. (1992) *Ultramicroscopy*, **42–44**, 989.
86. Snétivy, D. and Vancso, G.J. (1992) *Macromolecules*, **25**, 3320.

87. Snétivy, D., Rutledge, G.C. and Vancso, G.J. (1992) *ACS Polymer Preprints*, **33**, 786.
88. Snétivy, D., Guillet, J.E. and Vancso, G.J. (1992) *Polymer*, **34**, 429.
89. Lotz, B., Wittmann, J.-C., Stocker, W. *et al.* (1991) *Polymer Bulletin*, **26**, 209.
90. Wittman, J.-C. and Lotz, B. (1990) *Progr. Polym, Sci.* **15**, 900.
91. Magonov, S.N., Bar, G., Cantow, H.-J. *et al.* (1991) *Polym. Bull.* **26**, 223.
92. Eng, L.M., Fuchs, H., Buchholz, S. and Rabe, J.P. (1992) *Ultramicroscopy*, **42–44**, 1059.
93. Magonov, S.N., Kempf, S., Kimmig, M. and Cantow, H.-J. (1991) *Polymer Bulletin*, **26**, 715.
94. Bickmann, B., Kempf, S., Stocker, W. *et al. Polym. Bulletin* (in press).
95. Stroup, E.W., Pungor, A., Lea, A.S. *et al.* (1992) *ACS Polymer Preprints*, **33**, 749.
96. Hamada, E. and Kaneko, R. (1992) *Ultramicroscopy*, **42–44**, 184.
97. Wickramasinghe, H.K. (1989) *Sci. Am.* (Oct), 74.
98. Hansma, P.K., Drake, B., Marti, O. *et al.* (1989) *Science*, **243**, 64.
99. Uozumi, K. and Yamamuro, K. (1989) *Jap. J. Appl. Phys.* **28**, L1297.
100. Weaver, J.M.R., Walpita, L.M. and Wickramasinghe, H.K. (1989) *Nature*, **342**, 783.
101. Williams, C.C., Hough, W.P. and Ishton, S.A. (1989) *Appl. Phys. Lett.* **55**, 203.
102. Williams, C.C. and Wickramasinghe, H.K. (1991) *J. Vac. Sci. Technol. B*, **9**, 537.
103. Sidles, J.A. (1992) *Phys. Rev. Lett*, **68**, 1124.
104. Reddick, R.C., Warmack, R.J. and Ferrell, C. (1989) *Phys. Rev. B*, **39**, 767.
105. Courjon, D., Sarayeddine, K. and Spajer, M. (1989) *Optics Communications*, **71**, 23.

3

X-ray studies using synchrotron radiation

G. Ungar

3.1 INTRODUCTION

X-ray diffraction research on polymers has benefited considerably from synchrotron radiation. The high brilliance of synchrotron X-ray radiation has opened new possibilities, such as fast time-resolved scattering, simultaneous dynamic diffraction and DSC, light scattering or mechanical experiments, studies of single molecular layers by X-ray reflectivity, etc. In addition, a number of advances in X-ray instrumentation and method of analysis, initiated at national synchrotron facilities, have subsequently spread to many individual off-line laboratories. In this chapter we review the information on instrumentation useful to the synchrotron user, and also give a selection of examples of polymer application. The reader is also referred to reviews by Elsner *et al.* [1], Russell (specific to SAXS) [2] and by Gehrke [3].

3.2 INSTRUMENTATION

3.2.1 Synchrotron radiation sources

The core of a synchrotron radiation facility [4] is the **storage ring** around which particles circulate at velocities v close to the speed of light; velocity of the 2 GeV electrons at SRS Daresbury is such that the electron mass is 4000 times that of the rest mass. The storage rings, located at different national research facilities, are several tens of metres in diameter. Charged particles with high energy, usually electrons, circulate within a high-vacuum tube with alternating straight and bent sections. Bending of the electron trajectory is accomplished by **bending magnets** (see Fig. 3.1, top). The 30 m diameter SRS storage ring at Daresbury has 16 dipole electromagnets, with a magnetic field $B = 1.2$ T, which bend the 2 GeV

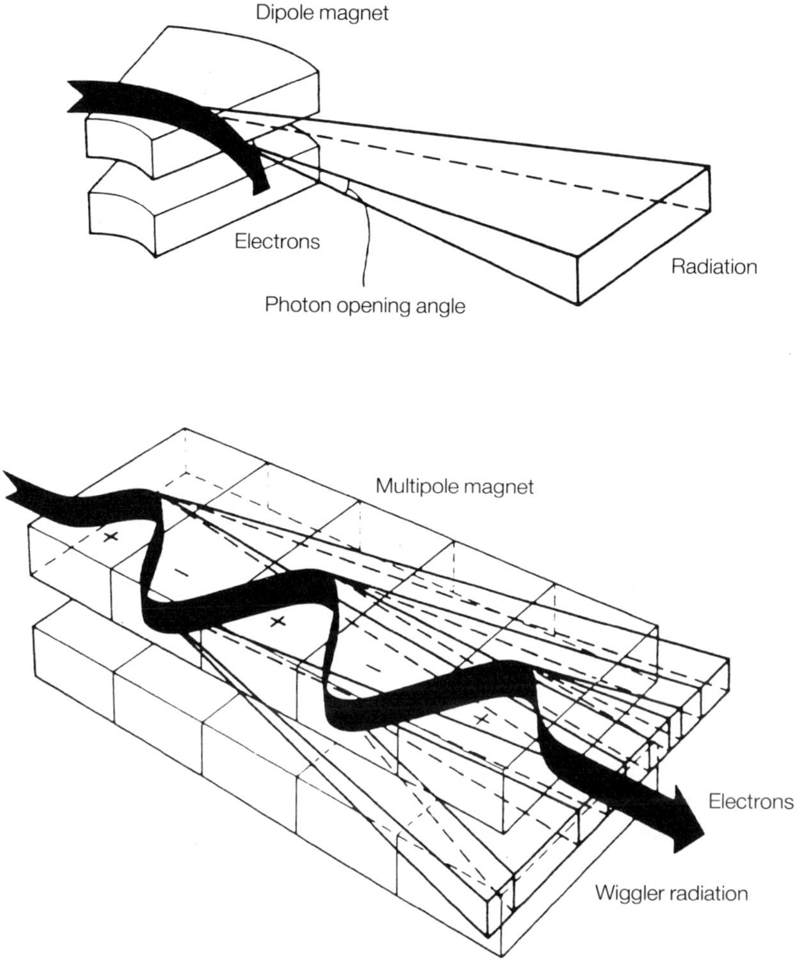

Fig. 3.1 Schematic view of an electron beam and synchrotron radiation emitted from a bending magnet (top) and a multipole wiggler (bottom). (From [4] by permission of Blackie and Sons.)

electron beam at a 5 m radius. As the electrons are deflected, electromagnetic radiation is emitted due to centripetal acceleration.

If the charged particles were low energy, the laws of classical electromagnetism would apply and the angular intensity spread of emitted radiation would describe a torus around the centripetal acceleration vector. Radiation of a single frequency would thus be emitted evenly both in the forward and backward direction, and both up and down. However, due to relativistic effects at high electron velocities, the radiation beam emitted from any point on the curved particle path is only confined to a narrow cone in the forward tangential direction. The **opening angle**, by which this self-collimation effect is measured, is inversely proportional to ν. For GeV electrons, such as those used in synchrotrons, the opening angle is typically 0.5 mrad (0.03°) or less. As a further consequence of relativistic effects, the radiation is white due to Doppler spread, with a high-energy cut-off.

Combining all the point sources on the curved section of the particle trajectory, each bending magnet produces a horizontally spread fan of synchrotron radiation which can be utilized by one or sometimes several beam-lines. The latter can be tens of metres long with several experimental stations along them. Synchrotron radiation spans the range from IR to hard X-ray; accordingly, the largest synchrotron user groups are the UV/VUV and X-ray research communities.

The electron (or positron) current is typically in the range between 50 and 400 mA, depending on the storage ring. The current gradually diminishes with time due to electron scattering on residual gas molecules. Thus the beam is periodically refilled, e.g. every 12 hours. Before being injected into the storage ring, electrons undergo a two-stage acceleration process, first in a **linear accelerator** and then in a circular **booster**. Final energy is reached in the storage ring itself. The electrons in a storage ring are confined to a thin beam of between several mm^2 and $0.1\,mm^2$, rendering radiation sources of high brilliance.

In contrast to the current, which gradually diminishes, particle energy lost through emitted radiation is continuously being replenished by application of radio frequency (rf) fields from klystron cavities disposed around the storage ring. One consequence of the application of rf field is that the particle current actually consists of discrete **bunches** spaced a few nanoseconds apart in the standard multi-bunch mode of operation. Single-bunch mode is sometimes employed, producing isolated short pulses of radiation of extremely high peak brilliance. These are primarily used for time-resolved spectroscopy experiments. The time-averaged current in the single-bunch mode is about an order of magnitude lower than in the multi-bunch mode, depending on the pulse separation. Although this may still provide a useful quasi-continuous X-ray source, significantly more intense than any rotating anode generator, single-bunch mode is rarely used for diffraction experiments. However, with the high source brilliance of the new synchrotrons and insertion devices, a recording of a whole X-ray diffraction pattern within a single bunch is becoming feasible. Thus, for example, it will be possible to condense $\sim 5 \times 10^{10}$ photons of white radiation in a single 50 ps bunch onto a small $0.2 \times 0.2\,mm^2$ specimen for Laue diffraction [5].

Beside bending magnets, synchrotron radiation is produced by the already mentioned **insertion devices**, of which the main types are wigglers and undulators. These are multipole magnets, inserted into the straight sections of the storage ring. They impose horizontal oscillation on the electron beam without causing it to change net direction. **Wigglers** are constructed in such a way that the radiation beams, emanating from each sharp bend caused by the high-field multipole magnet, are collinear; thus their intensities add up (Fig. 3.1 bottom). Wigglers can produce harder X-rays than bending magnets, and the wavelength distribution can be shifted, or tuned, by changing the magnetic field. A three-pole wiggler is sometimes called a wavelength shifter. The second SRS wiggler, for example, has a central superconducting magnet giving a 6 T field, and two return field of -1.2 T [6]. Current multipole wigglers consist of 20–30 lower field electromagnets, and are the best sources of intense white radiation. **Undulators** work on a similar principle; however, the electron beam is made to oscillate only within the opening angle of the radiation beam, so that permanent magnets suffice. Interference between electrons and the emitted beam causes the white radiation to split into a series of spectral bands, and the beam intensity within certain wavelength intervals $\Delta\lambda$ becomes very high indeed.

For a given magnetic field, the high-energy cut-off of synchrotron radiation, E_c, is proportional to the square of the energy of the electrons. It is thus clear that high electron energies are required in order to obtain both sufficiently high energy and efficient collimation of synchrotron radiation.

The most useful quantity for comparing intensities of different sources of synchrotron radiation is **spectral brilliance**, defined as the number of photons per second emitted into a mrad^2 solid angle through a 1 mm^2 cross-section within a 0.1% band-pass ($\Delta\lambda/\lambda = 0.001$). Fig. 3.2(a) compares wavelength dependencies of spectral brilliance (log scale!) for bending magnets (BM), wavelength shifters and multipole wigglers on the existing 2 GeV SRS storage ring and at the projected 3 GeV DIAMOND synchrotron. For further comparison, Fig. 3.2(b) shows the equivalent curve for a 34-pole undulator on beam-line 4 of the 6 GeV ESRF synchrotron at Grenoble [7]. For diffraction experiments that require the highest possible monochromatic X-ray flux compressed into a narrow beam, and do not need particularly short wavelengths, an undulator operating with high-energy electrons is the best radiation source available to date.

When making comparisons between X-ray intensities produced by synchrotrons and other sources, e.g. rotating anode tube generators, allowance has to be made for two characteristics of the latter. (1) Although the total amount of X-ray energy emitted by a tube is considerable, this is spread nearly isotropically; as will be shown in the next section, even with focusing optics only a small fraction can actually be utilized. (2) The bandwidth of characteristic emission lines from a metal target is in the

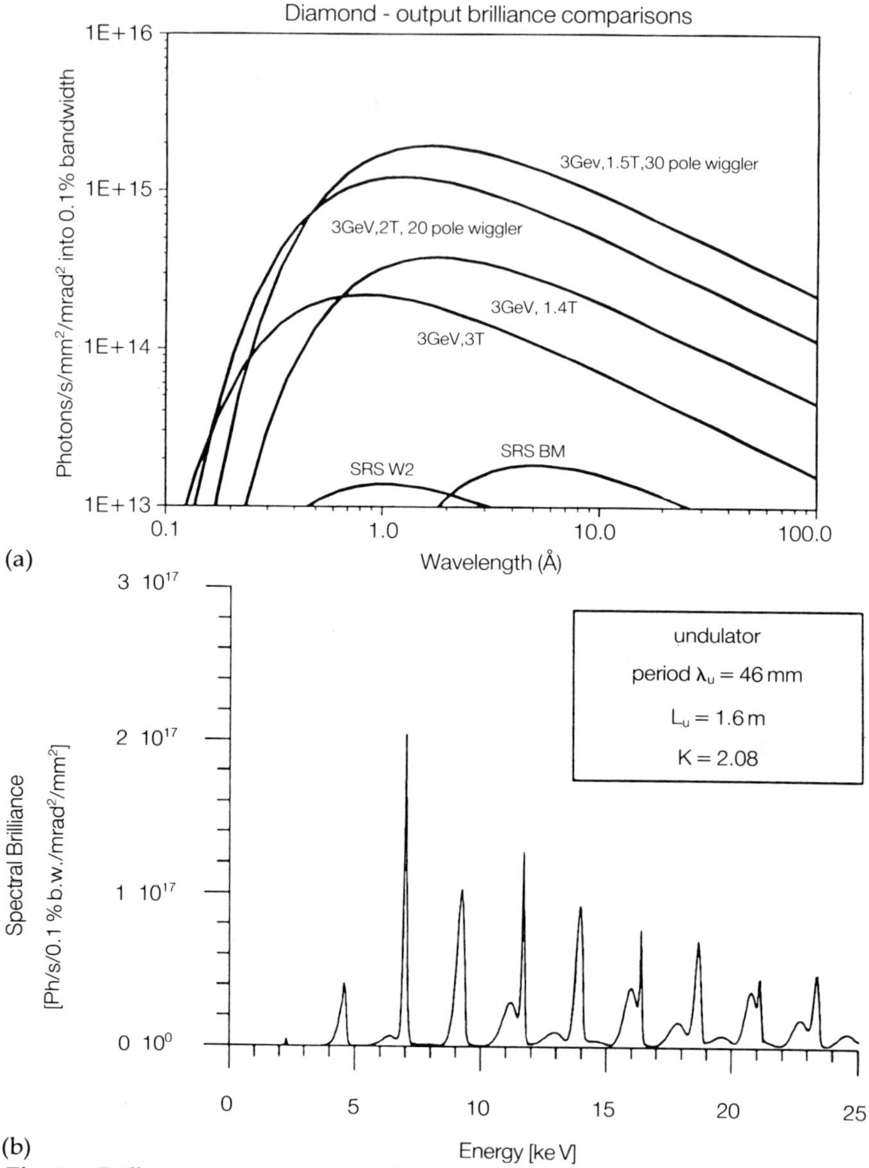

Fig. 3.2 Brilliance versus wavelength of synchrotron radiation produced by (a) a bending magnet (BM) and a wavelength shifter (W2) on the existing 2 GeV SRS storage ring at Daresbury, and by bending magnets and multipole wigglers on its projected 3 GeV replacement DIAMOND; (b) undulator on beam-line 4 of the 6 GeV ESRF synchrotron at Grenoble [7]. For comparison of abscissa scales note that the 1.54 Å wavelength (CuKα) corresponds to a photon energy of 8.26 keV. Reproduced from Bösecke, P., *European Synchrotron Radiation Facility User Meeting Information*, Grenoble, 1991.

range $\Delta\lambda/\lambda = 5\times10^{-4}$. For some experiments such a narrow band-pass may not be necessary, and while with the white synchrotron radiation increasing $\Delta\lambda$ means an increase in intensity, this does not apply to characteristic radiation. On the other hand, the white Bremsstrahlung from a tube has significantly lower brilliance. Quantitative comparisons [8] have shown brilliance of synchrotron radiation produced by bending magnets to be at least three orders of magnitude higher than that of characteristic radiation from a rotating anode. With new highly focused particle beams and the use of insertion devices (see later) this ratio is being further increased by several orders.

Table 3.1 Research synchrotron radiation facilities suitable for X-ray diffraction

Name	*Location*	*Electron energy (GeV)*	*Typical beam current (mA)*
SRS	Daresbury, UK	2	300
DCI	Orsay, France	1.8	300
NSLS X-ray	Brookhaven, USA	2.5	500
DORIS II	Hamburg, FRG	3.7–5.3	100
SPEAR	Stanford, USA	3.5	100
ALS	Lawrence Berkley, USA	1.5–1.9	400
ADONE	Frascati, Italy	1.5	100
ALADDIN	Wisconsin, USA	1	100
Photon Factory	Tsukuba, Japan	2.5	200
ESRF	Grenoble, France	6	100
APS	Argonne, USA	7	200
CHESS	Ithaca, USA	8	100
ELETTRA	Trieste, Italy	1.5–2	200
VEPP-3	Novosibirsk, Russia	2.2	100
BEPC	Beijing, China	2.6	100
VEPP-4	Novosibirsk, Russia	7	10

The subject of the present review is work carried out at research synchrotrons. A current list of these national facilities, with large-diameter storage rings and producing X-rays in the Ångstrom region, is given in Table 3.1. There are other synchrotrons, some generating lower and some much higher energy radiation. Also, in recent years compact synchrotrons have become available, designed for use in X-ray lithography. Their small size has been achieved through the use of superconducting bending magnets allowing a 180° bend with a radius of only 0.5 m [9].

3.2.2 X-ray optics

The radiation beam emerging, e.g. from an insertion device, although of narrow divergence, will have broadened somewhat before reaching the specimen or detector, as experimental hutches are located tens of metres down-beam. Thus focusing is required, as well as monochromatization. The latter is omitted if white radiation is to be used, i.e. for Laue diffraction or energy dispersive diffractometry. Otherwise monochromatization is accomplished by some form of crystal monochromator. Filters do not give a sufficiently narrow band-pass, but can be used as pre-monochromators to reduce the heat load of the full-spectrum beam on a more sensitive optical element. Multilayers may also be used for partial monochromatization (see later). Most stringent demands on X-ray optics are imposed by small-angle X-ray scattering (SAXS); it is essential to achieve small beam dimensions while maintaining high photon flux and low background scatter.

Traditionally, with sealed tube X-ray generators, slit rather than point collimation has often been used for SAXS in order to maximize the beam intensity. The popular Kratky camera [10] produces a beam of line cross-section (thin wide beam) with low background scatter. The disadvantage of this geometry is that it does not allow recording of two-dimensional SAXS patterns, and that it requires numerical 'de-smearing' of the scattering curve [11, 12] to correct for the beam width. However, with the intrinsically collimated synchrotron radiation of high brilliance, point focus geometry is mostly used instead. A modified Kratky camera, in conjunction with a linear position-sensitive detector, is in use on beamline X3A2 at NSLS [13]; however the beam dimensions are small enough for smearing effects to be negligible.

In order to obtain a **point focus**, i.e. the image of the synchrotron radiation source in the detector plane, both vertical and horizontal focusing is desirable. In the case of conventional X-ray tube sources double focusing is often achieved by two bent mirrors rotated by 90° (Franks optics) [14, 15]. However, since this arrangement does not provide effective monochromatization of white synchrotron radiation, one of the mirrors is replaced by a bent crystal monochromator (Huxley–Holmes optics) [16]. The widespread use of the Huxley–Holmes arrangement is also partly due to the convenient rejection by the mirror of high-energy photons, which would otherwise give rise to transmission of harmonics by the crystal monochromator. A typical optical arrangement for low-angle X-ray scattering of synchrotron radiation, using a one-dimensional position-sensitive detector, is shown in Fig. 3.3.

(a) Mirrors

X-ray mirrors work on the principle of external total reflection [17], which is possible because the refractive index n inside matter is somewhat less

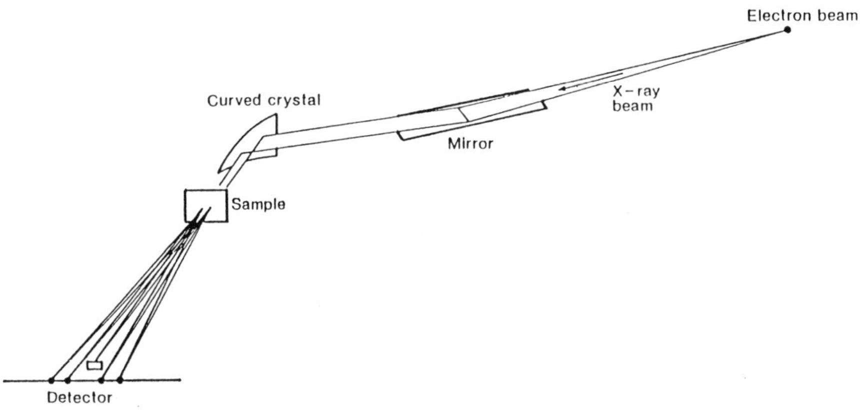

Fig. 3.3 Schematic diagram of the optical arrangement used for SAXS. A cylindrically bent mirror focuses the beam vertically and a bent triangular crystal provides horizontal focusing and monochromatization. A series of shutters and adjustable slits (not shown) is also present. Almost all of the beam-path is enclosed in a vacuum.

than 1 for X-rays. According to the Drude–Lorentz theory of dispersion, for high-frequency radiation the real part of n is given by:

$$n = 1 - \delta = 1 - \sum_r \frac{N_r e^2}{2\pi m_e(\nu^2 - \nu_r^2)} \tag{3.1}$$

Here, N_r represents the number of electrons per unit volume which vibrate with the natural frequency ν_r, e and m_e are electron charge and mass, and $\nu = c/\lambda$ is the frequency of incident radiation. Summing over all the different electrons and rearranging, one obtains for δ:

$$\delta = \frac{\lambda^2}{2\pi} r_e \rho_{el} \tag{3.2}$$

Here r_e is the electron radius and ρ_{el} is electron density. At grazing angles below a critical angle θ_c the beam is totally externally reflected, and, the critical angle below which total reflection occurs is given by

$$\sin\theta_c = \sqrt{2\delta} \tag{3.3}$$

Since for 1.5 Å X-rays the refractive index decrement δ is of the order of only 10^{-6}, θ_c is of the order of only 1° even for heavy elements.

If the mirror is elliptically bent, X-rays can be focused into a line (horizontal line in the case of vertical focusing as in Fig. 3.3). Accurate elastic bending of mirrors can be achieved (see later). X-ray mirrors are usually made of optically flat glass on which a thin film of heavy metal is deposited (gold, platinum, rhodium; nickel for CuKα radiation from an

X-ray tube). The high density metal coating serves to increase θ_c, which allows the mirror to be shorter while still maintaining a sufficient acceptance angle to capture the incident beam. Even so, being sometimes 30 metres away from the source down a divergent beam, mirror lengths are of the order of a metre. For comparison, in a conventional laboratory diffraction set-up the mirror can be close to the X-ray tube; hence it can capture a significant fraction of the useful X-ray beam while being only a few centimetres long.

Being usually the first optical element in the beam-path, a mirror is exposed to intense heat from the broad spectrum radiation, and it needs to be cooled. Liquid gallium is used at Daresbury while cryogenic coolants are employed at ESRF. Coarse pre-monochromatization is sometimes applied, as already mentioned. Although all X-ray optics for the synchrotron beam are kept under vacuum, chemical degradation under the action of heat and extensive irradiation limits the life of the optically flat mirror surface.

Some of the recent developments in X-ray optics include focusing by capillary arrays [18] and production of intense microbeams by single capillaries.

(b) Focusing monochromators

Crystal monochromators [19] work on the principle of Braggs law

$$n\lambda = 2d\sin\theta \qquad (3.4)$$

By changing the inclination angle θ of the diffracting planes with respect to the incident beam, different wavelengths are selected by the diffracted beam. Single crystals with strongly diffracting planes are used for monochromatization. The most common crystals presently used at synchrotrons are silicon and germanium, mainly utilizing the (111), and sometimes (220) planes. Beside technical factors affecting availability of large flat dislocation-free single crystals, as well as their mechanical, thermal and radiation stability, two important diffraction parameters of monochromator crystals are peak diffracting power (reflectivity) and Darwin width. **Darwin width** ω_0 is the intrinsic width of a Bragg reflection from a perfect single crystal and is small but finite. From Darwin's dynamical diffraction theory:

$$\omega_0 = \frac{4r_e d^2}{\pi V_c} PF\tan\theta \qquad (3.5)$$

where r_e is the classical electron radius, d is the interplanar spacing, V_c the unit cell volume, and P and F the polarization and structure factors, respectively. Thus, for a hypothetical perfectly parallel beam coming from an infinitesimal source, ω_0 determines the wavelength dispersion through the simple relationship

$$\Delta\lambda/\lambda = \omega_0/\tan\theta \tag{3.6}$$

In practice the optimum balance between acceptable spectral bandwidth, source size and diffracted beam intensity is sought for a given type of experiment by choosing the type of crystal and the particular diffraction plane. Thus, for example, the band-pass $\Delta\lambda/\lambda$ for Si(220) is 5.6×10^{-5}, for Si(111) 1.4×10^{-4}, and for Ge(111) 3.3×10^{-4}. Since for normal diffraction experiments on polymers the latter value is adequate, Ge(111) crystals are usually employed as monochromators. Since the full band-pass is utilized, the integrated diffracted intensity is the relevant reflectivity parameter. This is comparatively high for Ge(111), partly due to the relatively large ω_0. For high-resolution diffraction work Si(220) is more appropriate, although at a cost of reduced intensity.

Focusing in the horizontal direction is achieved by bending the monochromator crystal (cf. Fig. 3.3). **Bent monochromators** are standard on beam-lines dedicated for diffraction work, where the wavelength remains more or less fixed – note that changing λ requires rotation of the entire station on a 2θ arm around the monochromator axis. Provision for such rotation normally exists.

Bent crystal monochromators are based on the parafocusing principle of the Rowland circle, commonly used in reflection two-circle diffractometry. The crystal cut can be one of the three types: the Johann [20], the Johansson [21] and the asymmetric. With synchrotron beams, where the distance from the source to the crystal may be several tens of metres, asymmetrically cut crystals are appropriate. At the same time demagnification of the source is achieved with good focusing, although angular beam dispersion, and hence the band-pass, are increased.

For comparison, we note that asymmetrically cut bent crystal monochromators are often used with conventional X-ray generators in Guinier cameras with forward-scattering sample transmission geometry. In those cases, however, the crystal is turned around to give magnification of the source, in order that the necessarily large crystal-to-film focal length can be achieved while keeping the crystal close to the source.

The angle of incidence θ_i for a crystal monochromator is considerably larger than that for a mirror. The bent monochromator is therefore oriented so as to handle the larger horizontal divergence of the synchrotron beam, while the mirror is made to focus the beam vertically (Fig. 3.3).

Two methods are mainly used for cylindrical bending of optical flats (mirrors and monochromators). In the four-point bend method a pair of forces is applied at each end by means of four rollers in total. At synchrotron sources this method is mainly used for mirrors, while it is used for both mirrors and crystal monochromators on smaller scale laboratory equipment. Regarding the second method, it has been established that cylindrical bending of crystal monochromators is achieved if these are cut in the shape of an isosceles triangle (Fig. 3.3), with the base held fixed and a force exerted on the apex; this technique is now commonly used for large monochromators on synchrotron beam-lines.

(c) Nonfocusing monochromators

Variable wavelength beam-lines use **double crystal monochromators** [22]. The beam is diffracted twice, from two parallel crystals. In the process it is first deflected by 2θ and then by -2θ; thus the exit beam is parallel, though shifted, with respect to the incident beam. In order to select the exit beam wavelength both crystals are simultaneously rotated. Good spectral resolution is achieved since resolution after n Bragg reflections is that for a single reflection raised to the n-th power [23]. Incidentally, the same experimental arrangement, when used with monochromatic incident radiation, yields the rocking curve of the crystals.

The flat double monochromator provides no focusing, and the beam remains divergent on exit. For this reason vertical rather than horizontal diffraction is chosen for double monochromators on a synchrotron line, horizontal focusing being accomplished by a bent mirror. Alternatively, the beam is focused by two orthogonal mirrors additional to the monochromator (e.g. Microfocus Line at ESRF).

Most polymer studies are performed with X-ray wavelengths around 1.5 Å; this energy region usually provides a good balance between angular resolution and sample transmittance. A need for varying the wavelength may arise in polymer diffraction studies where a high-pressure cell or highly absorbing solvents are used (e.g. polyphosphoric acid), or where anomalous dispersion may be of interest (e.g. organo-metallic polymers or polymer–salt complexes).

For highest resolution, allowing small-angle scattering studies of structures on a micron scale, the appropriate diffractometer design is that by **Bonse and Hart** [24, 25]. Its key elements are two multiple reflection **channel-cut monochromators**, one in the incident beam and the other in the rotatable detector arm (Fig. 3.4). Each monochromator is a large, almost dislocation-free single crystal of silicon or germanium with a groove cut in it. This is done so as to allow the incident beam to be diffracted n (e.g. 5) times from e.g. the 220 planes before exiting the groove. In the process the single-crystal diffraction profile is raised to the n-th power, which practically eliminates the tails of the rocking curve. In other words, the beam exiting the groove is made almost perfectly parallel

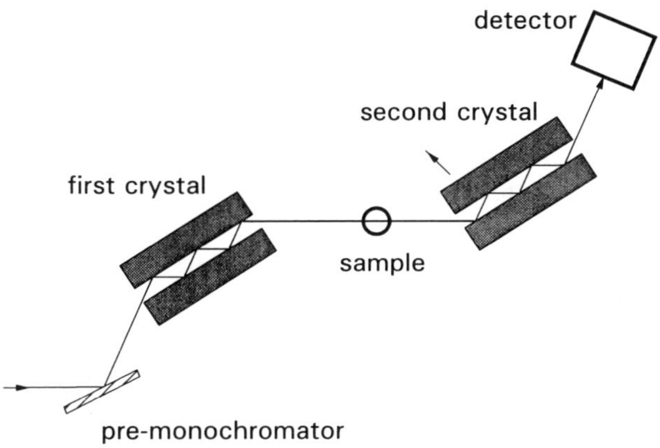

Fig. 3.4 Schematic diagram of the Bonse–Hart diffractometer for very small-angle scattering, containing two channel-cut crystal monochromators. The second crystal and the detector are located on a moving 2Θ arm.

by rejection of stray rays. As the walls of the groove belong to the same single crystal, identical orientation of diffracting planes is guaranteed.

It should be emphasised that it is the low-angular divergence rather than small-beam cross-section which is crucial for the operation of the Bonse–Hart diffractometer. The beam is in fact not particularly narrow and thus reasonable intensity is maintained. The analyser crystal ensures that only those photons diffracted within a narrow angular range are accepted, even though they might originate from widely separated areas of the specimen. Fig. 3.4 shows the arrangement with multiple diffraction in only one dimension; a design also exists with additional double-diffraction collimation in the other direction [23,25]. Bonse–Hart diffractometers have been constructed on synchrotron beam-lines at Daresbury, Hamburg and Stanford. A commercial version is now available for use with rotating anode generators [26].

In order to obtain a parallel tail-less monochromatic primary beam a single channel-cut crystal can be employed. However, if an analyser crystal is not used, then in order to obtain both good resolution and intensity the beam has to be compressed into a smaller cross-section. This can be achieved by cutting a tapered rather than parallel groove in the monochromator crystal [27]. The optical principle of such a **condensing-collimating channel cut** (CCCC) monochromator is shown in Fig. 3.20. Note that beam compression in CCCC monochromators is based on the long established principle of diffraction from asymmetrically cut crystal surfaces [28]. Interest in CCCC monochromators exists in the synchrotron community as a beam condensing alternative to focusing monochromators; due to single diffraction and optical abberations from imperfect geometry, the latter produce beam profiles with extended tails, unsuitable for high-resolution diffraction work using position-sensitive detectors. In principle CCCC monochromators can also be used with laboratory X-ray sources. However, high intensity at the exit of the monochromator can only be achieved if the incoming beam is already pre-collimated, as is the case with synchrotron radiation.

Within the context of non-focusing monochromators, it should be mentioned that transmission **Laue diffraction** on a perfect, thermally resistant single crystal such as diamond is being considered. Selection of wavelength is achieved by rotating the crystal, while compensatory counter-deflection of the beam, including focusing, can be achieved by a bent multilayer (see later). Such an arrangement is being tested at ESRF.

(d) Multilayers

Multilayers are recent additions to the limited range of X-ray optical elements. They consist of a stack of alternating layers of high and low electron density, such as Si and W, or C and W. The deposited layers are optically flat and the layer spacing is in the range 20 to 40 Å. Thus multilayers are artificial 'smectics', and their rocking curve width depends on the number of layers, which can be in the range of hundreds. Their acceptance angle is larger than that of mirrors, since they operate on the principle of Bragg diffraction. However, maximum θ is limited by the minimum layer spacing; multilayers with too small spacings are unstable due to interphase diffusion. Multilayers may potentially be used as pre-monochromators on synchrotron beam-lines, as well as in X-ray microscopy and lithography. They are also worth considering as mirror replacements for mirror-monochromator (Huxley–Holmes) SAXS cameras attached to conventional X-ray sources, when high intensity is more important than high resolution [29]. Multilayers are also being used in some time-resolved synchrotron experiments, e.g. involving diffuse scattering, where large bandwidth can be tolerated. Thus allowing energy resolution to drop to 10^{-2} provided an intensity gain of almost two orders of magnitude over the normal crystal monochromator set-up on beam-line X-20C at NSLS. This enabled recording of scattering patterns in milliseconds [30]. Multilayers are now available commercially.

(e) Optics for energy-dispersive diffraction

In energy-dispersive powder diffraction the specimen is exposed to white X-ray radiation while the detector receiving angle 2θ is kept fixed. The diffraction pattern is then reconstructed from the energy spectrum. The wide spectrum radiation is provided by a wiggler or wavelength shifter. For a review of the technique the reader is referred to Reference [31].

Energy-dispersive diffraction has an advantage over angle-dispersive diffraction for certain experiments such as those using high pressure and/or some temperature cells. The exit window of the cell can be small, since only a fixed narrow beam of scattered radiation is recorded. In this context, the energy-dispersive method is particularly useful for diffraction at large **q**-vectors (small *d*-spacings); hence it is less applicable to studies of polymers than to ceramics or metals.

Resolution Δd increases at higher X-ray energies E, since

$$\frac{\Delta d}{d} = \frac{\Delta E}{E} \tag{3.7}$$

Thus high-energy X-rays are used which, in turn, requires a lower diffraction angle. Experimental resolution in energy-dispersive diffraction relies critically on a high degree of collimation, both of the incident as well as the diffracted beam. Collimation of the latter is achieved at SRS by the beam passing between two nearly optically flat plates of molybdenum, 50 cm long and typically 0.1 mm apart [32]. A solid-state germanium or silicon detector is used.

3.2.3 Detectors

(a) General

Photographic film is the simplest and oldest way of recording X-ray diffraction patterns and is often used with synchrotron radiation. On the positive side are the capacity for recording two-dimensional diffraction patterns and high spatial resolution ($\geq 1\,\mu m$, depending on grain size). Among the numerous disadvantages are cumbersome film processing, the need for subsequent measurement of optical densities if intensities are to be quantified, low dynamic range (10^2–10^3), low signal-to-noise, non-linearity and lack of time resolution. The latter handicap can be partially redressed by using carousel-type mechanical film changers. Alternatively, streak cameras may be used, such as the Guinier–Lenné temperature-scanning powder camera, which have the film move continuously along one coordinate axis, diffraction angle being the other; this results in a two-dimensional record of time (or temperature) evolution of the powder pattern. Polaroid film is often useful for quick checking of sample or beam-stop alignment, or for observing the general appearance of a diffraction pattern when only a one-dimensional detector is used.

Most diffraction work at synchrotrons uses photon counting detectors. These are mainly **position-sensitive** (PSD), meaning that information is provided not only on the number and, possibly energy range, but also on the position of incident photons. A linear PSD gives only one position coordinate (i.e. scattering angle) along a narrow window, a quadrant PSD gives the radial coordinate on a circular segment, and an area detector gives two cartesian coordinates on a square or circular window. Position-sensitive detectors eliminate the need for angular scanning, thus increasing the recording speed and enabling the advantages of high synchrotron beam flux to be exploited in time-resolved experiments. However, angular scanning detector mode is still indispensable for high-resolution work, where the moving detector arm contains collimating slits or monochromators.

X-ray detectors have been discussed in some detail by Hendrix (general) [33], Russell (general) [2], Gruner (CCD and Vidicon) [34], Amemiya *et al.* (image plates) [35], Hall *et al.* (multiwire and IP) [36] and Cernik *et al.* (CCD and IP) [37].

There are **single photon counting** and **integrating** detectors. The former, e.g. proportional chambers, count and electronically process information on individual photons as they arrive, while in integrating detectors, e.g. CCD cameras or image plates, counts are accumulated and read subsequently. Only the former are suitable for true time-resolved studies, although read-out from CCD and Vidicon cameras can take place at TV rates, which is adequate for most studies. In contrast to integrating detectors, single photon counting detectors have the potential of energy resolution. However, they are more likely to have a lower count rate limit, due to the time it takes to electronically process and store individual photon signals.

Maximum detector counting rate is currently the limiting time factor for many synchrotron experiments, rather than limited beam flux. Attenuators are often used to cover all or part of the detector window, where scattered intensity is high. It must be also born in mind that, since the synchrotron beam is pulsed, the detector must be able to handle count rates an order of magnitude higher than those calculated on the basis of average continuous intensity.

(b) Gas-filled detectors

The simplest detector of this type is the **ionization chamber**. It consists of a noble gas filled chamber with two plate electrodes across which a moderate voltage is applied (several hundred V cm^{-1}). As X-ray photons are absorbed by the gas, photoelectrons are ejected with excess energy $E = h\nu - E_0$, E_0 being the ionization potential of the gas, e.g. 25 eV for argon. Photoelectrons produce further cations and electrons by inelastic collision of photons with gas, until their excess energy is used up. An 8 keV (1.5Å) photon thus produces about 300 ion-electron pairs, which separate and drift to the electrodes. Depending on the exact capacitance of the chamber, this drift results in an electrical signal in the microvolt range. Thus sensitivity of ionization chambers is low but the maximum count rates are high. Open ionization chambers, with air as the absorbing gas, are often used to monitor the intensity of traversing incoming or transmitted primary beam in synchrotron experiments.

Much higher gain and thus higher quantum efficiency of the detector is obtained as the voltage between electrodes is increased to several kV cm^{-1} in the **proportional chamber**. The electrons produced by inelastic collisions are accelerated sufficiently to cause further ionizations, resulting eventually in an **avalanche** following each photon absorption. The multiplication factor, given as the ratio between n_0, the number of primary electrons and n, the total number of electrons z centimetres downfield, is

$$\frac{n}{n_0} = \exp(\alpha z) \tag{3.8}$$

where α is the number of ion pairs formed per cm at the normal gas pressure.

The maximum applied voltage and, consequently maximum gain, is limited by the discharge threshold. Because of the gas amplification effect, proportional detectors can count single photons. They also provide limited energy resolution, since the number of ionizations, and hence the amplitude of the electrical pulse, is roughly proportional to the photon energy.

Positionally non-resolving proportional detectors, together with scintillation counters, are commonly used in scanning angle-dispersive diffractometers, including those on synchrotron beam-lines.

(c) One-dimensional proportional detectors

Proportional detectors can be made **position-sensitive**; at present one- and two-dimensional proportional PSDs are probably the most widely used detectors for polymer diffraction studies at synchrotrons. In many laboratories they are also employed with conventional sources. First the principle of operation of a one-dimensional (linear) PSD will be discussed, which will be followed by an extension to quadrant and two-dimensional detectors.

As with all proportional chambers, a linear PSD is essentially a capacitor: the electric signal on the electrodes arises as the charges, created by the photon, migrate within the electric field. Behind the entrance window and immersed in a highly absorbing noble gas in most linear PSDs is an anode wire or wire screen, with a series of connected metal strips as the cathode further behind. The electric field is high around the thin anode wires, but relatively low elsewhere. The photoelectron and the secondary electrons move toward the anode until they reach the narrow high-field region where they create the avalanche. The negative charge is quickly

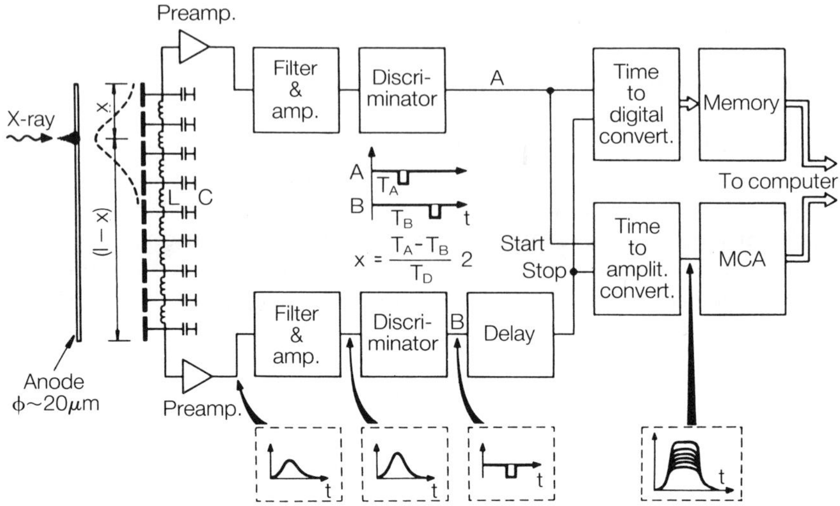

Fig. 3.5 Operational principle of a delay-line position-sensitive detector. (From [33], courtesy of Springer–Verlag.)

collected by the anode and, due to the capacitor effect, a signal is induced over several cathode strips with the centre directly opposite the location of the avalanche (Fig. 3.5).

Let l be the active length of the detector and x the distance from one of its ends to the avalanche. There are two principal ways of extracting the location x:

(a) by recording the **time difference** between the pulses travelling the distances x and $l–x$ and arriving at either end of the electrode (in Fig. 3.5 the cathode), and
(b) by recording the **charge division** between x and $l–x$ portions of the electrode.

Currently the most popular PSD system based on time difference between incoming pulses is the **delay line** detector [38, 39]. The cathode strips are connected via inductance-capacitance links (see Fig. 3.5), the speed of signal propagation being determined by the LC time constant. The signals from each end of the delay line are amplified, amplitude discriminated and additionally delayed by a fixed interval in order to make all delays positive. The time difference $T_A–T_B$ is then either passed through a time-to-amplitude converter (TAC) and digitized, or digitized directly. The encoded events are stored in a multichannel analyser (MCA) or histogramming memory. Direct time-to-digital conversion allows current delay line detectors to operate at around 1 MHz. While the processing time of a TAC–MCA combination used to be considerably longer than 1 μs, this is no longer the case.

Spatial resolution of all gas detectors is rather limited, typically not better than 200 μm for detectors with high count rate capability for use with synchrotron radiation. The limit is determined by a number of factors, the obvious one being the speed of time digitizer relative to the propagation time along the delay line. On the other hand, the delay line cannot be made too slow, as this would cause long dead time and limit the maximum count rate, a major consideration with detectors for synchrotron applications. Other critical parameters affecting spatial resolution are pre-amplifier noise, delay line impedance, and the total charge generated per event. The latter is in turn limited by maximum voltage before gas discharge, which is affected by gas purity and pressure and the design of the electrodes.

The alternative PSD design, also using the time delay concept (a) is based on **rise time** principle or RC delay [40]. Here a single resistive wire serves as the anode and itself collects and transmits the signal. The cathode is the detector housing. The wire is either carbon coated quartz or thin steel, and the speed of signal propagation towards its two ends is determined by its RC time constant. As with the detector speed, spatial resolution is also inversely proportional to the product RC. Thus, here again a compromise between count rate and resolution must be found. The main disadvantage of this detector is the critical dependence of detector performance on the thin anode wire, which is sensitive to mechanical damage and discharge corrosion.

The second principle (b) of locating the photon striking a position-sensitive detector is recording the division of the charge between x and l–x portions of the electrode. One way of realizing this is by having the cathode strips mutually connected not via LC links in a delay line, but via resistors in a transmission line. The charges accumulated at each end of the wire will be inversely proportional to the resistance, and hence distance, from the site of the avalanche. Since small charges have to be measured accurately, sophisticated and thus slow electronic processing is required, so that this method is not often used at synchrotrons.

An ingeniously simple alternative, using the same charge division concept, is the so called **backgammon** principle. Instead of being made up of lateral strips, the cathode consists of an even number of longitudinal intermeshed elongated triangles, stretching alternatively from end A to end B and vice versa, as on a backgammon board. The cloud of positive ions created by an avalanche deposits over several triangles, and since the width of a triangle is linearly related to the distance from its base, the ratio of total charges on two adjacent antiparallel triangles is equal to the ratio of the distances of the avalanche from detector ends.

Thus

$$\frac{Q_A}{Q_B}=\frac{x}{l-x} \tag{3.9}$$

The advantage of backgammon detectors is good stability and sturdiness of the hardware which can be produced with high mechanical accuracy. Otherwise accurate charge measurement is also required with this detector, which limits its count-rate capability and hence its use with synchrotron radiation.

(d) Two-dimensional proportional detectors

Gas-filled **two-dimensional** or **area** detectors work on the same principle as linear proportional PSDs. However, two sets of read-out cathodes are required, one being a plane of wires or strips aligned along X and the other along the Y direction. A plane of thin parallel anode wires is interposed half-way between the two cathode planes. The negative charge on the anode induces positive charges in both cathode planes. The cathodes in each plane are connected to their respective delay line or charge division resistance chain, providing, in turn, x and y coordinates of the avalanche. Beside this three-plane (cathode–anode–cathode) array, two-plane arrangements are also possible, with anodes as read-out electrodes.

Early multiwire area detectors suffered from spatial-resolution limitation in the direction perpendicular to the anode wires. This was because the design of the detector, as described earlier, makes it essentially a composite set of parallel one-dimensional detectors, with the high electric field having cylindrical symmetry around each anode wire. In such an arrangement electrons are diverted towards the anode wires and hence avalanches always occur at the wires themselves. This presents no problem for one-dimensional detectors, where only the positional coordinate along the anode is of interest. However, in area detectors of the type described, the coordinate scale perpendicular to the wires becomes quantized, with the period determined by anode wire spacing. At the time this spacing in practice could not be made less than 1 mm.

Modern area detectors use a screen of thin (e.g. ϕ 50 μm) closely spaced (e.g. 250 μm) wires instead of the widely spaced anodes. This results in a virtually homogeneous electric field, with negligible deflection of electron trajectory and a resulting detector resolution of *c.* $0.2 \times 0.2\,\text{mm}^2$, or 0.1% of the full detector linear length.

Continuous effort is being devoted to increasing the detector **count-rate capacity**. This is often the most important consideration for application with synchrotron radiation, particularly for fast time-resolved experiments on new high-flux sources such as ESRF or APS. Three factors determine maximum count rate: (a) the speed at which signals travel to detector ends, e.g. along the delay line, and are electronically processed, (b) the speed at which large volumes of data can be stored, and (c) the space charge effect limiting high **local** count rates.

Among the means of reducing detector dead time is segmentation, which shortens the signal path length of a delay line. Furthermore, two instead of one time-amplitude converter (TAC) can operate on one delay line in a time-sharing mode. This can reduce the dead time additional to that of the delay line itself, from 70 to 10 ns [41]. The current generation of 2-D proportional detectors at Daresbury achieve a maximum rate of 1.5 MHz for a detector with 4096×4096 pixels. Time resolution in the microsecond range is already realizable, which demands a fast high-volume data processing capability. The detector has 1 Gword of memory.

A detector may nominally be capable of handling a given overall count rate only if the photons are spread relatively evenly over the active area. In regions of intense diffraction peaks, or of the central scatter near the beam-stop, high local concentration of positive ions is generated since their drift velocity to the electrode is about 1000 times lower than that of electrons. This disparity may result in a significant positive **space charge** which distorts the local electric field and diverts incoming electrons. The consequence is a distorted diffraction profile. The problem is alleviated somewhat by reducing the gap between the anode and the cathode, but too small a gap leads to electric discharge.

In order to improve photon efficiency, negative potential is applied to a conductive inner surface layer of the entrance window. The **drift space** created between this screen and the front zero-potential cathode, with its uniform electric field, directs the primary electrons into the cathode–anode acceleration gap. Since most photons are absorbed in the drift space, efficiency and definition of the detector are improved.

The gases most often used in proportional detectors are argon or xenon, due to their high linear absorption coefficients for X-rays in the 5–10 keV region. Quantum efficiency of a 2 bar Xe-filled detector is typically 0.85. A 'quencher' polyatomic gas, like CO_2, is added to limit the extent of the avalanche and prevent unwanted fluorescence.

Flat gas-filled position-sensitive detectors are well suited for small-angle scattering experiments, with photons at normal incidence. However, in the wide-angle scattering region, where photons impinge at an angle, these detectors suffer from **parallax**. Photons are absorbed at different depths of the gas layer, the thickness of which is of the order of a centimetre. The position of the avalanche will depend on the depth at which a particular photon is absorbed. The use of pressurized gas can be advantageous, since it provides a thinner effective absorption layer, but there are drawbacks too. For suppression of parallax xenon is preferable to argon because of its higher absorbance. In order to eliminate parallax completely, an area detector with spherical drift chamber has been constructed and is being used at the LURE synchrotron [42] for wide-angle diffraction from proteins. The electric field gradient within the drift chamber is directed radially, the centre point coinciding with the position of the scattering sample.

For wide-angle powder diffraction, curved one-dimensional detectors are available which avoid the parallax problem. The commercial detectors, however, are not specifically designed for the high count rates encountered in synchrotron experiments.

Many-fold improvement in overall detector speed can be achieved by connecting each individual electrode wire or strip to its own electronic circuit, viz. amplifier-discriminator pair. One-dimensional versions of this detector type, the so-called **multiwire linear detector**, exist at LURE [43] and Daresbury. The Daresbury detector has an active length of 20 cm, with 200 anode wires placed at right angles to the long dimension of the window. It can operate at 1 MHz **per 1 mm pixel**, i.e. each pixel can take the full load of a whole delay-line detector, as far as signal processing is concerned, while still retaining linearity of 1% over five decades of count rates [44]. The drawback of this type of detector is a somewhat coarse spatial resolution of 1 mm, although interpolation techniques have been devised which can improve this figure.

The problem of space charge, of course, remains even in the case of multiwire linear counters. Solution may possibly be found in **Microstrip** [45] and **Microgap** [44] detectors. In the latter the high field gap between the anode wires and the solid Cu cathode plane is only 0.4 mm. The anodes are at ground potential and are used for the read-out. The positive ions are rapidly removed by the nearby solid negatively charged cathode. Development of an area detector on the same principle is in progress.

For **powder diffraction** experiments the use of a two-dimensional detector produces a large amount of redundant data, particularly in the multiframe time-resolved mode; regarding the position of the incident photon, the experimentalist is only interested in radial distance from the beam centre. On the other hand, the alternative detector is a linear PSD, but it only captures a small proportion of photons within a diffraction cone. An appropriate solution for small- and intermediate-angle powder diffraction is the **quadrant detector**. This is a multiwire detector with read-out strips, usually cathodes, in the shape of concentric arcs, centred around the undiffracted beam. The concentric strips are linked to a delay line or similar, as in the case of straight parallel strips. The active area of the detector is in the form of a circular segment, typically with a 20 cm radius and an apex angle of 60°. The anode wires in front are oriented parallel to the axis bisecting the sector in order that the angle between the wires and the underlying cathode strips is as close to normal as possible. As discussed above, the avalanches are localized on the anode wires, hence spatial resolution is degraded if the cathodes are not perpendicular to the anodes. This is the reason that a full circular detector, if constructed, must be sectorized [46].

(e) Photodiode arrays

Detectors based on solid-state devices are discussed next. A **photodiode array** (e.g. Reticon S) detector consists of a series of *p–n* junctions produced by diffusing *p*-regions into the surface of *n*-type silicon. A charge depletion region is created around each junction by reverse biasing, each diode becoming a small charged capacitor. The detector operates on the principle of partial discharge of these capacitors in the vicinity of electron-hole pairs produced by absorbed photons. The read-out process involves shift registers. The charges are connected in turn to a video output line and as recharging by reverse bias takes place, the amplified charging voltage is recorded. It should be mentioned that, operating in a non-integrating mode, photodiode detectors can be used as photon energy analysers for applications such as energy dispersive diffraction. Typical energy resolution for this purpose is of the order of $\Delta E/E = 0.003$.

The size of individual pixels in diode arrays is of the order of 2.5 mm $\times$ 25 μm, and the overall length of the array is 25 mm. Larger devices also exist, but there are no two-dimensional diode arrays. Operating in integrating mode they provide no energy discrimination. In fact diode arrays are sensitive to visible light and have to be shielded from it when used for detecting X-rays.

The important advantage over CCD detectors (see later) is that diode arrays can be exposed to direct X-rays, since they do not suffer from radiation damage.

Noise is always a prominent factor under consideration when solid-state devices are compared with other types of detectors. Apart from the intrinsic statistical fluctuations in incoming photon level ('shot noise', $\sqrt{N}$, where N is the total number of detected photons), there are two contributions from the detector. These are called 'dark current' and 'read-out noise'. **Dark current** has its origin in thermally induced electron-hole pair production. The high activation energy involved means that cooling is a very effective way of reducing dark current: the noise is halved for every 7°C reduction in temperature. However, cooling to sub-zero temperatures requires placing the detector in vacuum to prevent icing.

The second type of detector noise, the **read-out noise**, is the error in measuring the recharge of each pixel. For a given data collection time this noise increases with the read-out frequency.

Due to these sources of noise, dynamic range of diode array detectors does not exceed 10^4. The cumulative effect of continuous dark current makes these detectors less suited for measurements of weak scattered intensities, i.e. for experiments with weak sources or weak scatterers. Conversely, they are at their best when used as low-angle detectors on synchrotron beam lines. Their high spatial resolution allows construction of very compact SAXS set-ups.

(f) CCD and Vidicon cameras

These detectors are mainly used for recording X-rays indirectly, since direct exposure damages the semiconductor device. However, some work has been reported using direct exposure and it has been claimed that cumulative damage to a charge coupled device (CCD) caused by soft X-rays can be annealed subsequently [47].

In most experiments X-rays are first converted to visible light by a phosphor screen. The image is then reduced in size to match the dimension of the chip and, usually, it is also intensified. The electron-hole pairs produced in the silicon are read out either at video rates (5–10 MHz) using standard TV technology or, in the 'slow scan' mode at 5–20 kHz, by direct coupling to a computer.

Quantitative application of low-noise silicon Vidicon tubes has been pioneered by astrophysicists requiring detection of low light levels. Silicon Vidicon tubes and silicon intensified target (SIT) tubes (two-dimensional arrays of photodiodes) were employed as they became commercially available. These were later joined and largely replaced by charge coupled devices (CCD). The operation of all these systems was modified essentially in two ways. Slow-scan mode was introduced, enabling read-out with low-noise narrow band-pass amplifiers. Secondly, the devices were cooled to as low as −70°C by Peltier elements, which reduced dark current dramatically.

A CCD consists of a two-dimensional array of capacitive elements produced by depositing alternating metal oxide regions of two different thicknesses on to *p*-type silicon. Each element, or pixel, is typically 25 μm × 25 μm in size or less, and can be considered a potential well for charge storage. Since the depletion regions overlap, charges can be shifted and read out sequentially. Most commercial charge coupled devices contain arrays with several hundred pixels in each direction, the largest size thus far being 2048×2048 [48].

Unfortunately, the 25 μm pixel size of a CCD sensor is not matched by the spatial resolution of the phosphor, typically of the order of 80 μm (FWHH of the point spread function). The spread of the signal in the phosphor is due to scattering of light in the polycrystalline layer, the thickness of which must be sufficient to absorb the bulk of the X-rays. To match the resolution of the phosphor with that of the CCD sensor, as well

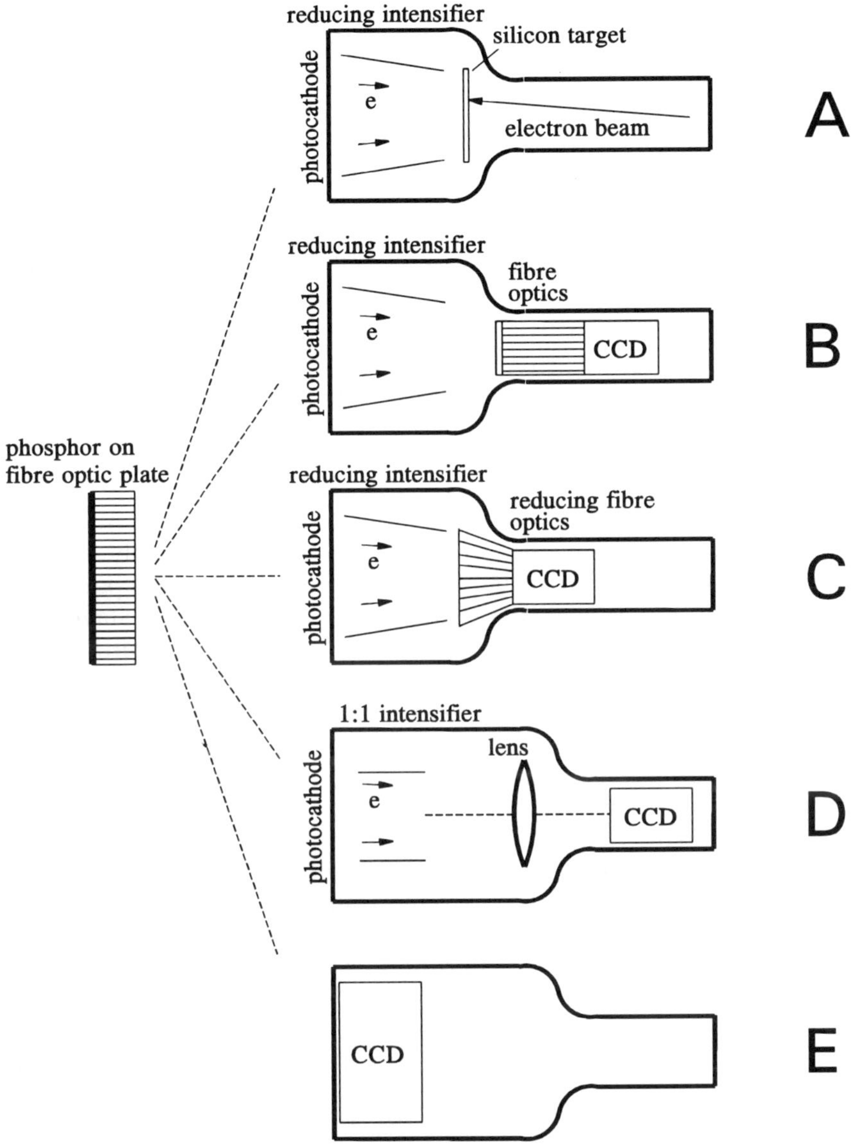

Fig. 3.6 Schematic representation of different 2-D detector arrangements using a CCD or silicon-intensified target with a scintillator at the front end. A: Vidicon detector [1]; B: reducing image intensifier coupled to CCD by parallel fibre optics; C: reducing intensifier coupled to CCD by tapered fibre optics; D: non-reducing intensifier with lens coupling to CCD (possibility of CCD being cooled by liquid N_2) [34]; E: direct parallel fibre optic coupling of scintillator to large 2048 × 2048 pixel CCD array [48].

as to fit the pattern on to the 1–2 cm chip size, the light image must be reduced. Typically a demagnification factor of at least 3:1 is appropriate.

Demagnification can be realized (a) by using a reducing (tapered) image intensifier, (b) by tapered fibre optics, or (c) by a lens. While SIT tubes have their own built-in intensifiers, CCD-based detectors for X-rays are built from individual components according to different designs [34]. The scintillator is usually coupled by coherent fibre optics to a microchannel plate or magnetically or electrostatically focused image intensifier. Some of the assemblies used or proposed with CCD and Vidicons are illustrated schematically in Fig. 3.6.

The aim of the design is a detector which would provide quantum-limited operation, good spatial resolution and an undistorted image [49]. Quantum-limited operation implies that the inherent input shot noise is above the total noise introduced by the detector. In practice this means that **at any stage** along the path of the signal there must be at least several statistically significant events (photons, electrons, electron-hole pairs) for every X-ray photon absorbed in the phosphor [34]. If the number of events per X-ray photon falls too low, subsequent gain merely amplifies the noise. There are several stages of gain and loss along the path of the signal, depending on detector configuration. The first gain stage is the scintillator, now made primarily of gadolinium oxysulphide, where several-hundred light photons are emitted for every 5–10 keV X-ray photon. However, even with efficient fibre optic coupling, this drops to only about 5–15 photoelectrons emitted from the photocathode of the photomultiplier [34]. The gain in the single- or multiple-stage image intensifier is adjusted so as to counter subsequent losses, notably those in optical image reduction.

Image reduction can be achieved either electro-optically by reducing intensifiers, which unfortunately often leads to distortion, or optically. Optically, the image can be reduced by lens or by tapered fibre optics. Lens coupling leads to less distortion but is very inefficient. The light efficiency of a lens, C, is given by

$$C = \left[\frac{M}{2f(1+M)}\right]^2 \qquad (3.10)$$

where M is the magnification and f the 'f-value' of the lens. From the above equation it follows that for a typical f=1.8 lens with demagnification $M = 0.3$, C is only 0.4%. Fibre optic coupling results in much smaller losses, although tapering also reduces efficiency: for parallel fibres ($M = 1$) C is 75%, while for $M = 0.3$ it is only 13%.

Clearly demagnification should be avoided, if possible. The large Tektronix 2048×2048 pixel CCD chip, measuring 55×55 mm^2, has been used by Eikenberry *et al.* [48] in an X-ray area detector without any image reduction. A phosphor screen of the above dimensions is connected to the CCD sensor directly via a parallel fibre optic plate (Fig. 3.6E). The low losses deemed the use of a photomultiplier unnecessary, which is beneficial in view of a number of problems often associated with these components. The detector showed excellent resolution and lack of distortion. Unfortunately, at the time of writing, large CCD arrays are still not readily available.

The problem of noise in CCD detectors can be greatly reduced by cooling. The current MPP (multiphase pinning) chips produce a dark current of several tens of electron per pixel per second at room temperature. Already water cooling reduces this figure noticeably. The device can be cooled to almost −70°C, which reduces the noise to about 1 electron per pixel per second. Freezing of silicon immersion oil used for fibre optic coupling precludes cooling to still lower temperatures. This is unfortunate since cooling the sensor by liquid nitrogen practically eliminates dark current altogether (*c.* 1 electron per pixel per hour). However, at such low noise levels even the inefficient direct lens coupling between phosphor and the CCD may become a viable alternative. In Fig. 3.6D a compromise solution is presented, which allows liquid N_2 cooling of the CCD but which also includes an image intensifier.

Vidicon detectors have been used on several synchrotron beam-lines in the past, and currently several CCD detectors are operational or are being tested. At Daresbury, for example, a CCD detector is used for Laue diffraction and for quick alignment and checking the quality of single crystals [50]. A TV detector and, occasionally a CCD camera, are also used for protein crystallography [49] and fibre diffraction. The advance of the application of CCD detectors to X-ray diffraction has not been as rapid as may have been predicted, because of strong competition from gas-filled detectors and image plates with superior signal-to-noise ratio (dynamic range), virtual absence of dark current and positional linearity. Considering the constant contributions to background noise from dark current and the read-out process, CCD detectors perform best (highest *S/N* ratio) with strong X-ray signals, i.e. with synchrotron radiation and strong scatterers. They are less suitable for detection of low intensities where long recording times are required.

(g) Image plate detectors

Image plates [35] have an active surface layer of photoluminescent amorphous barium europium halide (BaFBr:EuX). When exposed to photons in a wide energy range, including UV and X-rays, metastable colour centres form which decay very slowly. However, transition to the ground state is stimulated by laser irradiation. Thus 2 eV photons of a red helium–neon laser stimulate blue luminescence (4 eV) from the colour centres. The number of blue light quanta emitted is proportional to the number of X-ray photons absorbed.

After the plate is exposed to X-rays it is scanned, or 'read', by the laser. During laser scanning of an exposed image plate the emitted blue light is separated by a colour filter and its intensity is recorded and stored digitally. An efficient read-out system is essential, and where this is achieved an overall detector quantum efficiency of one can be approached. After scanning is completed, the plate is exposed to white light for some 20 seconds to wipe out the remaining colour centres. The plate is then ready for a new X-ray recording.

Image plates are currently produced by Fuji (Image Plates) and by Kodak (Storage Phosphor). Their main use is in low-dose medical radiography. Their entry into the X-ray diffraction field was primarily in protein crystallography, both on synchrotron beam-lines and on laboratory generators. Exposed image plates may be scanned off-line, in a batch flat-bed scanner, in which case they are treated essentially like photographic plates. Alternatively, the detector encompasses an on-line scanner, which eliminates the need for handling the plates and adds to the reproducibility of the detector. Scanning can be line-by-line (rectangular or square plates) or in a record-player mode (circular plates) with polar coordinates being converted to Cartesian during the scanning process.

The dead time between exposures for current commercial on-line scanners is 90 seconds for a one-plate detector (18 cm diameter circular plate, $150 \times 150\ \mu m^2$ pixels) and 30 seconds for a two-square plate detector; however, the plate read-out in the latter case is actually longer (a $20 \times 20\ cm^2$ plate is read within eight minutes with a $100 \times 100\ \mu m^2$ pixel resolution). The fraction of unread (residual) colour centres decreases as $\exp(-It_p)$, where I and t_p are the laser light intensity and the illumination time per pixel, respectively. It is clear that for highly efficient and fast read-out considerable laser power (up to several tens of mW) needs to be used.

The advantages of image plates are high dynamic range (10^5), low noise, high detective quantum efficiency ($\geq 80\%$ for $\lambda = 1–2$ Å) and linearity and virtually unlimited count-rate capability. The latter property is particularly useful for synchrotron application, but is neutralized to a large extent by the lack of time resolution. Like with other integrating detectors there is no energy resolution, which is not a problem with monochromatized radiation. Spatial resolution is determined by the size of the laser beam and the scanner optics, but there is a lower limit determined by scattering of luminescent light in the halide layer. Generally the overall resolution is considered to be $150 \times 150\ \mu m^2$ FWHH (full width at half height of the point spread function). Unlike CCD detectors, image plates can in principle be very large; 30 cm diameter

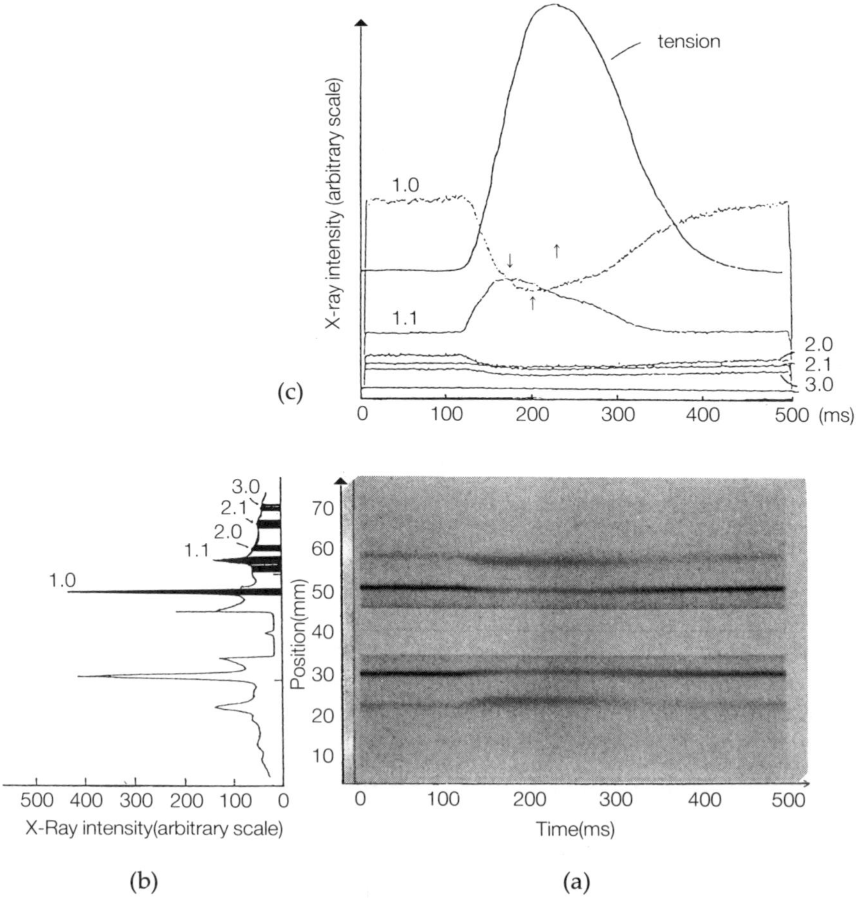

Fig. 3.7 (a) Time-resolved pattern of equatorial low-angle X-ray diffraction from frog muscle during twitch contraction. The data were obtained using a streak camera with an image plate. The trigger pulse to stimulate the muscle was applied at 100 ms. (b) Equatorial intensity profile near $t = 0$. (c) Time profile of observed reflection intensities. Reproduced from Allinson *et al.*, *Phase Transitions*, **39**, 145, 1992 and from Amemiya *et al.*, *Rev. Sci. Instr.*, **60**, 1552, 1989.

circular or 20×20 cm square plates with 100–150 μm pixel size are the current state of the art in commercial image-plate detectors with on-line read-out.

It has been reported [36] that the number of colour centres decreases by 10 per cent upon storage of the image plate for one hour after the exposure. This initial decay of latent image is followed by a much slower decay. This behaviour may to some extent affect linearity of image plate response if read-out is delayed for variable times, or if long and variable exposures are used.

Various methods have been devised to enable image plate recording with increased time resolution for dynamic synchrotron radiation experiments. Two methods have been reported by Amemiya and co-workers [51]. One is based on a slide-projector-type mechanism for rapid transport of plates from one stack, to the exposure position, and finally to the second stack. Up to 40 two-dimensional exposures with 0.3 second time resolution can be achieved in this way. The other method is based on the streak-camera design (see section on photographic film detection). It works as a 20 cm long **linear** detector, with the plate moving continuously in the direction normal to the linear detector slit. Time resolution of 23 μs has been claimed, with the total duration of the scan of up to 46 ms. Figure 3.7 shows the result of a low-angle diffraction experiment on frog muscle during twitch contraction, using the streak-camera technique. The whole experiment takes 0.5 s.

A further example of a streak camera, this time using a cylindrical image plate for wide-angle diffraction, is shown in Fig. 3.20.

3.2.4 Sample cells

Ancillary equipment for controlling sample environment is usually custom designed for specific experiments. Certain heating or cooling stages are commercially available, or else temperature stages designed for microscopy or spectroscopy can be adapted.

(a) T-jump cells

In today's experiments using synchrotron radiation, time resolution of temperature-induced processes is restricted not by the limited photon counts, but rather by the limited rate of change in specimen temperature. A piston-driven T-jump assembly has been in use at DORIS [52, 53, 54], and a dual chamber device at NSLS [55, 56], in order to attain the desired temperature rapidly. In the latter device the chambers are kept at preset temperatures T_1 and T_2; when required, the small sample cell is pneumatically driven from T_1 into the X-ray beam at T_2 (Fig. 3.8.). Both the above devices take about 20 s for temperature equilibration with T-jumps of 10–20 K. The device at Hamburg is placed within the vacuum beam-pipe. A system containing a pre-calibrated air blower and equipped with a remotely operated gas flow deflector has been used at Daresbury for crystallization and annealing studies of small free-standing samples of polyethylene [57] and long chain *n*-alkanes.

Internal heating by means of an infrared laser has been employed for investigations of kinetics of phase transitions on millisecond scale [58].

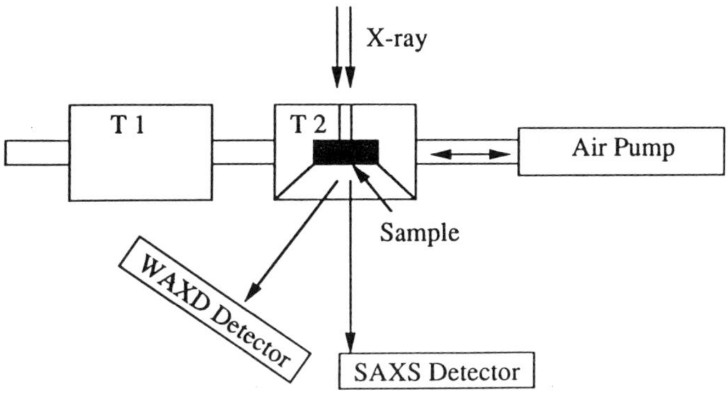

Fig. 3.8 Simultaneous SAXS and WAXS experimental set-up using a dual chamber temperature jump cell (after Song *et al.* [55]).

A 2 ms pulse from an erbium laser produced a 10–12°C temperature rise. This enabled a kinetic study of the transition from the gel phase to lamellar and from lamellar to inverse hexagonal phase in a lyotropic non-polymeric liquid crystal. It is expected that even faster kinetics could be followed using *Q*-switched lasers. Pulsed laser systems are very promising for rapid heating, provided the sample is IR transparent.

(b) Stretching frames

Molecular rearrangements in polymers during mechanical deformation can be studied *in situ* by time-resolved diffraction using synchrotron radiation [59]. Heatable stretching devices have been constructed for this purpose, having the facility for recording stress–strain data in parallel with scattering patterns [60]. Some examples of *in situ* deformation studies are given in the section on applications.

(c) Magnetic cells

Work on liquid-crystal polymers often requires the application of a magnetic field to obtain molecular orientation. At the same time the specimen must normally be kept at an elevated temperature. Magnetic fields are achieved either by electro-magnets or permanent magnets. Compact high-remanence permanent magnets made of materials such as cobalt–samarium or iron–neodymium–boron are capable of producing fields in excess of 1 Tesla if the pole gap is small. Correct design of the yoke and placement of magnets are essential for minimizing losses. Fields up to 1.5 Tesla can be achieved with conical pole pieces made from suitable materials such as FeCo. Cobalt–samarium magnets can be used

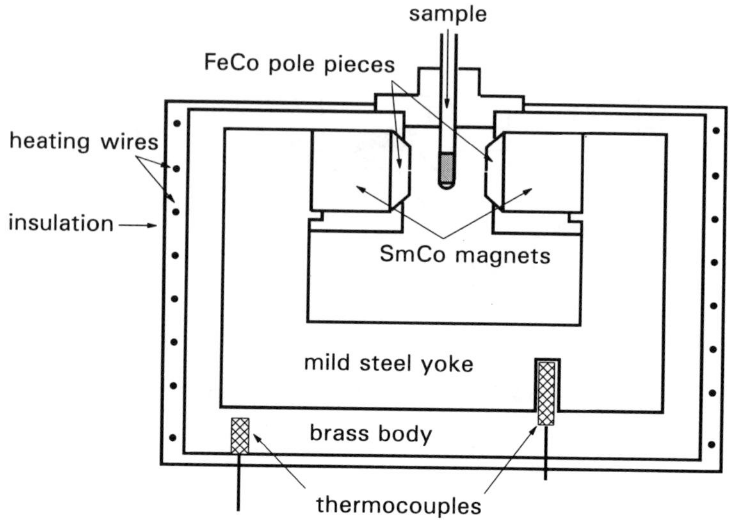

Fig. 3.9 Compact temperature-controlled X-ray cell with SmCo permanent magnets for alignment of polymer liquid crystals. The pole gap is adjustable (mechanism not shown) [61].

within a temperature cell. The advantage of this is that pole pieces can be brought as close as the specimen and beam sizes permit.

Compared to cobalt–samarium magnets, iron–neodymium–boron magnets are somewhat more powerful at room temperature but are restricted to temperatures up to *c.* 100°C. Figure 3.9 shows schematically a compact temperature controlled X-ray cell with small SmCo magnets and a variable pole gap achieving a field of up to 1.1 T [61].

A magnetic cell with high temperature stability but lower field has been described in reference 62. Large permanent magnet pieces are placed outside the heated cell, which consists of an outer and an inner shell. The inner compartment is held at c. 3 K higher temperature than the outer, and the temperature gradient over the entire irradiated volume of the sample held in a beryllium cuvette was reportedly less than 2 mK.

Molecular alignment can sometimes be achieved using an off-line magnetic field. During transfer and the diffraction experiment itself, orientation may be preserved by cooling into the glassy or the smectic state, provided such treatment is allowed. A high-field superconducting magnet can be applied in these cases.

(d) Shear cells

Studying diffraction on polymers and liquid crystal subjected to shear flow has recently become an area of interest. Synchrotron radiation

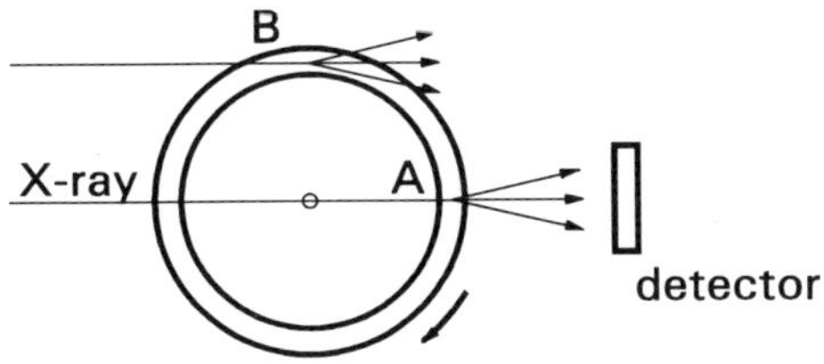

Fig. 3.10 Couette-type shear cell used for studies of shear-induced orientation and transitions in smectic liquid crystals. The static and the rotating cylinders are made of Lexan. (After Safinya *et al.* [64].)

provides sufficient intensity to perform time-resolved studies of orientational relaxation upon cessation of flow and of other transient effects. A heated cell with **parallel rotating plates** made of boron nitride has been used for investigating flow-induced orientation effects and relaxation dynamics in highly viscous nematic solutions of rigid-rod polymers [63].

A **Couette shear cell** consisting of two concentric Lexan cylinders has been used for studying effects of flow on the nematic (N)–smectic-A (S_A) transition in a thermotropic low-molar-mass liquid crystal [64]. The X-ray path was either through the centre (A in Fig. 3.10) or through the edge of the cell (B in Fig. 3.10), which could be temperature controlled to high accuracy. The beam was thus parallel to the velocity gradient in A and to the velocity itself in B. Effects such as critical slowing down of smectic-like fluctuations under non-equilibrium steady-state conditions could be observed.

These or similar shear cells may potentially also be employed in studies of orientability of thermotropic polymers, flow-induced crystallization, etc.

(e) Pressure cells

For X-ray diffraction studies of polymers at high pressure ($p \leq 10$ kbar) diamond anvil cells are used most often. A better control and a more uniform distribution of pressure can be achieved with a cell recently designed by Hikosaka [65] (see Fig. 3.11). The cell can be used for both optical microscopy and X-ray diffraction. The specimen is sandwiched between two flat 1 mm thick diamond windows. The pressure imposed by compressed nitrogen gas on the inner piston is amplified and transmitted to the windows via a conical high-speed steel anvil. The sides of the windows are sealed by PTFE rings compressed by the outer piston. The uniform hydraulic pressure obtained in the sample compartment can be accurately controlled by nitrogen gas. This cell has so far been used for studies of polymers using a rotating anode X-ray source [66]. However, it can equally be ported to a synchrotron beam-line. Furthermore, the

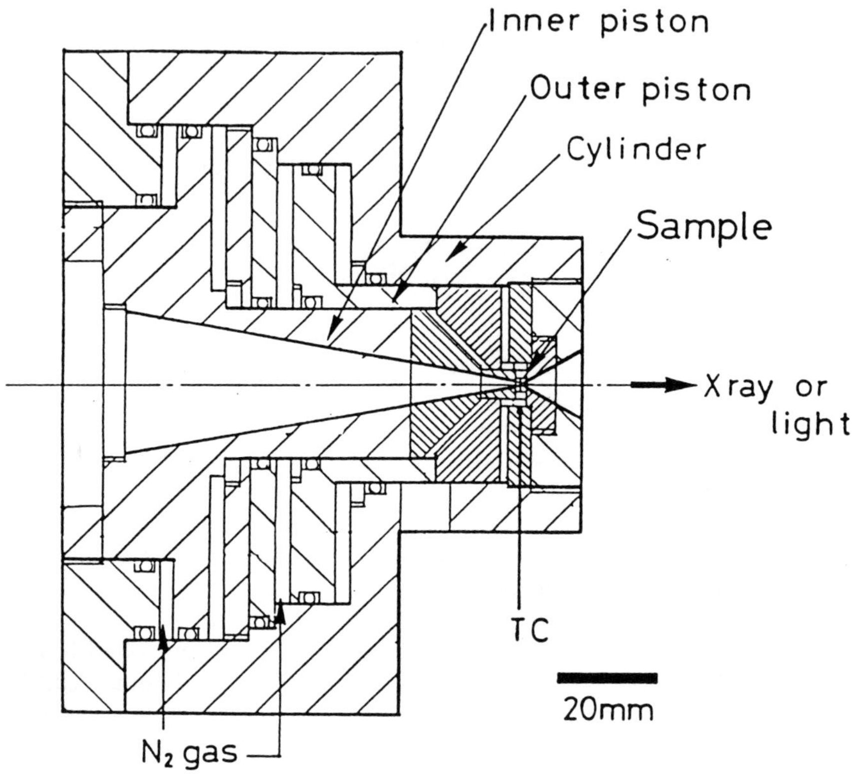

Fig. 3.11 Section through the heated pressure cell with flat diamond windows for optical microscopy and X-ray diffraction. (From [65] by permission of Marcel Dekker.)

efficient PTFE sealing system is likely to make it suitable also for less viscous low-molar-mass liquids.

As already mentioned, energy-dispersive diffraction technique may be advantageous, e.g. in very high-pressure experiments. Such experiments require an intense white X-ray beam which includes high-energy radiation; thus synchrotron radiation, preferably produced by a wiggler, is necessary. Even for angle-dispersive experiments with diamond pressure cells it is advantageous to use shorter wavelength and thus more penetrative radiation; this is likely to be found on a wiggler or wavelength shifter beam-line.

(f) *Other* in situ *cells*

A number of experiments have been performed where the change in scattering pattern was monitored while a chemical or physical process was taking place *in situ*. To investigate the complex interplay between microphase separation and crystallization in segmented polyurethanes and their role in industrial processes such as reaction injection moulding (RIM), a small-scale RIM machine [67] fitted with a Kapton-windowed mould has been placed into the synchrotron X-ray beam [68]. Both real-time WAXS and SAXS intensities were recorded following the filling of the mould.

3.2.5 Miscellaneous

Between the source and the exit slit, the synchrotron beam-line is evacuated to avoid beam attenuation and chemical degradation of optical elements. As usual for SAXS and some WAXS experiments, a vacuum pipe is positioned between the specimen and the detector to minimize air scatter. The beam-stop is typically stepper motor driven for easy alignment, and can be part of the vacuum chamber. Detector and sample stage are usually also moveable by remote control.

A thin semi-transparent beam-stop is sometimes used (e.g. at Hamburg). Thus the attenuated primary beam is recorded together with the scattering curve, which is used both for normalizing the intensities and locating the primary beam. Correction for changes in beam intensity, mainly due to decay in storage ring current, can also be made by means of an ionization chamber, positioned either before or after the specimen. Ionization chambers are also very useful for alignment purposes, as well as for measuring absorption in the specimen for purposes such as determination of absolute scattered intensities.

3.3 APPLICATIONS

Small-angle X-ray scattering (SAXS) has most benefited from the application of synchrotron radiation. This is because, unlike conventional sources, the synchrotron produces a highly collimated beam in the first place, so that its further collimation required for SAXS does not reduce beam intensity significantly. Excellent collimation means further that high-resolution WAXS studies of line profiles can be performed, an opportunity not as much exploited in the polymer field as, for example, in the field of low-molar liquid crystals.

3.3.1 Weak scatterers and microdiffraction

The high brilliance of synchrotron radiation enables execution of microdiffraction experiments, in which scattering is confined to a small irradiated volume. Scattering volume may be small for a variety of reasons. For example, a thin single-strand polymer fibre may be investigated, as described later in the example of Kevlar. Alternatively, very small amounts of material might be available, which is often the case with biological samples. Further, beam size must sometimes be reduced to fit a small sample cell. An example is the study of orientation of liquid crystals in an electric field where, in order to limit the operating voltage to a reasonable level, the electrode gap is kept to only 100 μm [69]. The beam must be confined to the gap and avoid the electrodes. High brilliance of the beam is especially necessary if dynamic effects are investigated. The constraint of small irradiated volume also applies to diffraction studies from thin surface layers, including evanescent wave scattering experiments described at the end of this chapter.

As an example of microdiffraction one can take the experiment by Li *et al.* [70] who determined the lateral contraction of the crystal unit cell of poly(*p*-phenylene terephtalamide) (Kevlar) fibre under longitudinal tension. Since the strain at break of Kevlar fibres is very limited, and measurements are confined to only a fraction of a per cent in macroscopic strain ϵ, measurements had to be carried out on a single strand rather than a bundle of fibres. In this way problems with slippage or straightening of misaligned strands were avoided.

In some cases SAXS intensities can be extremely weak and require the use of synchrotron radiation in order to observe any scattering at all. Thus, in the example of high-modulus polyethylene fibres, the long-period scattering intensity gradually vanishes as the draw ratio exceeds 40 or so. This is due to increasing density of the non-crystalline phase, leading to an overall decrease in electron density fluctuations $\Delta\rho$ [see next section, equations (3.11) and (3.12)]. Synchrotron SAXS experiments enabled van Aerle and Braam [71] to observe long-period scattering up to a draw ratio of 40 in the real-time mode during stretching.

Variations in electron density can also be very small in amorphous polymer blends or block and graft co-polymers. At the other extreme, SAXS intensity may be very low if the crystallinity (X) is very high. If X is 95% or 94%, as in extended-chain or solution-grown once-folded-chain crystals of long *n*-alkanes (see Table 3.2), the scattering intensity is low because of the low value of the factor $X(1-X)$ in equation (3.12).

3.3.2 Time-resolved studies

(a) SAXS and semicrystalline polymers

Before describing some of the results of time-resolved experiments on crystallizable polymers, we shall briefly mention some of the different kinds of information on lamellar crystal morphology obtainable from SAXS experiments. The simplest and most frequent usage of SAXS is in determining the long period L from the diffraction peak position. Provided lamellar stacks are large and regular, the Bragg equation can be applied as a good approximation. In case of unoriented specimens the Lorentz correction must be applied, which in the case of SAXS amounts to multiplying the measured intensities by q^2, to obtain $i(q)$. Here q is the scattering wave vector $(4\pi\sin\theta)/\lambda$. The scattering peak profile and the intensity ratio between different diffraction orders contain further information about the size of the coherently scattering units (size of lamellar stacks), about the distribution of lamellar thicknesses, crystallinity, etc. [72]. Although the distribution of lamellar thicknesses can be narrow, there is no true long-range order. A stack of lamellar polymer crystals is sometimes referred to as a one-dimensional paracrystal [73]. There is no driving force for a stack to regularize as there is for a true three-dimensional crystal. This poor regularity is manifested as a general lack of higher-order diffraction peaks.

One frequently used property of the scattering function is that, irrespective of how the scattered intensity is distributed over the reciprocal space, the total scattered intensity per unit-volume of the specimen is determined entirely by the mean square electron density fluctuation in the sample, $<\Delta\rho^2>$. The so called 'SAXS invariant' [74] is defined as

$$4\pi \int_0^\infty I(q)q^2 dq = Q = <\Delta\rho^2> \qquad (3.11)$$

where I is scattered intensity in electron units. For a two-phase system, such as a semicrystalline polymer with crystalline and amorphous densities ρ_c and ρ_a, Q is given by

$$Q = X(1 - X)(\Delta\rho)^2 \qquad (3.12)$$

where X is the volume crystallinity and $\Delta\rho = \rho_c - \rho_a$.

Further, the analysis of diffraction intensities, either relative or absolute, can provide information about the electron density profiles. The example of a method [12] for determining density fluctuations between comparatively regular alternating crystalline and non-crystalline layers will be given further below in connection with chain-folded long *n*-alkanes. Another method of analysing lamellar morphology by SAXS, which is more appropriate for less regular structures, is via the correlation function.

The correlation function in SAXS is the equivalent of the Patterson function used widely in single crystal structure determination, as it is the self-convolution of electron density in real space. It is fortunate that in the case of morphology based on large stacked lamellae, typical of semicrystalline polymers, only the one-dimensional correlation function, $\gamma_1(x)$, need be considered. Here x is the dimension normal to the lamellae. γ_1 is obtained by Fourier transformation of the scattering function:

$$\gamma_1(x) = \frac{\int_0^\infty i(q)\cos(2\pi xq)dq}{\int_0^\infty i(q)dq} \qquad (3.13)$$

The 'observed' correlation function thus derived is then compared with the calculated correlation function based on a specifically chosen model. If the model electron density function is $\eta(\xi)$, with η being the deviation from the mean density $<\rho>$, then γ_1 is calculated as

$$\gamma_1(x) = \frac{\int_0^\infty \eta(\xi)\eta(x-\xi)d\xi}{\int_0^\infty \eta^2(\xi)d\xi} \tag{3.14}$$

Specific procedures for dealing with the ideal two-phase lamellar model structures, in which density can assume only two values ρ_c (crystalline) and ρ_a (non-crystalline), have been developed by Vonk and Kortleve [75]. The 'transition layer' of thickness E in which ρ changes linearly between ρ_c and ρ_a, has later been added to the model [76]. Using the second derivative of γ_1 Ruland [77] has shown that distributions of crystalline and amorphous layer thicknesses can be obtained separately in cases where first three diffraction orders are resolved.

Measurement of absolute SAXS intensity is not often undertaken but it can be very useful (see example later). The recommended method is using the standard sample of low-density polyethylene Lupolen, obtainable from Anton Paar AG, Graz, Austria. The scattered intensity at a particular low angle has been calibrated as a fraction of the primary beam flux by the method of Kratky [78]. It is less well known that the parameters of this sample have been determined for both slit collimation (Kratky-type cameras) [78] and for pinhole collimation [79], the latter being effectively the SAXS geometry mostly used with synchrotron radiation. The definite advantage of using a well-monochromatized beam for absolute intensity measurements is that the irradiated mass of the specimen per unit area, the necessary normalization parameter, can be determined accurately by measuring the transmission of the monochromatic primary beam. With the knowledge of chemical composition and mass absorption coefficients, the Lambert–Beer law can be applied.

(b) Crystallization, annealing and melting

A significant contribution of time-resolved synchrotron radiation SAXS to the field of polymer crystallization was the finding [80] that in isothermal melt crystallization the long period of **initially formed lamellae** can be only half as large as the long period observed subsequently. Isothermal lamellar thickening occurs during crystallization and is particularly pronounced for larger supercoolings, where the initial lamellae are highly metastable. The 'self-seeded' crystallization procedure employed in these experiments gave rise to simultaneous appearance of a large number of primary lamellae, whose transient diffraction signal could therefore be detected. When the primary L thus obtained is plotted against supercooling, the previously missing continuity is restored between melt and solution crystallization, as predicted by theory. For example, melt-grown primary lamellar stacks with $L = 16$ nm were observed, where the smallest L measured previously on mature melt-crystallized samples was of the order of 25 nm.

Much information on the recent **chain-folded long alkanes** comes from synchrotron SAXS experiments. These perfectly monodisperse model polymers offer new insights into crystallization and morphology of long-chain molecules. Thus it was found that alkane chains longer than $C_{150}H_{302}$, and including the longest $C_{390}H_{782}$, fold on crystallization with the fold length tending to assume values very close to L_0/n, where L_0 is the contour length and n is an integer [81]. The existence of these 'integer folds' (IF) was established from a combination of L-values from SAXS and Raman LAM spectroscopy. For some melt-crystallized samples chain tilt was observed, similar to that in shorter alkanes and polyethylene.

The thin intercrystalline layer in n-alkanes has been investigated by measuring absolute diffraction intensities [82]. For this purpose the method of Strobl [12] was applied, whereby the extrapolated limiting structure factor of the zeroth diffraction order, B_0, gives the total electron deficiency per unit area of the non-crystalline layer, κ. The extrapolation is performed by fitting the measured structure factors of several diffraction orders (B_1, B_2, . . .) to an even (usually second) order polynomial. The curvature gives the parameter σ^2, which is the second moment of the electron density profile and can be regarded as the one-dimensional analogue of the radius of gyration. Note that this method is only applicable if $t \ll L$, where t, is the thickness of the non-crystalline layer, and if t is fairly uniform throughout the sample. The absolute SAXS intensity is a very sensitive probe for fold layer thickness t: to the first approximation the extrapolated zero order intensity increases with t^2.

Table 3.2 lists the non-crystalline layer parameters of two integer forms, E (extended chain) and F2 (folded in two), for n-alkanes $C_{198}H_{398}$ and $C_{246}H_{494}$, as obtained from the absolute intensity measurements [82]. While the parameters κ and σ^2 are obtained directly, t_1, t_2 and the non-crystalline fraction t_1/L are based on the assumption of a discrete two-phase model, i.e. of a rectangular density profile. Data for a melt-crystallized sample of high-density polyethylene are also included for comparison. The thinness of the fold layer in n-alkanes, particularly in the solution crystallized sample, indicates the high crystallinity and, accordingly, 'tightness' of chain folds. Viewing the non-crystalline layer as consisting of two fold layers belonging to the two adjoining lamellae, the thickness of each fold layer is $t/2$. $t_1/2$ values are 7Å for melt-quenched, and only 4Å for solution-grown once-folded chain crystals (corresponding values for $t_2/2$ are 8.5Å and 5.5Å). The small fold-layer thickness,

Table 3.2 Intercrystalline layer parameters for integer forms of long *n*-alkanes from absolute SAXS intensities [82]

	E melt crystd ($C_{198}H_{398}$, $C_{246}H_{494}$)	*F2 quenched from melt ($C_{198}H_{398}$, $C_{246}H_{494}$)*	*F2 solution crystd ($C_{198}H_{398}$)*	*Linear polyethylene, melt crystd*
electron deficiency per unit area of interlayer, κ (eÅ^{-2})	0.66	0.73	0.40	2.4
average thickness of non-crystalline layer from κ, t_1(Å) (assuming amorphous density = 0.85 g cm^{-3})	13	14.5	8	48
non-crystalline fraction, t_1/L	5.5%	10.5%	6.3%	21%
2nd moment of density deficiency profile across non-crystalline layer, σ^2(Å^2)		25	11	
thickness of non-crystalline layer, from σ^2, t_2(Å) (assuming two-phase model)		17	11	

particularly in the case of solution-grown crystals, suggests that a large proportion of folds are 'tight' and adjacently re-entrant.

Although in their final form mature alkane crystals have integer fold lengths, real-time SAXS using synchrotron radiation has revealed that melt crystallization of folded, and in some cases extended chains, proceeds via a transient non-integer folded form (NIF) [83]. The NIF form has a long period intermediate between those of the extended (E) and the folded-in-two (F2) forms. The NIF form is characterized by only a single

Fig. 3.12 Series of consecutive SAXS curves recorded during melt-crystallization of *n*-alkane $C_{246}H_{494}$ at $T_c = 122.3°$ C (a) and $T_c = 121.1°$ C (b). T-jump from 135° C to T_c was effected in seven seconds. Narrowly spaced curves were recorded for 12 seconds, with 12 seconds pause, while widely spaced curves were recorded consecutively for 60 seconds. The values of 25.7, 12.9 and 8.6 nm correspond to layer thickness of extended chain $C_{246}H_{494}$ tilted at 35° to the layer normal. (From [84] by permission of Butterworth–Heinemann.)

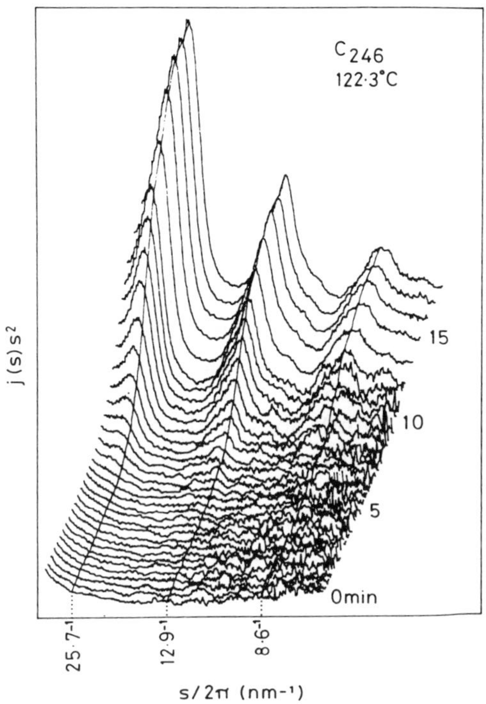
C246
122·3°C
j(s)s²
15
10
5
0min
25·7-1
12·9-1
8·6-1
s/2π (nm-1)

(a)

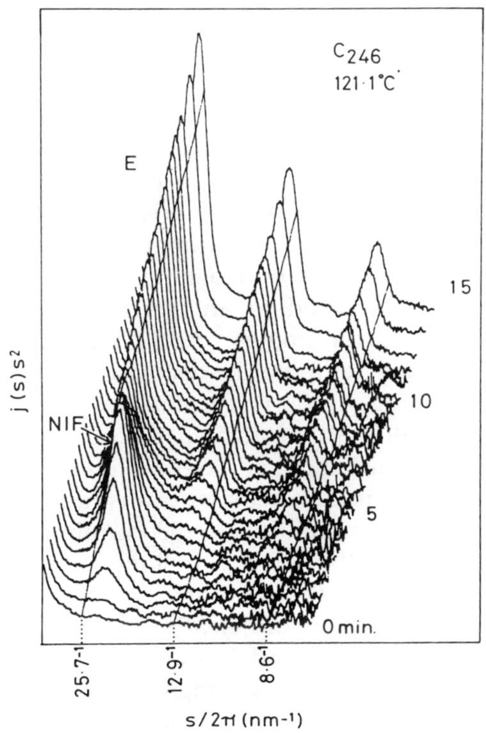
C246
121·1°C
E
NIF
j(s)s²
15
10
5
0 min.
25·7-1
12·9-1
8·6-1
s/2π (nm-1)

(b)

intense SAXS peak, with a very weak second order seen occasionally. This indicates a comparatively large κ and σ^2 parameters. At lower crystallization temperatures T_c, NIF transforms subsequently into F2, while at higher T_c it transforms into E.

An illustrative example of the role of NIF in crystallization of $C_{246}H_{494}$ is presented in Fig. 3.12 [84]. Figure 12(a) and (b) show, respectively, the development of SAXS at $T_c = 122.3$°C and 121.1°C. Rapid T-jump was achieved by a dual-heated helium gas flow system. At 122.3°C crystallization occurs directly in the E form: three diffraction orders of the 25.7 nm periodicity are seen, corresponding to the extended chains tilted at the usual angle of 35°. At 121.1°C the first crystals observed are considerably thinner (20 nm), show only one diffraction order and are typical of the NIF form. However, NIF$\rightarrow$E transformation soon takes place and once the extended-chain substrate is created, further crystallization ($t > 6$ minutes) occurs directly in the E form.

Recent time-resolved SAXS studies on sharp low-molecular-weight fractions of poly(ethylene oxide) (PEO) show a behaviour similar to that of long alkanes [85]. In addition to the NIF and a number of IF forms, the slower crystallization of PEO enabled observation of a second NIF form with L-value intermediate between those of F2 and F3.

The thickening process during **annealing** of lamellar crystals of conventional polymers has been the subject of several time-resolved SAXS studies using both synchrotron [86, 87] and rotating anode radiation [88]. These studies have addressed the long standing question as to whether crystal thickening occurs via solid-state chain diffusion or through partial melting–recrystallization. Grubb *et al.* [86] used a 2-D Vidicon detector to observe changes in the SAXS pattern of single-crystal polyethylene mats. The characteristic sharp SAXS arcs were found to rapidly spread towards lower angles, producing a continuous radial streak about 10 seconds after reaching the annealing temperature T_A. This eventually condensed to an intense spot at longer times, moving gradually to lower angles. Additional WAXS experiments showed that crystallinity temporarily decreases in the early stages of annealing if fast heating to T_A is employed. This suggested partial melting. Spells and Hill [87] investigated the effect of heating rate on subsequent isothermal thickening at different temperatures T_A, and also observed sections of annealed mats by transmission electron microscopy. Both slow (10°C/min) and fast (~500°C/min) heating rates were used. Figure 3.13 shows the change in long period L with time at $T_A = 117$°C after fast heating (○), and at $T_A = 122$°C after slow heating (×). Three characteristic stages are clearly distinguished: (i) induction period, (ii) rapid increase in L and (iii) logarithmic increase in L with time. It was concluded that partial melting/recrystallization played a part in all annealings in case of the high heating rate, whereas for the low

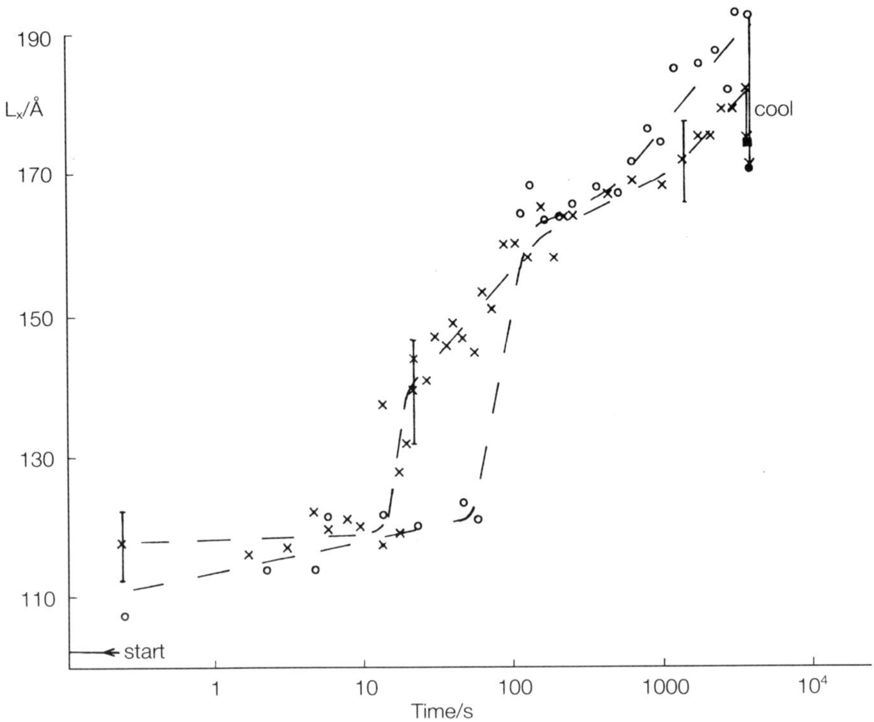

Fig. 3.13 Variation of X-ray long period of polyethylene single crystals with annealing time at $T_A = 117°C$ after fast heating at $\sim 500°C\,min^{-1}$ (○) and at $T_A = 122°C$ after slow heating at $10°C\,min^{-1}$ (×). (From [87], by permission of Butterworth–Heinemann.)

heating rate it only occurred to a significant extent above T_A = 123°C [87].

Crystallization and melting of PEEK was studied by both low- and wide-angle scattering [56, 89, 90]. These experiments were designed to help understand structure formation in this typical representative of aromatic thermoplastics and resolve the question of the origin of double melting endotherms observed by DSC in cold crystallized PEEK [91, 92]. The model based on bimodal lamellar thickness distribution seems to have been upheld [56, 89]. Time-resolved SAXS data recorded during isothermal crystallization of PEEK were analysed in terms of the correlation function $\gamma(r)$ [89]. The methodology is comparable to that applied to poly(ethylene terephtalate) by Santa Cruz *et al.* [93]. Figure 3.14 shows time development of $\gamma(r)$ during isothermal crystallization at 230°C. An increase in amplitude and a shift to smaller r of the first maximum are immediately apparent. The position of the first maximum,

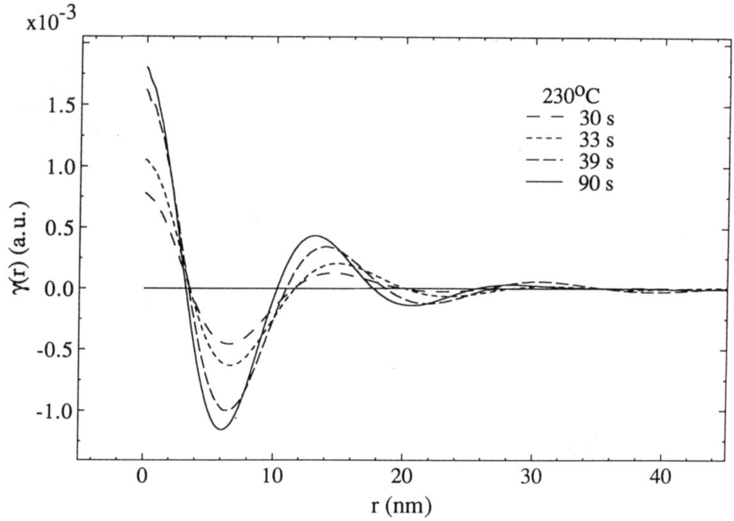

Fig. 3.14 Correlation function $\gamma(r)$ of PEEK as a function of time of isothermal crystallization at 230° C. (From [89] by permission of Butterworth–Heinemann.)

L_C^M was shown to be a more accurate measure of the long period than the value derived from the Bragg law. Furthermore, by using $\gamma(r)$, linear crystallinity X_{CL} within lamellar stacks could be determined separately from the bulk crystallinity measured by WAXS. The former was derived from the relation [93]

$$X_{CL}(1 - X_{CL})L_C^M = B \tag{3.15}$$

where B is the position of the first zero crossing of $\gamma(r)$.

Isothermal crystallization experiments such as those in Fig. 3.14 have shown that the long period decreases with crystallization time, but that lamellar thickness $l_C = X_{CL}L_C^M$ remains constant. By comparing linear crystallinity X_{CL} and the overall bulk crystallinity X_{WAXS} primary crystallization (growth of spherulites) could be differentiated from secondary crystallization (insertion of additional thin lamellae). Since the spherulitic volume fraction X_{WAXS}/X_{CL} continued increasing through a considerable part of the experiment, using X_{WAXS} values in place of X_{CL} would have led to erroneous results for lamellar thickness. Formation of secondary lamellae was considered responsible for the decrease in long period during isothermal crystallization. Regarding the melting process, the low-temperature DSC endotherm was attributed to melting of the thin secondary lamellae.

SAXS intensity changes during **melting and recrystallization of poly(ethylene terephtalate)** (PET) have been analysed by Gehrke *et al.* [54].

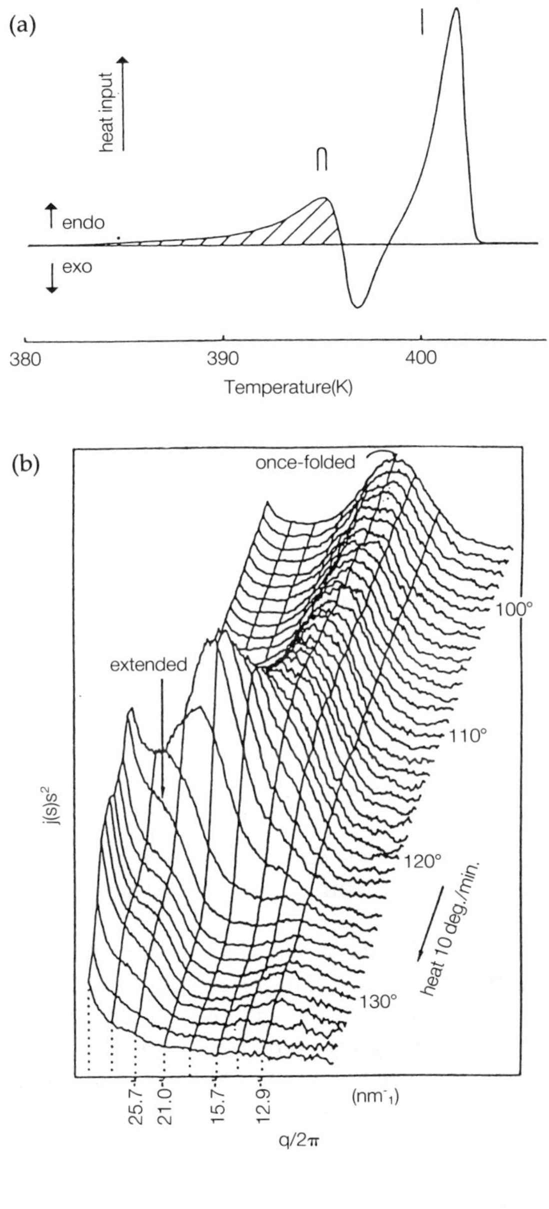

Fig. 3.15 (a) DSC melting thermogram of once-folded chain crystals of alkane n-$C_{246}H_{494}$; shaded area marks surface pre-melting. (b) Corresponding synchrotron SAXS traces. Heating rate 10° C min^{-1}. (From [94] by permission of American Chemical Society.)

Three different effects could be distinguished: (a) changes in $\Delta\rho$, the difference between densities of the crystalline and amorphous phases, (b) partial melting, and (c) recrystallization. The change in the SAXS invariant Q could be fully accounted for by the change in overall crystallinity X_{WAXS} as measured from the wide-angle diffraction. It was hence concluded that partial melting and recrystallization took place within the spherulites and that there was no need for invoking a change in the overall spherulitic volume fraction.

Because of their uniform lamellar thickness, integer-folded long n-alkanes provide clean systems for studying **pre-melting and recrystallization** phenomena. The DSC thermogram in Fig. 3.15(a) shows the complex melting behaviour of crystals of n-$C_{246}H_{494}$ with chains folded-in-two. Figure 3.15(b) shows the series of SAXS curves recorded during a heating run at the same rate of 10 °C min^{-1} as that used in Fig. 3.15(a) [94]. It is clear from the comparison of these two figures that the first endotherm corresponds to melting of the chain-folded crystals, the subsequent exotherm to recrystallization of the more stable chain-extended crystals, and the final endotherm to melting of the latter.

The interesting feature of Fig. 3.15(a) is the extended low-temperature end of the first (shaded) endotherm. The pre-melting is reversible as long as the crystals are not heated above 122–123°C, and provided they are not kept close to melting point long enough for chain-extension to occur through solid-state diffusion. The intensification and the shift to larger spacings of the 'once-folded' SAXS peak show that pre-melting involves an increase in thickness of the non-crystalline interlamellar layers [compare equation (3.12)]. This 'surface melting' consumes a significant fraction of the heat of fusion of the 15 nm thick lamellar crystals. However, it is not until lamellar cores melt fully that recrystallization in the extended-chain form can commence; hence the abrupt onset of the exotherm at 123°C.

Melt crystallization of **blends of low-density (LDPE) and high-density polyethylene (HDPE)** was investigated by Song *et al.* [55] using real-time SAXS. Slow cooling (0.3°C min^{-1}) as well as stepwise temperature-lowering experiments were made, which resulted in HDPE/LDPE phase separations on different spatial scales. In a comprehensive study of crystallization and melting of blends of deuterated high-density polyethylene (DHDPE) and hydrogenous linear low-density polyethylene (LLDPE) Tashiro *et al.* [95] employed time-resolved SAXS and WAXS and compared the results with those of low-angle light scattering and IR spectroscopy. This provided a broad view, on a scale from Ångstroms to microns, of the state of phase separation, which could potentially also be extended by neutron scattering. It emerged that mixing occurred on the scale of individual lamellae or small lamellar stacks.

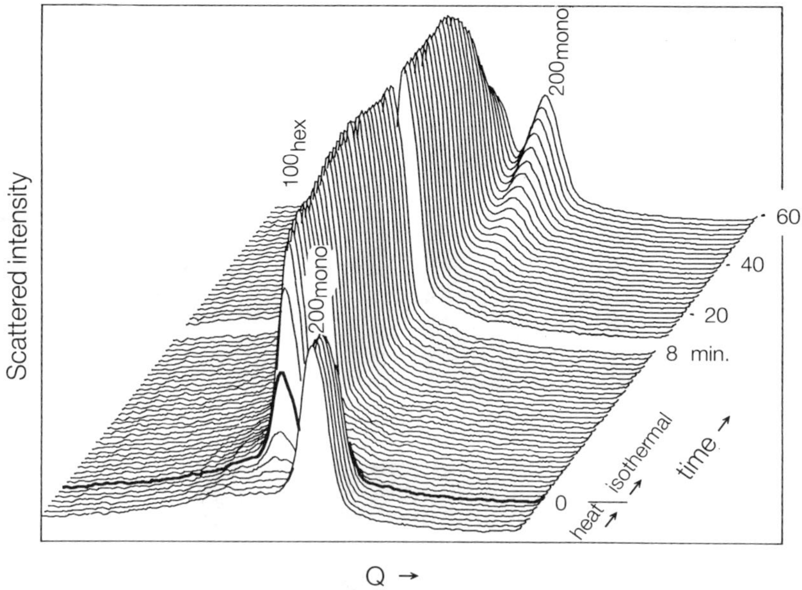

Fig. 3.16 WAXS experiment showing isothermal phase reversal in 1,4-*t*-polybutadiene. The angular range of the main monoclinic (200_{mono}) and hexagonal (100_{hex}) reflections is shown. The single crystal mat is heated at 3° C min^{-1} to 68.5° C (bold curve), after which the temperature remained constant within ±0.1° C. Separation between time-frames is 15 s for the first 37 curves, and 2 minutes for the last 25 curves. Monoclinic → hexagonal transition is seen at the beginning and the reverse hexagonal → monoclinic transition towards the end of the run.

(c) Phase transitions in crystalline and liquid crystalline polymers

The phenomenon of polymorphic phase transitions was one of the first in synthetic polymers to be studied by synchrotron radiation, using VEPP-3 in Novosibirsk [96]. Different crystal modifications in **polypropylene** were studied. More recently the effect of nucleation agents on the formation of the β (hexagonal) phase and on its transformation to the stable α (monoclinic) phase was investigated [97].

The well-known ability of the high-pressure 'hexagonal phase' in polyethylene to produce extended-chain crystals was found to be a general property of the conformationally disordered **2-dimensional hexagonal phase** (also described as columnar liquid crystal [98]) in a number of polymers. Time-resolved synchrotron SAXS experiments on polyphosphazenes [99] and 1,4-*t*-polybutadiene (PBD) [100] have shown

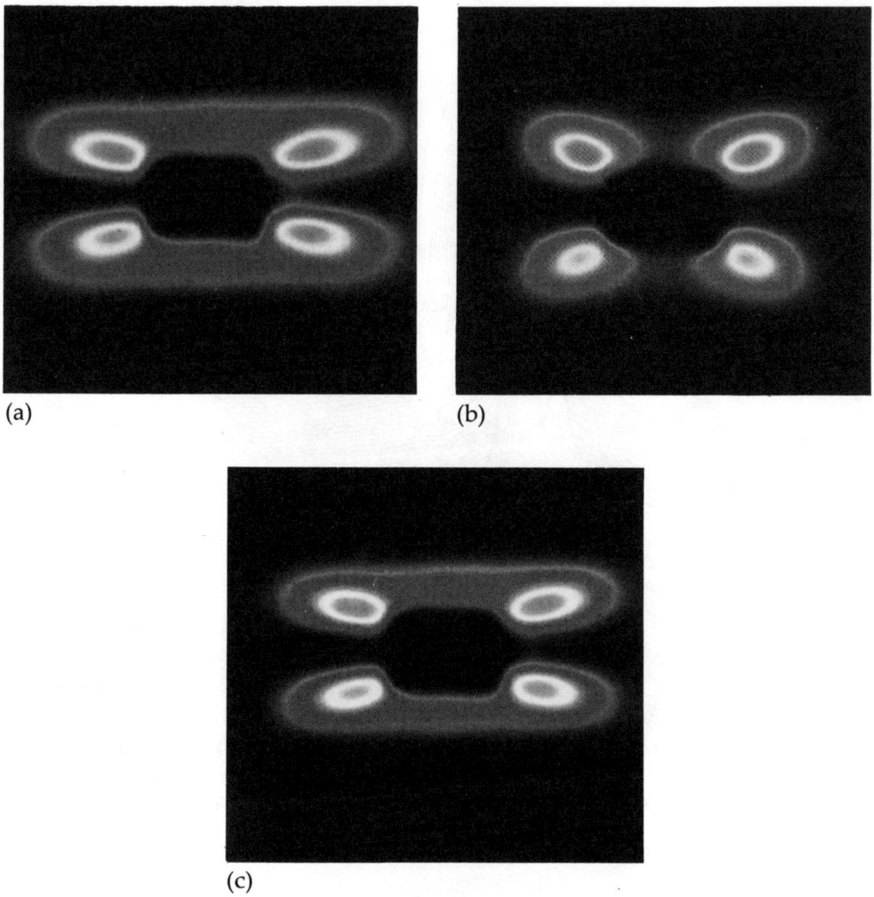

Fig. 3.17 Reversible change in the SAXS pattern of Hostalen GUR during cyclic tensile deformation at 100° C. (a) $\lambda_1 = 5.0$, (b) $\lambda_2 = 4.0$, (c) $\lambda_3 = 5.0$. (From [102] by permission of International Union of Crystallography.)

that once the crystal → columnar transition has occurred, lamellar thickening takes place at great speed, even at temperatures well below the isotropization transition.

As a consequence of lamellar thickening in the columnar phase, the crystal-columnar transition in PBD was found to be isothermally 'self-reversing'. The thin (10 nm) solution-grown crystals of PBD were first brought to 68.5°C, i.e. just above the temperature T_t at which the monoclinic-hexagonal transition occurred – see Figure 3.16. The sample was then held isothermally at 68.5°C, and the newly formed hexagonal phase reverted spontaneously to the monoclinic crystal state. Time-resolved SAXS and post-mortem electron microscopy showed that the lamellae had thickened several-fold in the process. The phase reversal can be explained by the fact that T_t increases with lamellar thickness due to the higher surface energy of the crystal compared with that of the columnar phase. Hence the T_t had increased well above 68.5°C during isothermal annealing, causing recrystallization seen in Fig. 3.16.

Time-resolved SAXS and WAXS studies during cycling through the ferro-electric–para-electric transition in **vinylidene fluoride-trifluoro-ethylene co-polymer (VDF-TFE)** have also been made [101]. The para-electric phase is, in fact, structurally of the same type as the hexagonal phases in polyphosphazenes and in PBD, i.e. columnar, with only 2-D positional order. The starting sample had a long period of 40 nm, and this changed on first heating through the transition to 50 nm. Subsequent traversing the transition temperature range had little effect on L. Changes in crystallinity and hysteresis effects were also investigated.

More examples of real-time phase transition studies are presented in the section on simultaneous techniques.

(d) Polymer deformation

Wilke and Bratrich [102] studied **restructuring in polyethylene during drawing** at different temperatures. The transformation from lamellar to microfibrillar morphology was investigated in detail. An interesting reversible change was recorded in the four-point SAXS pattern of high-molecular-weight Hostalen GUR (Fig. 3.17). At 100°C and draw ratios λ up to *c.* 4 a four-point pattern is observed. This was attributed to the formation of the 'V-structure', schematically represented in Fig. 3.18(b). With further drawing at $\lambda = 5.0$, layer line streaks develop across the meridian, indicating narrowing of microfibrillar elements through slippage [Fig. 3.18(c)]. On release of stress and redrawing to $\lambda = 4.0$ the four-point pattern reappears [Fig. 3.17(c)], indicating that the narrow microfibrillar elements had slipped back into line as shown in Fig. 3.18(b). This unusual reversibility is attributed to the high degree of chain entanglement or physical cross-linking between crystalline elements, present in the high-molecular-weight polymer.

Kinetics of crystallization of uncross-linked polyisobutylene rubber **following rapid stretching** was monitored by measuring wide-angle reflection intensities [103]. Draw ratios of 300–800% were achieved in a fraction of a second, and the half-time for crystallization was 40 s. Slower stretching did not produce crystallization. On stress release crystallinity disappeared instantly.

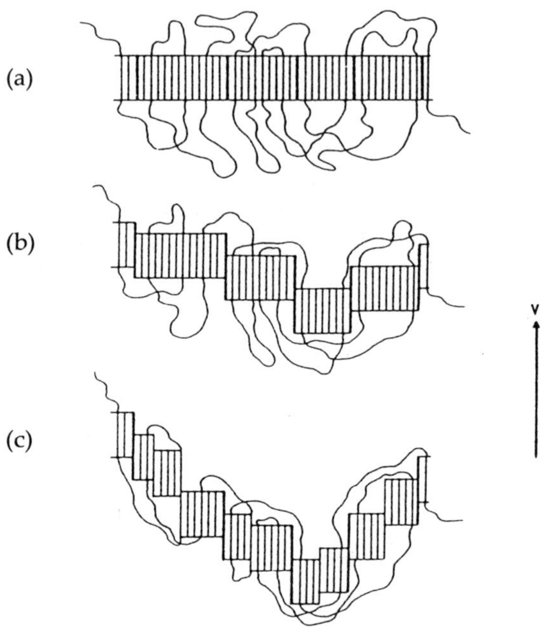

Fig. 3.18 Schematic diagram of the V-structure occurring during drawing of high-molecular-weight polyethylene (cf. Fig. 3.17). (From [102] by permission of International Union of Crystallography.)

High strain rate deformation of **rubber-modified plastics** has been studied at CHESS by real time SAXS [104] using a linear photodiode array detector. Transmitted, as well as scattered intensities from high-impact polystyrene and ABS co-polymers were investigated and **craze formation** was distinguished from the general microcavitation process. The average diameter of the bridging microfibrils was found to be 14 nm, irrespective of strain rate. Contrary to some established theories, a large extent of plastic deformation was found to precede craze formation. The increase in volume, normally observed on deformation of this type of materials, was attributed largely to cavitation of the rubber particles, rather than to craze formation. Reduction in the size of rubber particles in high-impact polystyrene from 2.8 μm to 1.2 μm increased the lag time for crazing from 19 ms to 25 ms, indicating the higher effectiveness of smaller particles.

(e) Microphase separation

As most polymer pairs are immiscible, phase separation is a typical feature of polymer blends. In the few polymer pairs that do show miscibility, an upper (UCST) or a lower critical solution temperature

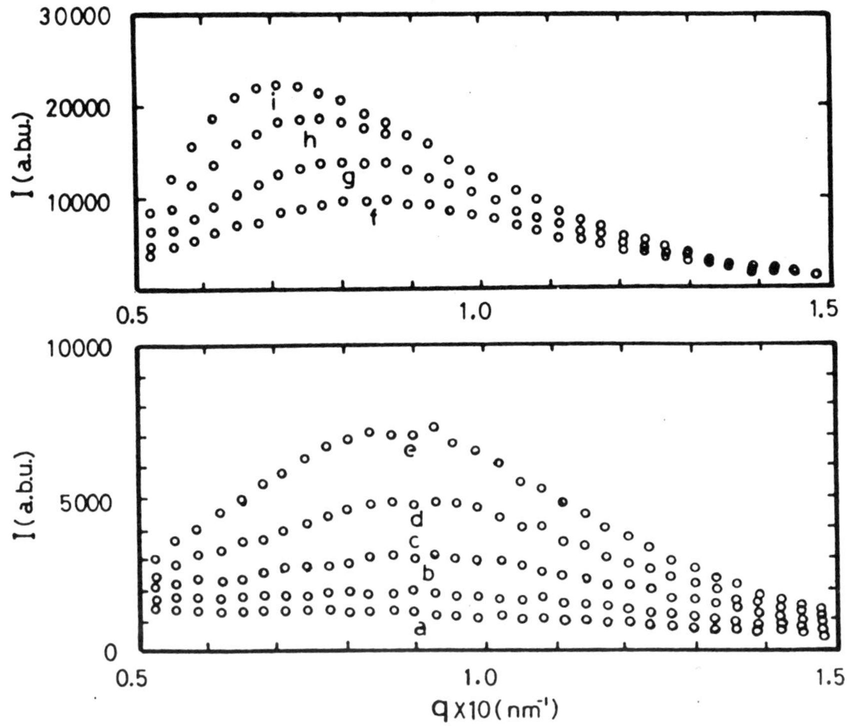

Fig. 3.19 Time-resolved SAXS profiles of a 60/40 PMMA-SAN blend following T-jump to 200° C (a) after 35 s; (b) 47 s; (c) 59 s; (d) 71 s; (e) 83 s; (f) 95 s; (g) 119 s; (h) 137 s; (i) 161 s. (From [108] by permission of Polymer Society of Japan.)

(LCST) exists. On heating through LCST or cooling through UCST **spinodal decomposition** occurs. This takes place through coherent concentration fluctuations, their amplitudes increasing with time. The scale of these fluctuations is usually in the region of the wavelength of light, but in the early stage the scale is several tens of nanometres, i.e. suitable for investigation by SAXS. Several time-resolved studies, using synchrotron radiation, have so far been carried out [105, 106, 107, 108]. Figure 3.19 shows the time evolution of SAXS following a temperature increase through LCST of a blend of PMMA and styrene-acrylonitrile (SAN) co-polymer [108]. As concentration fluctuations increase, scattered intensity increases continuously with time. However, the domain size remains constant in the early period of spinodal decomposition, as indicated by the constant position of the scattering maximum, curves a–e. The increase starts only later, curves f–i, to continue into the visible light scattering range further on. Quantitative comparisons with existing theories of spinodal decomposition showed certain discrepancies which could point to failures of theories when fluctuations are not much greater than the radius of gyration of the polymer [109].

The facility for *in situ* recording of X-ray scattering during the **reaction injection moulding (RIM)** process at the Stanford synchrotron has already been mentioned. The apparatus has been employed in studies of microphase separation and crystallization during the fast polyurethane production process [68]. Real-time WAXS and SAXS intensities were recorded following the filling of the mould. The former gave information on crystallization and the latter on microphase separation. No crystallization occurred in the miscibility region, whereas below a limiting temperature crystallization was found to be induced by a microphase-separated amorphous precursor.

Binary blends of poly(ϵ-caprolactone) (PCL) and polystyrene oligomer show an UCST in the same temperature region as the crystal melting point of PCL. Time-resolved SAXS experiments [110] showed that **crystallization and phase separation** occur simultaneously. A model was proposed in which two distinct phases coexist at low temperatures: one amorphous and rich in polystyrene, and the other semicrystalline with lamellar PCL crystals separated by mixed amorphous layers.

A study of **block co-polymers** of PCL and polystyrene [111] showed that at 36/64 composition a mesophase is formed immediately upon quenching to 28.7°C; this is characterized by a very sharp diffraction peak at 10 nm. However, crystallization of PCL segments takes place subsequently, giving rise to a broader SAXS peak centred around 18 nm in place of the sharp mesophase peak. The nature of the mesophase was not established as only the first diffraction peak was observed, but it was clear that in this case the subsequently developed semicrystalline morphology, of the usual lamellar type, was not determined by that of the precursor mesophase.

3.3.3 Simultaneous techniques

With the advent of time-resolved X-ray diffraction, it became desirable to simultaneously perform two or more experiments on the same specimen. For studies of polymers one clear candidate for such integration is simultaneous recording of **small-** *and* **wide-angle scattering**. Such experimental arrangements, with different combinations of one- and two-dimensional detectors, are currently operational at several synchrotron sources [56, 112, 113].

A proposed design for a combined high-resolution WAXS and SAXS camera [114] for time-resolved synchrotron experiments is depicted in Fig. 3.20. The cylindrical WAXS camera contains a translating imaging plate drum, while the SAXS pattern is recorded by a 2-D position-sensitive detector. In this way an extended wide-angle range could be

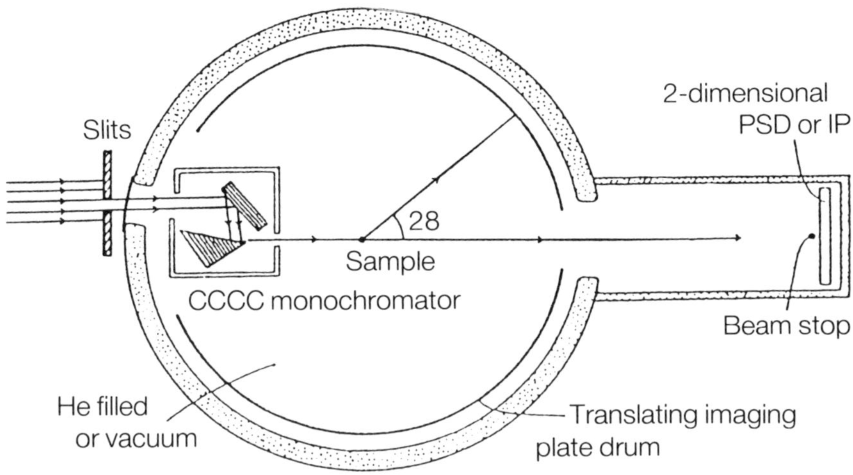

Fig. 3.20 Schematic drawing of the design for a high-resolution WAXS powder diffractometer and SAXS system for time-resolved experiments [114]. The size of the channel cut monochromator is exaggerated. Camera radius is 300 mm. (By permission of the American Institute of Physics.)

covered, the parallax problem eliminated, and intense diffraction lines recorded at high speeds due to unlimited counting rate capability of the imaging plates.

Combined WAXS–SAXS studies of *in situ* melt crystallization of several polymeric systems have been performed using the apparatus at DESY. In the case of poly(ethylene terephtalate) (PET) it was reported that long-period density fluctuations, observed by SAXS, appeared well in advance of any crystallinity detectable from WAXS [115]. The suggested interpretation of this finding was that critical density fluctuations, of the type found in spinodal decomposition, took place first in this single component system, resulting in a lamellar crystal morphology which is not caused by the usual effects of polymer crystallization kinetics. More detailed studies along these lines have recently been carried out using the Franks SAXS camera with a rotating anode source at Kyoto University [116].

A further addition to the SAXS–WAXS system at DESY is small-angle **light scattering** which can be recorded simultaneously with the X-ray data. The whole set-up is depicted schematically in Fig. 3.21. SAXS is recorded with a linear or area proportional counter C. WAXS is captured on a phosphor screen S_c and mirrored (M_1) on to the objective lens O of a Vidicon camera (V). At the same time visible light from laser L passes through the sample (S) in the reverse direction, is polarized (PS) and reflected on to a CCD camera. This triple instrument has been used for

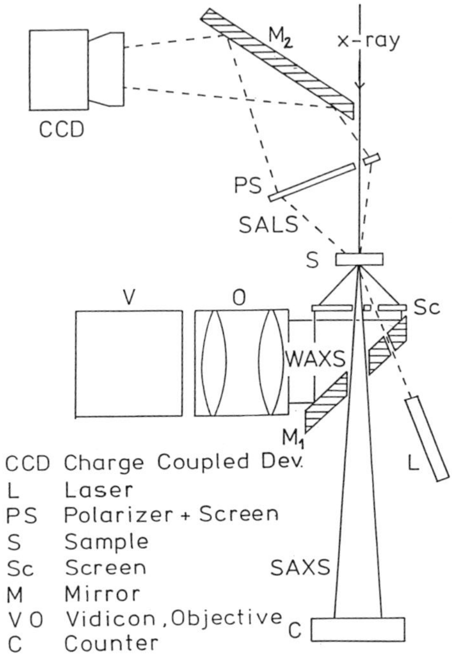

Fig. 3.21 Schematic diagram showing the facility for simultaneous recordings of SAXS, WAXS and small-angle light scattering (SALS) at DESY [117].

studies of primary and secondary crystallization of polymers like PET, poly(ethylene naphtoate) (PEN) and VDF-TFE co-polymer where, respectively, spherulitic, ellipsoid and rod-like morphology could be distinguished from light-scattering patterns [117].

The second example of simultaneous time-resolved techniques is the combination of SAXS with **differential scanning calorimetry** (DSC) [118]. While the specimen is heated or cooled, and the diffractograms collected using synchrotron radiation, the heat flow into or out of the specimen is also being recorded. Thus a uniquely defined one-to-one correspondence is established between the structural changes as revealed by diffraction, and thermal events such as endotherms, exotherms or changes in heat capacity. The technique has been applied to monitoring the complex melting process in solution-crystallized polyethylene and to studies on segmented polyurethanes.

The simultaneous X-ray diffraction/DSC technique (XDDSC) has subsequently been modified by selecting a more X-ray transparent material for specimen cells and has been applied to studies of phase behaviour in a number of molecular and liquid-crystal systems, both polymeric and non-polymeric [119]. WAXS as well as SAXS ranges were

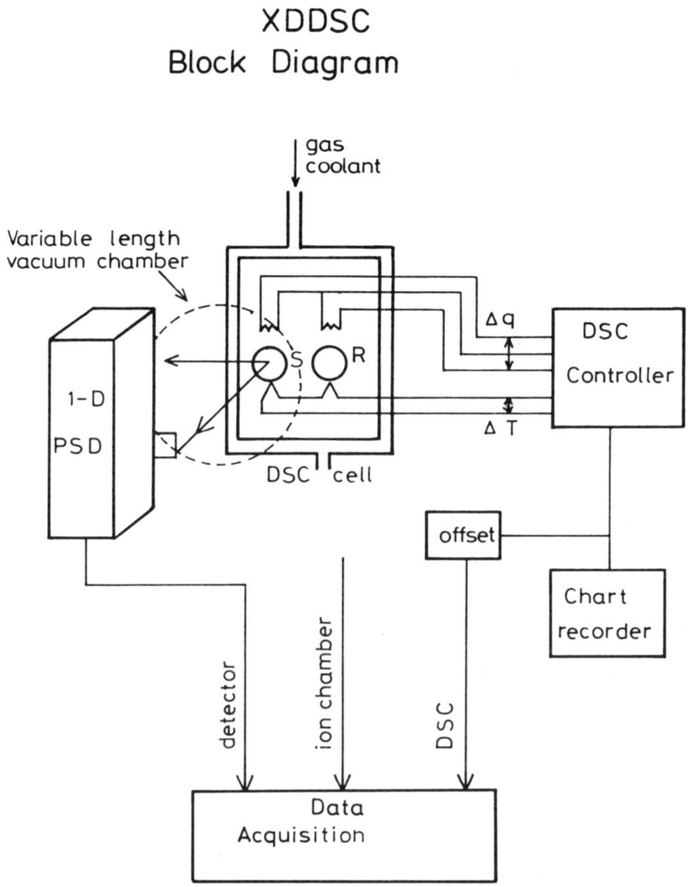

Fig. 3.22 Principle of operation of the apparatus for simultaneous scanning calorimetry and X-ray diffraction (XDDSC). S = sample, R = reference, 1-D PSD = linear position-sensitive detector. (From [119] by permission of Gordon and Breech Publishers.)

covered. The material used for sample pans and cuvettes is either boron nitride or graphite, provided the scattering angles of interest are not very low. Both materials are highly transparent to X-rays, have good thermal conductivity and can be easily machined. The use of sample pans made of these materials enables the achievement of a ten-fold increase in transmitted X-ray intensity over aluminium pans. The scheme of the experimental XDDSC arrangement is presented in Fig. 3.22.

The advantage of XDDSC is most prominent where metastable states are involved and the behaviour of the specimen is not fully reproducible. A good example is **melting of gel-drawn polyethylene**. When gel-drawn

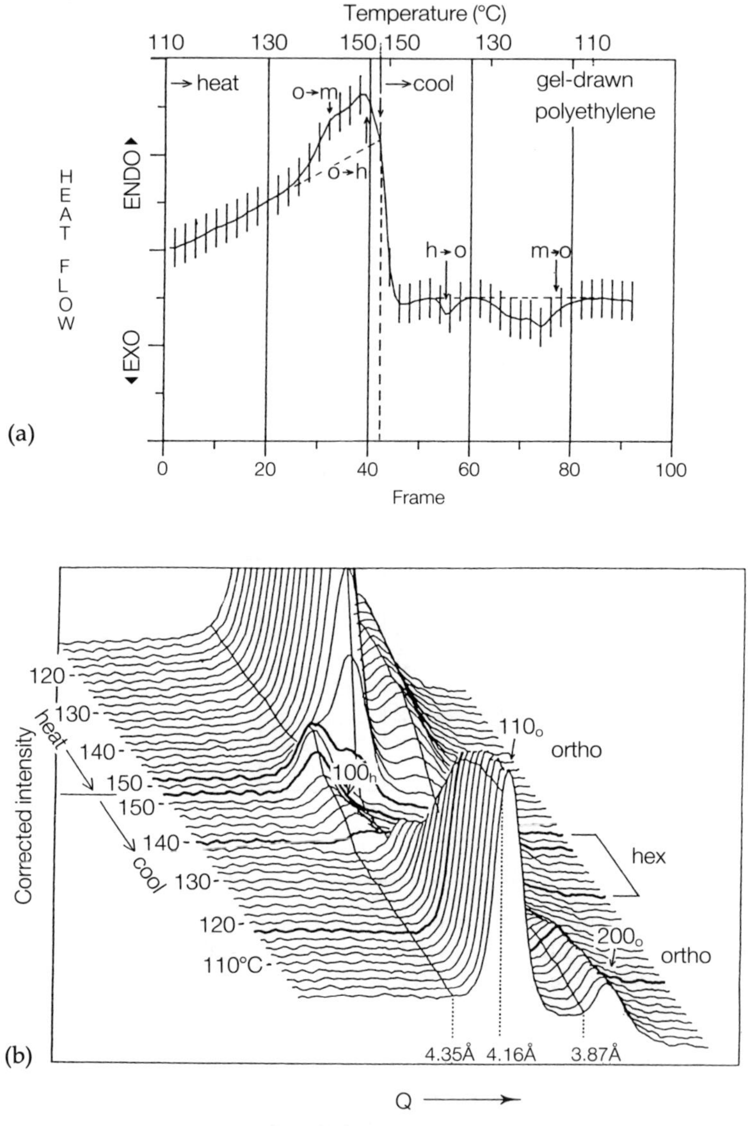

Fig. 3.23 Heating and cooling XDDSC cycle for an embedded gel-drawn high modulus polyethylene tape (a) thermogram; (b) simultaneously recorded diffractograms showing the strongest equatorial reflections: orthorhombic (o) 100 and 200 and hexagonal (h) 100 (Bragg spacings are indicated at the bottom). Heating (5° C min^{-1}) was stopped and cooling (−5° C min^{-1}) started at 152° C, i.e when part of the sample was in the melt (m) and part was in the hexagonal phase. Bold curves refer to the arrow positions in the thermogram. (From [120] by permission of American Chemical Society.)

fibres of ultra-high molecular weight polyethylene (UHMWPE) are mechanically constrained to prevent shrinkage, transition into the hexagonal columnar mesophase occurs around the normal melting temperature of orthorhombic crystals. Final isotropization takes place 20°C higher. The precise isotropization temperature and the mesophase fraction will depend on how well the particular specimen is constrained. Figure 3.23 shows a XDDSC scan [(a): thermogram; (b): diffractograms] of epoxy-embedded tape consisting of 5°C/min heating to the hexagonal phase (152°C) immediately followed by cooling at the same rate [120]. Each endo- and exotherm in the complex thermogram can be uniquely assigned here. Note in particular the small sharp cooling exotherm near 140°C which, for the first time, shows the columnar→crystal transition, uniquely fast and reversible in polyethylene.

XDDSC is particularly useful in identifying **monotropic liquid-crystal phases** in polymers. A monotropic phase is metastable and appears only on cooling. Polarized microscope identification is usually hampered as the texture is too fine grained, whereas by the time X-ray diffraction is recorded using a conventional source the sample would have already changed phase. The next example shows how the existence of a metastable nematic phase was confirmed in the main-chain polyether MBPE-9 [121]. Although the annealed polymer undergoes only direct crystal→isotropic melting on heating, an intermediate birefringent phase was found on cooling. The thermogram and accompanying diffractograms recorded in a cooling XDDSC scan are shown in Fig. 3.24. As can be seen, there is very little change in the scattering pattern between the isotropic melt and the intermediate phase, both showing only a diffuse scattering maximum. Hence the intermediate phase is identified as nematic. Such X-ray identification depends critically on the certainty that the relevant diffractograms had indeed been recorded between the two exothermic events and not, for example, while the sample was still isotropic.

Combining X-ray diffraction with other types of dynamic measurements is possible and has been explored to a certain extent. Recording diffraction patterns during mechanical deformation, with **simultaneously monitoring stress–strain behaviour** [60], has already been mentioned.

3.3.4 Surface techniques

(a) X-ray reflectivity

Structure of air–polymer or polymer–polymer interfaces has been studied increasingly actively in recent years by measuring reflectivity of neutrons, and to some extent also of X-rays. The refractive index n for X-rays (and

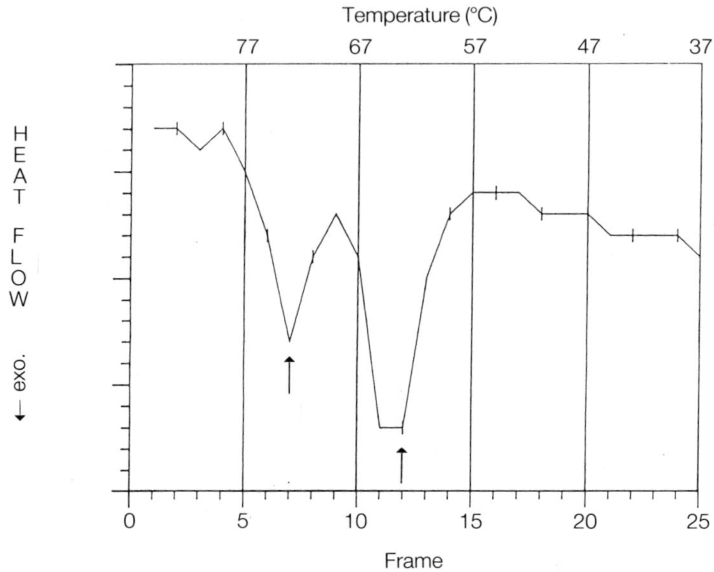

(a)

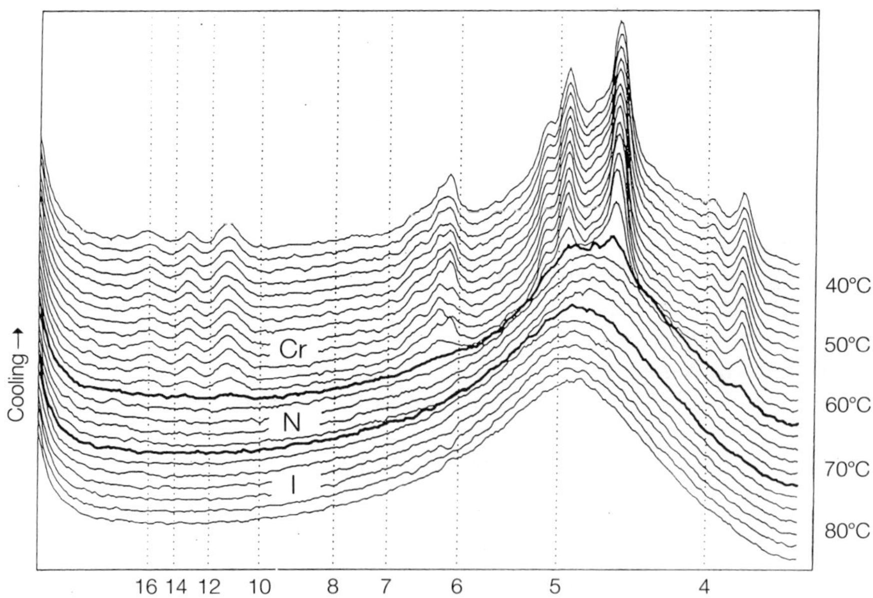

(b)

Fig. 3.24 Simultaneous X-ray/DSC (XDDSC) cooling scan (2° C min^{-1}) for polymer MBPE-9 (a) thermogram; (b) diffractograms (one every 2° C). Bold diffractograms correspond to arrowed positions on the thermogram. Cr = crystal, N = nematic, I = isotropic. (From [121] by permission of American Chemical Society.)

neutrons) in matter is somewhat less than 1, in contrast to n for visible light [see equation (3.1)]. Thus, as already mentioned in describing X-ray mirrors, **external** total reflection occurs when the beam impinges on a flat surface below the critical angle θ_c [Equation (3.3)]. Fresnel reflectivity R_F at an interface between two materials, denoted 1 and 2, is [122]:

$$R_F = \left(\frac{\mathbf{k}_{z_1} - \mathbf{k}_{z_2}}{\mathbf{k}_{z_1} + \mathbf{k}_{z_2}}\right)^2 \tag{3.16}$$

where

$$\mathbf{k}_{z_i} = (2\pi n_i/\lambda)\sin\theta_i \tag{3.17}$$

Here $\mathbf{k}_{z_i}$ is the wave vector component along z, i.e. normal to the interface, and θ_i is the angle between z and the propagation direction in medium i. For many applications reflected intensity R can be expressed as a product of R_F and the square of the Fourier transform of the **gradient** of electron density along the surface normal.

If there is more than one interface, e.g. if a thin layer of polymer A is deposited on a polymer substrate B, interference of the two reflected beams produces a modulated curve of reflectivity versus angle. For flat and parallel interfaces, such as may be obtained, e.g. by spin coating, **specular reflection** may occur on both interfaces, provided refractive index decrements are sufficiently large at each interface. A series of interference fringes is then obtained (Fig. 3.25), the fringe spacing giving an accurate measure of the thickness of the polymer A layer, irrespective of its density. More importantly, the complete experimental reflectivity function can be compared with calculated curves based on assumed electron density depth profiles [123]. Thus the averaged electron density projection on to the surface normal is determined. For example, interdiffusion of polymers A and B across the interface may be modelled and studied experimentally, which is of relevance to problems such as adhesion, welding etc.

Optically flat interfaces are required for specular reflection. The unavoidable residual surface roughness can be included in the model through a smearing error function of variance σ. In the case of neutron reflectivity, contrast between the two polymers can be high if one polymer is hydrogenous and the other deuterated. On the other hand, X-ray refractive indices of organic polymers are usually very similar, and by far the largest contrast is that between polymer and air. However, presence of heavy atoms in polymer B can enable X-ray reflectivity studies, such as has been performed on the system polystyrene–poly(bromostyrene) (PS–PBrS) [124, 125].

Figure 3.25 shows X-ray reflectivity curves of a PS–PBrS bilayer before (solid line) and after (dashed line) annealing at 136°C, i.e. between the glass transitions of the two polymers. Initial thicknesses of PS and PBrS

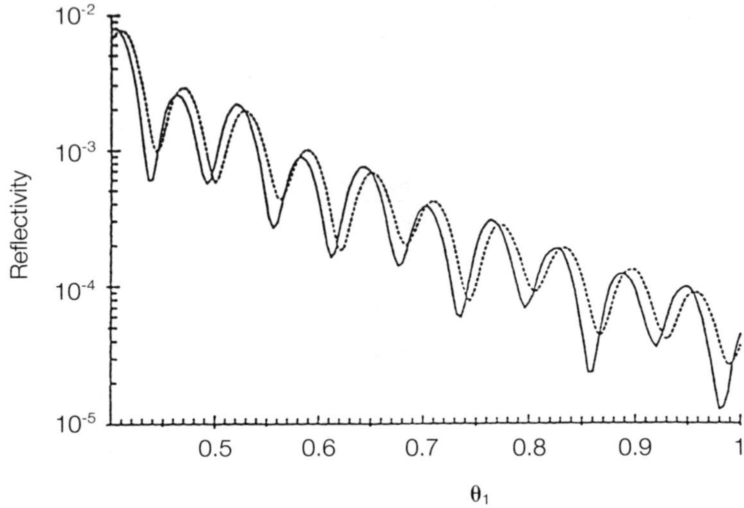

Fig. 3.25 Measured X-ray reflectivity curves from a PS–PBrS bilayer before (solid line) and after annealing at 136° C for 180 min (dashed line). (From [124] by permission of American Chemical Society.)

layers were 40 nm and 31 nm, respectively. The observed reduction in modulation amplitude after annealing is the result of some intermixing. Model calculation showed that the change can be accounted for by the formation of a partially mixed interlayer of tan profile from which the Flory–Huggins interaction parameter χ could be obtained.

Structures of spread monolayers on water surface have been determined by fitting squared Fourier transforms of model electron density profiles to observed X-ray reflectivity curves [126]. Langmuir–Blodgett film multilayers on solid substrate can be studied by the same technique. Since the total film thickness consists of an integer number of molecular layers, the interference fringes due to film thickness now become subsidiary maxima in an interference function dominated by a series of strong harmonics of the basic layer repeat. Ferro-electric and antiferroelectric L–B films of side-chain S_{C^*} liquid-crystal polymers, deposited by repeated (head-to-tail) or alternate reversed (head-to-head) layer deposition have been studied in this way [127]. In principle, lamellar phases of block co-polymers give similar patterns as L–B films, although the degree of regularity is usually lower.

While specular reflection gives information on the electron density profile normal to the surface, **diffuse reflectivity** provides an insight into the in-plane density fluctuations [128]. Thus surface roughening upon heat annealing of an initially smooth surface of polystyrene-PMMA diblock co-polymer has been investigated by Sirota and co-workers [129]

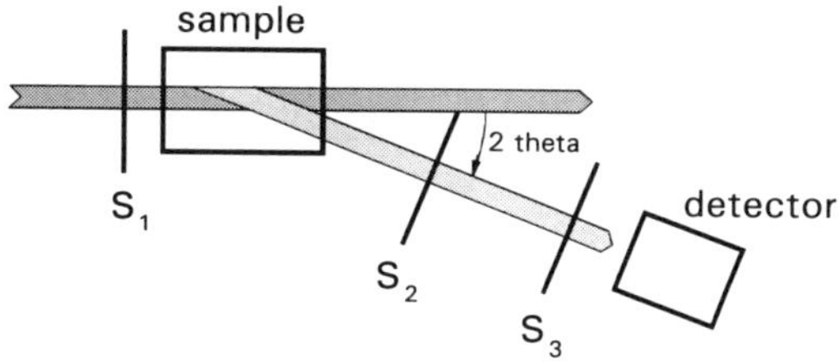

Fig. 3.26 Top view of the geometry for surface X-ray reflection–diffraction measurements of in-plane order. S_1, S_2 and S_3 are slits. The beam, normally wide but thin, is monochromatized by a double crystal and then focused by a mirror.

by observing diffuse rather than specular reflection profiles, i.e. by introducing a finite off-set between the incidence angle and the detector acceptance angle. Initially 7.5 bilayers of PS-*co*-PMMA were exactly covering the surface. However, the experiment revealed that thermal expansion on annealing leads to formation of islands of one additional bilayer on the top surface.

(b) Evanescent wave scattering

Marra, Eisenberger and Cho [130] pioneered the **diffraction–reflection**, or grazing incidence diffraction technique for studying the structure of thin surface layers. The incidence angle is kept fixed in this case and is below the critical angle θ_c in order to maximize the intensity diffracted from the surface layer. Since total reflection occurs, no beam escapes into the sample interior but there is instead a component of energy flow parallel to the surface (evanescent wave). For grazing angles of incidence less than θ_c the exponentially decaying penetration of evanescent wave in organic matter is typically of the order of 5 nm. The diffraction which this produces thus gives information on surface structure. As θ_c is approached and exceeded the penetration depth of the evanescent wave increases sharply and the technique ceases to be surface sensitive.

The reflection–diffraction technique has originally been applied to epitaxially grown metal layers on semiconductor substrate. However it has found application in studies of **lateral order** in organic monolayer films on water surface and in studies of two-dimensional order in thin films in general. It has also been used recently for studying surface ordering in solid polymer films [131, 132]. For such applications diffraction **out of the plane of reflection** is recorded.

Figure 3.26 shows schematically the top view of the experimental geometry for reflection–diffraction. A more detailed description is given, for example, in reference 133. By rotating the sample around the vertical axis one can determine whether the irradiated area is a single 2-D crystal or a 2-D powder, and whether there is preferential in-plane orientation. A comparatively wide beam is usually needed to cover sufficient area for obtaining reasonable intensities, which means that experimental angular resolution is not very high. This is not a problem in the case of floating monomolecular layers since intrinsic diffraction profiles are usually broad. Phase transitions, including hexatic ordering, have been observed with changing surface pressure [133]. The undulator-generated high brilliance beam for surface diffraction at ESRF will be only 1 mm wide and will thus enable high-resolution experiments.

Studying surface diffraction from spin-coated films of an aromatic polyimide, Russell *et al.* [134] found that although the film interior was non-crystalline (or liquid crystalline), an approximately 80 Å deep layer at the air–polymer interface gave characteristic crystalline Bragg diffraction peaks. The depth profile of crystallinity could be determined by taking scans at different $\mathbf{q}_z$ elevation, with $\mathbf{z}$ being normal to the surface.

The properties associated with total reflection of X-rays have further given rise to a method of studying polymer adsorption from solution. **Evanescent wave fluorescence** relies on a grazing incidence beam inducing X-ray fluorescence in the thin layer of polymer adsorbed on a flat solid–liquid interface. The polymer must be tagged, i.e. have atoms in which fluorescence can be induced. Thus, e.g. Mo Kα radiation from a sealed tube has been used to excite 13.5 keV fluorescence from Br K-edge [135]. Fluorescence is recorded with an energy-dispersive detector tuned into the fluorescent wavelength. Since changing the angle of incidence changes the penetration depth of the evanescent wave, one can obtain a detailed concentration profile on the nanometre scale. In the particular example quoted it was the solvent rather than the polymer which contained Br atoms, providing negative contrast. Alternatively polymers such as poly(bromostyrene) could be used in a variety of non-brominated solvents. As a further possibility the authors [135] suggest silicon-containing polymers which could be excited by Cr radiation. It would clearly be advantageous to use synchrotron radiation from, e.g. a wiggler, the wavelength of which could be chosen to suit the composition of the polymer or solvent under study.

3.4 CONCLUSION AND FUTURE PROSPECTS

Synchrotron radiation has augmented the power of X-ray diffraction generally, with polymer science being a major beneficiary. Intrinsic collimation and high brilliance make synchrotron sources especially well suited for small-angle scattering, an important technique in polymer research. Due to steady improvements in X-ray optics, a number of SAXS stations today offer resolution of several-hundred nanometres without the need to resort to special methods such as Bonse–Hart. Thus accurate intensity extrapolations to $q = 0$ can now be made, and phenomena like extended chain crystallization or near-micron scale phase separation can, in principle, be investigated. Time-resolved diffraction, both at wide and small angles, will undoubtedly remain a major area of application of synchrotron radiation. This is now increasingly being combined with other simultaneous experiments, like DSC and light scattering and, furthermore, WAXS and SAXS data are now often being collected concurrently. The development of high count rate area detectors with large-size memory and fast data transfer has enabled 2-D data collection with millisecond time resolution. This potential has been used to great advantage in studies of, e.g. muscle contraction, but has not yet been exploited in the area of synthetic polymers. Crystallization at high undercooling, fast phase transitions, switching and similar processes are some of the candidates for high time resolution experiments. With small source size and sophisticated X-ray optics, including capillary focusing, the beam size is being reduced to the order of a micron on selected experimental stations. This offers prospects for X-ray diffraction on polymer-single crystals using, for example, Laue diffraction. Radiation damage becomes a serious consideration under such circumstances, but not as serious as in biological systems, where Laue microdiffraction programmes are already under way. Finally, surface and thin-film studies by both reflectivity and grazing incidence diffraction are likely to continue increasing their share of synchrotron X-ray research on polymers, particularly in view of the growing area of organized molecular films for electronic and optical applications.

REFERENCES

1. Elsner, G., Riekel, C. and Zachmann, H.G. (1985) *Adv. Polymer Sci.* **67**, 1.
2. Russell, T.P. (1991) in *Handbook on Synchrotron Radiation*, Vol. 3, (Eds G.S. Brown and D.E. Moncton), North Holland, Amsterdam.
3. Gehrke, R. (1989) *Topics in Current Chemistry*, Springer–Verlag, **151**, 111.
4. Greaves, G.N. and Catlow, C.R.A. (1990) in *Applications of Synchrotron Radiation*, Blackie and Sons, Glasgow, p 1.
5. Wulff, M. (1991) European Synchrotron Radiation Facility User Meeting Information, Grenoble, p 35.
6. Ross, J.S.H. and Smith, K.D. (1992) SRS Users Meeting, Daresbury.
7. Bösecke, P. (1991) European Synchrotron Radiation Facility User Meeting Information, Grenoble, p 47.
8. Kunz, C. (1979) *Synchrotron Radiation Topics in Current Physics*, Springer–Verlag, Berlin.
9. Wilson, M.N., Smith, A.I.C., Kempson, V.C. *et al.* (1990) *Microelec. Engn.* **11**, 225.
10. Kratky, O. and Glatter, O. (1982) *Small-Angle X-Ray Scattering*, Academic Press, London.
11. Vonk, C.G. (1971) *J. Appl. Crystallogr.* **4**, 340.

12. Strobl, G.R. (1972) *Kolloid Z. Z. Polym.* **250**, 1039.
13. Chu, B., Wu, D. and Wu, C. (1987) *Rev. Sci. Instr.* **58**, 1158.
14. Franks, A.B. (1958) *J. Appl. Phys.* **9**, 349.
15. Hayashi, H., Hamada, F., Suehiro, S. *et al.* (1988) *J. Appl. Cryst.* **21**, 330.
16. Huxley, H.E. and Brown, W. (1967) *J. Mol. Biol.* **30**, 383.
17. Compton, A.H. (1923) *Phil. Mag.* **45**, 1121.
18. Chapman, H.N., Nugent, K.A. and Wilkins, S.W. (1991) *Rev. Sci. Instrum.* **62**, 1542.
19. Matsushita, T. and Hashizume, H. (1983) in *Handbook of Synchrotron Radiation*, Vol. 1A, (Eds D.E. Eastman and Y. Farge), North Holland, Amsterdam.
20. Johann, H.H. (1931) *Z. Phys.* **69**, 185; Warren, B.E. (1950) *Rev. Sci. Instr.* **21**, 102.
21. DuMond, J.W.M. and Kirkpatrick, H.A. (1930) *Rev. Sci. Instr.* **1**, 88; Johansson, T. (1932) *Naturwissenschaften*, **20**, 159.
22. Beaumont, J.H. and Hart, M. (1974) *J. Phys. E, Sci. Instr.* **7**, 823.
23. Bonse, U. and Hart, M. (1965) *Appl. Phys. Lett.* **7**, 238.
24. Bonse, U. and Hart, M. (1966) *Z. Phys.* **189**, 151.
25. Bonse, U. and Hart, M. (1967) in *Small Angle X-Ray Scattering*, (ed. H. Brumberger), Gordon and Breach, New York.
26. Matsuoka, H., Kakigami, K., Ise, N. *et al.* (1991) *Rigaku Journal*, **8**, 22.
27. Wilkins, S.W. and Stevenson, A.W. (1988) *Nucl. Instr. Methods*, **222**, 121.
28. Fankuchen, I. (1937) *Nature*, **139**, 193.
29. Riekel, C. and Suortti, P. (1991) *J. Appl. Cryst.* **24**, 549.
30. Stevenson, G.B., Ludwig, K.F., Jordan-Sweet, J.L. *et al.* (1989) *Rev. Sci. Instr.* **60**, 1537.
31. Buras, B. and Gerward, L. (1989) *Prog. Cryst. Growth and Charact.* **18**, 93.
32. Häusermann, D. and Barnes, P. (1992) *Phase Transitions*, **39**, 99.
33. Hendrix, J. (1985) *Adv. Polym. Sci.* **67**, 59.
34. Gruner, S.M. (1989) *Rev. Sci. Instr.* **60**, 1545.
35. Amemiya, Y., Matsushita, J., Nakagawa, A. *et al.* (1988) *Nucl. Instr. Meth.* **A246**, 572.
36. Hall, C.J., Lewis, R.A., Parker, B. and Worgan, J. (1991) *Nucl. Instr. Meth.* **A310**, 215.
37. Cernik, R.J., Clark, S.M., Deacon, A.M. *et al.* (1992) *Phase Transitions*, **39**, 187.
38. Grove, R., Lee, K., Perez-Mendez, V. and Sperinde, J. (1970) *Nucl. Instr. Meth.* **89**, 257.
39. Gabriel, A. (1977) *Rev. Sci. Instr.* **48**, 1303.
40. Borowski, C.J. and Kopp, M.K. (1975) *Rev. Sci. Instr.* **40**, 951.
41. Lewis, R.A., Fore, N.S., Helsby, W. *et al.* (1992) *Rev. Sci. Instr.* **63**, 642.
42. Kahn, R., Fourme, R., Bosshard, R. *et al.* (1989) *Rev. Sci. Instr.* **60**, 1568.
43. Kahn, R., Fourme, R., Bosshard, R. *et al.* (1982) *Nucl. Instr, Meth.* **201**, 203.
44. Lewis, R.A. (personal communication).
45. Oed, A. (1988) *Nucl. Instr. Meth.* **A263**, 351.
46. Hendrix, J. (1982) *EMBL Research Report*.
47. Magorrian, B.G. and Allinson, N.M. (1988) *Nucl. Instr. Meth.* **A273**, 599.
48. Eikenberry, E.F., Tate, M.W. Belmonte, A.L. *et al.* (1991) *IEEE Trans. Nucl. Sci.* **38**, 110.
49. Strauss, M.G., Naday, I., Sherman, J.S. *et al.* (1987) *IEEE Trans. Nucl. Sci.* **34**, 389.
50. Allinson, N.M., Carr, P.D., Colapietro, M. *et al.* (1992) *Phase Transitions*, **39**, 145.

51. Amemiya, Y., Kishimoto, S., Matsushita, T. *et al.* (1989) *Rev. Sci. Instr.* **60**, 1552.
52. Zeitz, R., Heuer, J. and Riekel, C. (unpublished).
53. Clements, J., Zachmann, H.G. and Ward, I.M. (1988) *Polymer*, **29**, 1929.
54. Gehrke, R., Riekel, C. and Zachmann, H.G. (1989) *Polymer*, **30**, 1582.
55. Song, H.H., Wu, D.Q., Chu, B. *et al.* (1990) *Macromolecules*, **23**, 2380.
56. Hsiao, B.S., Gardner, K.H., Wu, D.Q. *et al.* (1992) *Amer. Chem. Soc. Polymer. Prept.* **33**, 265.
57. Spells, S.J. (unpublished).
58. Kriechbaum, M., Rapp, G., Hendrix, J. and Laggner, P. (1989) *Rev. Sci. Instr.* **60**, 2541.
59. Wilke, W. and Bratrich, M. (1991) *J. Appl. Cryst.* **24**, 645.
60. van Aerle, N.A.J.M. and Braam, A.W.M. (1988) *J. Appl. Cryst.* **121**, 106.
61. Abramic, D. and Ungar, G. (unpublished).
62. Chan, K.K., Pershan, P.S., Sorenson, L.B. and Hardouin, F. (1986) *Phys. Rev. A*, **34**, 231.
63. Odell, J.A., Keller, A., Atkins, E.D.T. *et al.* (1989) *Mat. Res. Soc. Symp. Proc.* **134**, 223; Odell, J.A., Ungar, G. and Feijoo, J.L. (1993) *J. Polym. Sci., Polym. Phys. Ed*, **31**, 141.
64. Safinya, C.R., Sirota, E.B. and Plano, R.J. (1991) *Phys. Rev. Lett.* **66**, 1986.
65. Hikosaka, M., Rastogi, S., Keller, A. and Kawabata, H. (1992) *J. Macromol. Sci. – Phys.* **31**, 87.
66. Rastogi, S., Hikosaka, M., Kawabata, H. and Keller, A. (1991) *Macromolecules*, **24**, 6384.
67. Yang, W.P. and Macosko, C.W. (1989) *Makromol. Chem.*, Makromol. Symposium, **25**, 23.
68. Koberstein, J.T., Yu, C.C., Galambos, A.F. *et al.* (1990) *Polym. Prepts., Amer. Chem. Soc.* **31**, 110; Ryan, A.J., Willkomm, W.R., Bergstrom, T.B. *et al.* (1991) *Macromolecules*, **24**, 2883.
69. Bunning, T.J., McNamee, S.G., Klei, H.E. *et al.* (1992) *Amer. Chem. Soc., Polym. Prepts*, **33**, 315.
70. Li, Y.G., Wu, C. and Chu, B. (1991) *J. Polym. Sci., Pt. B: Polym. Phys. Ed.* **29**, 1309.
71. van Aerle, N.A.J.M. and Braam, A.W.M. (1989) *Colloid Polym. Sci.* **267**, 323.
72. Vonk, C.G. (1982) in *Small-Angle X-Ray Scattering*, (ed. O. Glatter and O. Kratky), Academic Press, London.
73. Hosemann, R. and Bagchi, S.N. (1962) *Direct Analysis of Diffraction by Matter*, North Holland, Amsterdam.
74. Fischer, E.W., Goddar, H. and Schmidt, G.F. (1967) *J. Polymer Sci. B*, **5**, 619.
75. Vonk, C.G. and Kortleve, G. (1967) *Kolloid Z. Z. Polym.* **220**, 19.
76. Vonk, C.G. (1973) *J. Appl. Cryst.* **6**, 81.
77. Ruland, W. (1977) *Colloid Polym. Sci.* **255**, 417.
78. Kratky, O., Pilz, I. and Schmitz, P.J. (1966) *Colloid Interface Sci.* **21**, 24.
79. Pilz, I. and Kratky, O. (1967) *Colloid Interface Sci.* **24**, 211.
80. Barham, P.J. and Keller, A. (1989) *J. Polym. Sci., Pt. B: Polym. Phys. Ed.* **27**, 1029.
81. Ungar, G., Stejny, J., Keller, A. *et al.* (1985) M.C., *Science* 1985, **229**, 386.
82. Ungar, G. and Rastogi, S. (unpublished).
83. Ungar, G. and Keller, A. (1986) *Polymer* **27**, 1835.
84. Ungar, G. and Keller, A. (1987) *Polymer* **28**, 1899.
85. Cheng, S.Z.D., Chen, J., Zhang, A. *et al.* (1992) *Polymer* **33**, 1140.
86. Grubb, D.T., Liu, J.H., Caffrey, M. and Bilderbeck, D.H. (1984) *J. Polym. Sci., Polym. Phys. Ed.* **22**, 367; Grubb, D.T. and Liu, J.H. (1985) *J. Appl. Phys.* **58**, 2822.

87. Spells, S.J. and Hill, M.J. (1991) *Polymer* **32**, 2716.
88. Kawaguchi, A., Ichida, T., Murukami, S. and Katayama, K. (1984) *Colloid Polym. Sci.* **262**, 597; Ichida, T., Tsuji, M., Murukami, S. *et al.* (1985) *Colloid Polym. Sci.* **263**, 293.
89. Hsiao, B.S., Gardner, K.H., Wu, D.Q., and Chu, B. (1993) *Polymer*, **34**, 3996.
90. Wang, J., Alvarez, M., Zhang, W.J. *et al.* (1992) *Macromolecules*, **25**, 6943.
91. Blundell, D.J. (1987) *Polymer*, **28**, 2248.
92. Bassett, D.C., Olley, R.J. and Al Raheil, I.A. (1988) *Polymer*, **29**, 1745.
93. Santa Cruz, C., Stribeck, N., Zachmann, H.G. and Balta-Calleja, F.J. (1991) *Macromolecules*, **24**, 5980.
94. Ungar, G. (1990) *Amer. Chem. Soc., Polym. Prepts*, **31**, 108.
95. Tashiro, K., Satkowski, M.M., Stein, R.S. *et al.* (1992) *Macromolecules*, **25**, 1809.
96. Forgacs, P. *et al.* (1980) *J. Polymer Sci., Polym. Phys. Ed.* **18**, 2155; Forgacs, P., Tolochko, B.P. and Sheromov, M.A. (1981) *Polym. Bull.* **6**, 127.
97. Garbarczyk, J., Starzynski, T. and Paukszta, D. (1989) *Polymer Commun.* **30**, 153.
98. Ungar, G. (1992) *Prog. Colloid Polym. Sci.* **87**, 53; (1993) *Polymer*, **34**, 2050.
99. Magill, J.H. and Riekel, C. (1986) *Makromol. Chem., Rapid. Commun.* **7**, 287.
100. Rastogi, S. and Ungar, G. (1992) *Macromolecules*, **25**, 1445.
101. Bourgaux-Leonard, C., Legrand, J.F., Renault, A. and Delzenne, P. (1991) *Polymer*, **32**, 597.
102. Wilke, W. and Bratrich, M. (1991) *J. Appl. Cryst.* **24**, 645.
103. Koch, M.H.J., Bordas, J., Schöla, E. and Broecker, H.C. (1979) *Polym. Bull.* **1**, 709.
104. Bubeck, R.A., Blazy, J.A., Kramer, E.J. *et al.* (1986) *Polymer Commun.* **27**, 357; Bubeck, R.A., Buckley, D.J., Kramer, E.J. and Brown, H.R. (1990) *Amer. Chem. Soc., Polym. Prepts*, **31**, 116.
105. Hill, R.G., Tomlins, P.E. and Higgins, J.S. (1985) *Macromolecules*, **18**, 2555.
106. Russell, T.P., Hadziioannou, G. and Warburton, W. (1985) *Macromolecules*, **18**, 2555.
107. Meier, H. and Strobl, G.R. (1987) *Macromolecules*, **20**, 649.
108. Ikawa, K., Uemura, A., Hosoda, S. *et al.* (1991) *Polymer J.* **23**, 1291.
109. Binder, K. (1983) *J. Chem. Phys.* **79**, 6387.
110. Nojima, S., Kato, K., Ono, M. and Ashida, T. (1992) *Macromolecules*, **25**, 1922.
111. Nojima, S., Kato, K., Yamamoto, S. and Ashida, T. (1992) *Macromolecules*, **25**, 2237.
112. Bark, M., Schulze, C. and Zachmann, H.G. (1990) *Amer. Chem. Soc., Polym. Prepts.* **31**, 106.
113. Laggner, P. and Mio, H. (1992) *Nucl. Instr. Methods*, **A323**, 86.
114. Barnea, Z., Clapp, R., Creagh, D.C. *et al.* (1989) *Rev. Sci. Instr.* **60**, 2537.
115. Zachmann, H.G. (1991) presented at the Europhysics Conference 'Solidification of Polymers', Stockholm.
116. Imai, M., Mori, K., Mizukami, T. *et al.* (1992) *Polymer*, **33**, 4451; 4457.
117. Zachmann, H.G. and Wutz, C. (1993) in *Crystallization of Polymers* (Ed. M. Dosière) Nato ASI Series, Kluwer, Dordrecht.
118. Russell, T.P. and Koberstein, J.T. (1985) *J. Polym. Sci., Polym, Phys. Ed.* **23**, 1109.
119. Ungar, G. and Feijoo, J.L. (1990) *Mol. Cryst. Liq. Cryst.* **180B**, 281.
120. Feijoo, J.L. and Ungar, G. (1990) *Amer. Chem. Soc., Polym. Prepts*, **31**, 139.
121. Ungar, G., Feijoo, J.L., Keller, A. *et al.* (1990) *Macromolecules*, **23**, 3411.
122. Lekner, J. (1987) *Theory of Reflection*, M. Nijhoff, Dordrecht.

123. Stamm, M. (1991) *J. Appl. Cryst.* **24**, 651.
124. Stamm, M., Reiter, G. Hüttenbach, S. and Foster, M. (1990) *Amer. Chem. Soc., Polym. Prepts*, **31**, 73.
125. Hüttenbach, S., Stamm, M., Reiter, G. and Foster, M. (1991) *Langmuir*, **7**, 2438.
126. Grundy, M.J., Richardson, R.M., Roser, S.J. *et al.* (1988) *Thin Solid Films*, **159**, 43.
127. Thibodeaux, A.F., Geer, R., Qadri, S. *et al.* (1992) *Amer. Chem. Soc., Polym. Prepts*, **33**, 1220.
128. Sinha, S.K., Sirota, E.B., Garoff, S. and Stanley, H.B. (1988) *Phys. Rev. B*, **38**, 2297.
129. Sirota, E.B., Hughes, G.J., Sinha, S.K. *et al.* (1990) *Amer. Chem. Soc., Polym. Prepts*, **31**, 81.
130. Marra, W.C., Eisenberger, P. and Cho, A.Y. (1979) *J. Appl. Phys.* **50**, 6927.
131. Russell, T.P. and Toney, M.F. (1990) *Phys. Rev. Lett.* **66**, 1181; Russell, T.P., Toney, M.F. and Factor, B.J. *Amer. Chem. Soc., Polym. Prepts*, **33**, 259.
132. Martin, D.C., Berger, L.L. and Gardner, K.H. (1991) *Macromolecules*, **24**, 3921.
133. Helm, C.A., Möhwald, H., Kjaer, K. and Als-Nielsen, J. (1987) *Biophys. J.* **52**, 381.
134. Russell, T.P., Tomey, M.F. and Factor, B.J. (1992) *Amer. Chem. Soc., Polym. Prepts*, **33**, 259.
135. Barton, S.W. and Rondelez, F. (1990) *Amer. Chem. Soc., Polym. Prepts*, **31**, 92.

4

Neutron-scattering techniques

A.R. Rennie

4.1 INTRODUCTION

Neutrons have been available to scientists for about sixty years but it is only in the last 25 years that they have been widely exploited as a tool in condensed matter research. This is largely due to the advent of high-flux beam reactors such as those at the Institut Laue Langevin in Grenoble, France, and at Brookhaven National Laboratory in New York. It is not feasible to review all the literature relating to solid polymers even given the finite number of laboratories equipped for such investigations. Reference will be made particularly to those papers that describe advances in the technique and to some review articles on specific applications. The description of theory will be limited to definition of the terminology and a few of the most basic equations.

There is considerable overlap, both in theory and practice, in the development of neutron scattering for the study of solid polymers and progress in other experimental techniques. First, the theory of scattering is largely identical for light, X-rays and neutrons and many models that have been developed for interpretation of data from one type of experiment can be transferred with little or no modification. Secondly we should observe that many experiments are made on polymer solutions, molten polymers or other non-polymeric materials. The methods and theory are normally identical and some reference to these experiments will be made where they indicate interesting potential for future studies on solid polymers or shed light on processes that influence final properties of solids. For example, the morphology of solid polymers is often dominated by kinetic processes in the melt.

The next section will provide a brief introduction to the theory of neutron scattering. It may be useful to mention some of the principle texts and reviews that provide a fuller treatment of this subject: the theory of neutron scattering is described in detail by Lovesey [1] and more concisely by Squires [2]. Neither of these authors devotes much attention to small-angle scattering, which is the most widely used application of neutrons in polymer science, but one book [3] and several reviews [4, 5, 6] are available which provide a good introduction to the subject for polymer scientists. The present description of the general theory will be limited to a discussion of the principles and sufficient introduction to the terminology to provide a guide through the literature. The review will then emphasize the distinctive features and recent developments in the application of neutrons to studies of solid polymers.

Neutrons are often used as just one of several different experimental tools to investigate a given problem in polymer science. Many publications although presenting some data on neutron scattering are not primarily concerned with this technique. In the scope of a short review it is not possible to give the full context to the contributions that neutrons have made to studies of solid polymers. After the introduction to theory and a description of the instrumentation, the main categories of neutron experiment will be considered in turn and recent, significant work will be highlighted.

4.2 PRINCIPLES OF NEUTRON SCATTERING

4.2.1 General remarks

A simple description of scattering can be found in many text books such as that by Born [7]. An idealized scattering experiment is shown in Fig. 4.1. An incident wave with defined wavelength and direction is allowed to fall on the sample. The scattered wave is measured and the intensity is recorded as a function of its angle and energy. It is normal to use vector notation to describe this process relating the incident wavelength λ_i and energy E_i to a wave vector $\mathbf{k}_i$ (momentum $\mathbf{p}_i$) with the de Broglie relationships:

$$E_i = h\nu_i\,,\ \mathbf{p}_i = \hbar\mathbf{k}_i \tag{4.1}$$

$$|\mathbf{k}_i| = 2\pi/\lambda_i \tag{4.2}$$

where ν_i is the frequency of the wave. Planck's constant h has the value 6.63×10^{-34} J s, $\hbar$ is equal to $h/2\pi$. The usual relationship of energy to velocity v or momentum and mass m also applies:

$$E_i = 1/2\, mv_i^2 = 1/2|\mathbf{p}|_i^2/m \tag{4.3}$$

Similar relations can be written for the scattered wave and we note that there is in general a change of momentum $\mathbf{q}$ which is expressed as:

$$\mathbf{q} = \mathbf{k}_s - \mathbf{k}_i \tag{4.4}$$

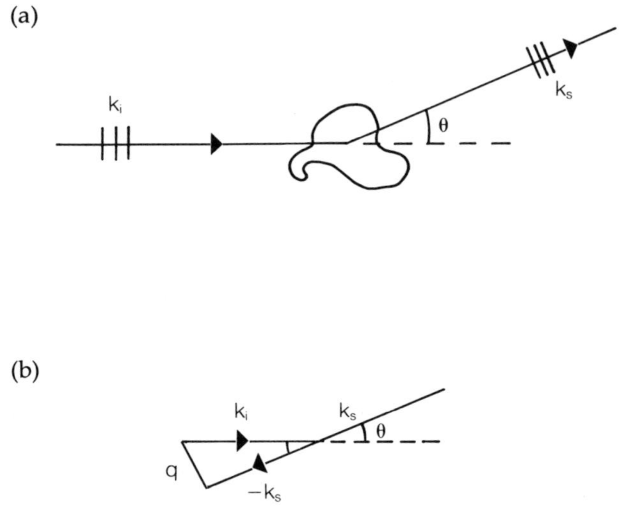

Fig. 4.1 Schematic diagram (a) of a scattering experiment showing an incident wave with vector $\mathbf{k}_i$ scattered at angle Θ to give a wave with vector $\mathbf{k}_s$. The vector diagram (b) shows the result for $\mathbf{q} = \mathbf{k}_i - \mathbf{k}_s$.

In the case of no energy transfer (elastic scattering) this can be reduced to the simple equation:

$$q = |\mathbf{q}| = (4\pi/\lambda).\sin(\theta/2) \tag{4.5}$$

where θ is the scattering angle. More generally there may be changes in both momentum and energy which can be written as:

$$E = h(\nu_s - \nu_i) = \hbar(\omega_s - \omega_i) \tag{4.6}$$

with ω the angular frequency of the wave ($2\pi\nu$). The theory of scattering is devoted to the calculation of scattered intensity I_s as a function of the incident intensity I_i and the variables $\mathbf{q}$ and E for a given sample. Within certain limits this can be written in the form:

$$I_s = I_i \,.\, \mathrm{d}\Sigma(\mathbf{q},E)/\mathrm{d}\mathbf{q}\,\mathrm{d}E \tag{4.7}$$

where the differential scattering cross-section $\mathrm{d}\Sigma/\mathrm{d}\mathbf{q}\ \mathrm{d}E$ depends on the spatial and temporal pair correlations of the distribution and magnitude of scattering centres in the sample as a Fourier transform. It would be inappropriate to develop this theory in detail but it can be found in standard references [1, 2]. The inverse relationship between the momentum transfer or scattering angle and the distance scale that is probed in an experiment arises from the properties of the Fourier transform. Measurements at small momentum transfer provide information about structure or correlations at large distances (roughly $2\pi/\mathbf{q}$). Similarly, data

Table 4.1 Scattering lengths of selected elements*

Element	*Scattering length* $\times 10^{-15} m$	*Incoherent cross-section* $\times 10^{-28} m^2$
1H	-3.74	79.9
2H(D)	6.67	2.0
C	6.65	0.001
N	9.36	0.49
O	5.81	0.0009
Si	4.15	0.015
S	2.85	0.007

* Values from Sears [8]. Where not specifically indicated the values are for the natural isotopic abundance of the element.

from measurements of small energy transfer concern long time scale motion in a sample.

The scattering amplitude for light depends on differences in the refractive index, for X-rays on electron density but for neutrons is a nuclear property known as the neutron-scattering length. For most purposes this can be assumed to be constant. However, there is some variation with wavelength and, more importantly, the spin of both the neutron and the nucleus.

The neutron has two distinct advantages when compared with other particles or waves in scattering experiments: it has a large mass and thus measurements of energy transfer in quasi-elastic or inelastic scattering experiments are straightforward. Further the interaction of neutrons with matter occurs via nuclear forces and these are not correlated with atomic number: indeed there can be large variations between isotopes. This is of great importance in the study of organic polymers where there are frequently no heavy nuclei present to provide contrast for electrons or photons. Indeed, the possibility of isotopically labelling molecules means that the configuration and size of individual polymers in a system that is chemically homogeneous can be investigated. The scattering lengths of some selected elements are listed in Table 4.1.

We can distinguish two categories of experiments, those in which there is no energy transfer (or the transfer is negligible) and for which measurements are made only as a function of momentum transfer and those other experiments (quasi-elastic and inelastic scattering) for which energy transfer (even if very small) is important. Numerically the first category is by far the largest and includes small-angle scattering which is the most-widespread application of neutrons to the study of polymers. There are considerable simplifications to the more general theory which can be made in this special case. Because of its relative importance some emphasis will be made of this topic both as regards theory and the practical aspects of the neutron experiments.

It is useful to introduce the idea of coherence. Interference between scattered waves to give information about pair correlation functions can only occur if the incident wave is coherent (of determined phase difference) at both scattering centres. This constraint is well known for light scattering but with neutrons a further complication arises – it is possible that a neutron will be scattered with a change of spin. A loss of spin coherence will exclude interference and give rise to scattering that depends only on the Fourier transform of the single-particle correlation function of the scattering lengths. Spin-incoherent scattering will give no direct structural information; the elastic component gives rise to an isotropic background. The probabilities of incoherent scattering are known [8] and are also included in Table 4.1. As the inelastic incoherent scattering contains information about the single-particle time correlation function, it is frequently measured in experiments on the motion in polymers. In elastic scattering experiments, this component of the scattering gives rise to a background that must be subtracted to obtain the structural information. Experiments can be made with polarized neutron beams and these can provide valuable extra information about coherent intensity as well as specific information on magnetic structures. These experiments are comparatively rare and, unless specifically mentioned, measurements discussed here can be assumed to involve randomly polarized beams.

4.2.2 Small-angle neutron scattering

As small-angle neutron scattering (SANS) is by far the largest use of neutrons in polymer science, it is appropriate to give some discussion of the features that distinguish it from most diffraction experiments. These aspects are not unique to neutrons and are described in books on X-ray small-angle scattering [9, 10]. The simplest application is the measurement of forward scattering from isolated objects such as polymer molecules. The angular variation of the scattering is dominated by the correlations of scattering amplitude within a single particle rather than any correlation between discrete particles. The Fourier transforms of certain simple geometrical shapes can be derived analytically. For example the transform of a Gaussian segment density distribution appropriate to a randomly coiled polymer is relatively straightforward. This was first calculated for the case of light scattering by Debye [11] and is of the simple form:

$$P(\mathrm{q}) = (2/x^2)\,\{e^{-x} - (1 - x)\} \tag{4.8}$$

where $x = q^2R_g^2$ and R_g is the radius of gyration of the molecule. The expression has to be multiplied by a constant which depends on the contrast to give the differential scattering cross-section. The model takes a simple form in the limit of small x

$$1/P(q) = 1 + x^2/3 \tag{4.9}$$

or in terms of intensity $I(q)$ and molecular mass, M_w at low concentration c

$$1/I(q) = (K/cM_w)\{1 + x^2/3\} \tag{4.10}$$

where the constant K is given by the product [5, 12]:

$$K = N_a\rho_m^2/(\rho_p - \rho_s)^2 \tag{4.11}$$

The quantity ρ_m is the normal mass density of the polymer, N_a Avogadro's number and the quantity in parentheses the scattering length density difference of polymer and solvent (or equally that of two differently labelled polymer species). This gives rise to the common representation of scattering data as a 'Zimm plot' of the reciprocal of scattered intensity versus q^2 [13]. The slope gives a direct measure of the radius of gyration and the intercept, the weight-averaged molecular mass. Standard texts on light scattering [14] describe this in more detail and in particular the use of measurements at a series of concentrations to extract the virial coefficient and limiting values at zero concentration.

There are some special features of neutron scattering that have given rise to extensions of this simple model. The technique is often applied to bulk samples or mixtures of two polymeric species. The concentration is then not small. Although early experiments were always made with low concentrations of one labelled species, it has now been recognized that this is not necessary if there is no thermodynamic interaction between the two species. The theory known as the random phase approximation (or RPA) [15, 16] has been experimentally verified for various D/H polymer mixtures [17, 18]. This is of enormous practical advantage in permitting measurements at a single high concentration at which there is a large scattered intensity to provide data on the unperturbed dimensions of polymers in the bulk. The theory has a simple extension to the case of a polymer blend with two chemically different but miscible species (1 and 2) which may interact with a Flory–Huggins parameter χ:

$$1/I(q) = 1/P_1(q) + 1/P_2(q) - 2^{\chi} \tag{4.12}$$

where P_1 and P_2 are the form factors as defined by equation 4.8 for species 1 and 2 respectively. This is the starting point for many studies of polymer blends which can, of course, be followed into regions of immiscibility. It is out-of-place to present a full theory of scattering from multi-component systems but reviews of this can be found elsewhere [9, 10]. It is largely identical to the theory of X-ray scattering.

Many models for the form factor of different objects have been developed and with the application of numerical methods it is readily possible to calculate the scattering from most simple geometrical forms and even composite structures. Such methods are described in the book edited by Kratky and Glatter [10]. A few cases of special relevance to polymer science are worthy of attention. The scattering from branched or star-like polymers has been described by Benoit [19] and used in a series of investigations [20, 21, 22, 23, 24] which, however, mostly concern melts or solutions. Block co-polymers have been treated in a number of papers [25, 26, 27]. Polydispersity can readily be included in numerical calculations, although it should be stressed that scattering methods on their own are rarely a good choice to obtain information about molecular-weight distributions. The mass obtained from a scattering experiment by evaluation of the zero-angle scattered intensity is always the weight average.

Models for the local structure of polymers have been described and these will normally relate to the persistence length of the chain. The scattering pattern for such models as the worm-like chain of Kratky and Porod has been calculated [28] and can be tested by scattering experiments. These experiments rarely fall in the range of small-angle scattering and great care is needed to avoid artefacts arising from transmission and sample geometry.

For many systems the simple Debye theory (equation 4.8) provides a description that is accurate to within a few per cent of measured intensity over the entire range of momentum transfer. Simple theory has proved of such success in analysing SANS data that many papers, particularly those concerned with relative changes in molecular dimensions between samples, do not extent beyond linear regression in a Zimm analysis. In practice this can be applied over a range that is significantly larger than $R_g q < 1$. The linear region in Zimm plots may extend as large as 2.5. This has been recognized, and small empirical corrections can be applied to linear fitting over an extended range of momentum transfer in the manner described by Ullman [29]. It should be stressed, however, that fitting of even the simplest theories (e.g. that of Debye [11]) will normally provide a better agreement at large momentum transfer.

Finally, before leaving the theory of small-angle scattering, it is useful to note that a particularly simple treatment of multiple scattering has been developed by Schelten and Schmatz [30]. This relies on the simple slab geometry used in most SANS experiments and permits a straightforward numerical evaluation of the effects of multiple scattering.

4.2.3 Diffraction and other scattering experiments

This section will be rather briefer and more general than the previous description of SANS. In part this reflects the relative importance in terms of the number of experiments reported in the literature. A secondary reason is that entirely adequate descriptions of both the theory and experimental technique can be found in reviews and monographs. For the most part the experiments do not emphasize polymeric properties and the methods are then identical to those for studies of low-molecular-weight species.

(a) Diffraction and wide-angle scattering

The simplest application of wide-angle scattering is in diffraction studies where there is coherent scattering from a regular crystalline lattice. The theory for neutrons can be reproduced from that for X-rays with only minor modification. The contrast arises from the nuclear scattering lengths. A simplification occurs in that no angular-dependent polarization factors are required for most experiments. The theory can be found in texts which describe diffraction of X-rays [31, 32]. The particular advantages of neutron diffraction are described in the book by Bacon [33] which refers largely to ionic crystals and materials of low molecular weight. There are several distinct advantages for neutron studies of polymer crystals. Neither X-rays or electrons are scattered well by hydrogen but replacement with deuterium provides a large coherent cross-section for neutron studies. As the presence of hydrogen is important in many polymers this can prove of decisive importance even given the need to prepare perdeuterated polymers. In this regard it should be stressed that although the ^1H hydrogen nucleus does scatter neutrons, this process is largely incoherent and gives a high background with only a small coherent signal containing structural information. Diffraction measurements on fully protonated polymers with neutrons are therefore rarely useful even to obtain information on the location of other nuclei. If deuterated samples of sufficient size are available, then neutron diffraction can usefully extend the information available about the positions of low-atomic-number nuclei in a crystal as well as giving data about the thermal motion in the crystal (which can be extended by further studies with inelastic scattering as described below).

There is little point in reproducing the standard treatments of Bragg diffraction and techniques for analysis of data and model refinement. Most of the usual procedures for single crystals, fibres and powders have been applied to neutron data. In most cases a good idea of the symmetry and structure has been obtained from preliminary measurements with X-rays and the problem is one of refinement of structural model parameters. This can be made in a variety of ways and even used for powder diffraction data using the method of Rietveld [34]. It should be stressed that the methods applied to other materials are not fundamentally different from those used in studies of polymers. The unit cells are often rather large, which may cause some difficulties. The constraint of chemical connectivity and repetition along a polymer chain may provide some simplification. Neutron diffraction has been applied particularly to studies of polymers of biological importance, for which hydrogen bonding may be of prime importance. Neutrons also have the advantage of generating less damage than an X-ray beam.

A variety of other studies are possible using wide-angle scattering; these include structures of glassy materials as well as studies of the local conformation of individual polymer chains. In general, the methods of analysis in terms of partial structure factors and radial distribution functions are analogous to those used for studies of simple liquids. The interested reader is referred to the review by Enderby and Neilson [35].

(b) *Reflection*

It is only comparatively recently that neutron reflectivity has been exploited as a tool for the analysis of the structure of interfaces. The phenomena of reflection of cold neutrons are largely the same as those for other types of particles or waves. Below a critical angle total reflection will occur, while at larger angles the reflectivity of an interface will decay sharply depending on the refractive index difference. It is often convenient to perform the calculations in manners analogous to those employed in optics. It has been shown [2] that a neutron refractive index, n can be defined as:

$$n = 1 - (N(z)\lambda^2 b/2\pi \tag{4.13}$$

where λ is the wavelength and $N(z)$ the number density of nuclei of scattering length b. The properties of neutrons are the same as for the optical s wave (electric vector polarized perpendicular to the surface). The usual methods of multilayer optics can then be used to calculate exactly the reflectivity of any given profile using the matrix methods described by Born and Wolf [36] or Abelés [37]. Further details of these calculations can be found in the texts by Lekner [38], Heavens [39] and Sears [40]. These methods have been widely used in model fitting of neutron reflectivity data [41] although an alternative approach based on the kinetic theory of scattering can provide a more direct insight into the variation of reflectivity with composition and structure of surface layers. In the case of neutron reflectivity it is often convenient, as in scattering experiments, to report the results in terms of $|\mathbf{q}|$ $(= 4\pi\sin\theta/\lambda)$, the momentum transfer normal to the surface, rather than the angle of incidence/reflection as measurements are often made at different wavelengths.

The results of kinetic theory can be summarized briefly [42] as stating that the reflectivity $R(\mathrm{q})$ as a function of momentum transfer q depends on the Fourier transform $N(q)$ of the density distribution of nuclei perpendicular to the interface $N(z)$ according to:

$$R(q) = 16\frac{\pi^2}{q^2}b^2|N(q)|^2 \qquad (4.14)$$

Although straightforward, this expression is an approximation that breaks down in the region of total reflection and, because it involves a Fourier transform, is not particularly convenient for numerical calculations. The advantage is that the variation of reflectivity can often be directly estimated and the theory of Fourier transforms can be used to provide certain simplifications for some distributions of different nuclei of particular symmetry. This approach has been exploited by Thomas *et al.* [43] to provide simple analysis of the distribution of different parts of diblock molecules.

(c) *Inelastic scattering*

A further particular advantage of neutron scattering is that it can be used to give information about dynamic processes in samples. If the approximation made in equation 4.5 that there is no energy transfer to or from the neutron is not valid, a further field of investigation is open. Any energy change, gain or loss, corresponds to a loss or gain in the energy of the sample. The experiment can then be considered as a kind of microscopic dynamic test in which the response of the sample is measured at different incident angles and energies. The analysis of this process in terms of simple elastic modes in the sample (such as phonons) or diffusive processes is found in the standard text books [1, 2, 44].

There are several features of polymers that give rise to interesting variations in the relaxation time with length scale. In particular the properties of polymer melts and solutions are described by the models of Rouse [45] and Zimm [46] which include the entropic elastic force and in the case of the second model, hydrodynamic effects of draining of solvent molecules through the polymer. These models, and the scattering laws associated with them, are described in the book of Doi and Edwards [47]. The reptation model of snake-like motion in polymer melts and concentrated solutions is also discussed. This may be of relevance to the mobility of elastomers but otherwise affects properties of solids in only a secondary manner. The behaviour of solid polymers is substantially similar to that of other organic solids. Crystalline polymers will show local diffusive motion as well as correlated lattice vibrations (phonons). Amorphous polymers have been used as prototype materials for the investigation of the glass transition. It is of particular interest to observe the onset of large-scale motion from the purely local vibrations that are possible in a glass. The wide range of distance and energy scales accessible to neutrons is of advantage in these experiments.

4.3 INSTRUMENTATION

4.3.1 Sources and general features of neutron instruments

The characteristic properties of polymers are that the molecules are large (typically 1–100 nm) and tend to move slowly. Measurements on polymers are simplified if a long wavelength is selected. Elastic scattering experiments can be performed at larger angles and the fractional energy changes that need to be measured in inelastic or quasi-elastic scattering experiments are bigger. The properties of the neutron and the source have a significant bearing on the design of diffractometers and spectrometers that give rise to differences between the instruments and those used for similar experiments with light, X-rays or electrons. Perhaps the most significant is the small flux at the source and consequent lower flux that reaches the sample (rarely more than 10^6 neutrons cm^{-2} s^{-1} and often much less if there is fine collimation or good resolution in wavelength). Instruments are therefore designed to accommodate large samples, often with lateral dimensions of a few cm^2, although the optimum thickness is usually of the order of 1 mm. In order to achieve high angular resolution, the instruments will often then become very large; the D11 SANS instrument discussed below has an overall length of 80 m. The second distinguishing feature of neutrons is the dispersion relation. The neutron is heavy and moves at finite low velocities (a wavelength of 1 nm corresponds to a velocity of about 400 m s^{-1}). This has two consequences. Firstly mechanical methods can be used to monochromate neutron beams, for example, two separated, fixed phase choppers can define a wavelength to a precision of about 1%. Secondly the ease of measurement of energy or wavelength by timing neutrons over a known distance has prompted many experiments that involve the measurement of energy transfer (inelastic scattering) and the design of novel instruments for elastic scattering based on 'white' beams exploiting the wavelength distribution on pulsed sources [48, 49]. It can be seen from equation 4.5 that equivalent information can be obtained by measuring elastic scattering at a fixed angle and different wavelengths as compared with conventional scans of angle at fixed wavelength.

There are two principle classes of neutron sources that are exploited for scattering experiments: nuclear reactors and accelerator-based spallation sources. The wavelength distribution of the neutrons that emerge from the moderator of a reactor is dependent on the temperature of the moderator. The thermal distribution will peak at about 0.15 nm for moderators at normal ambient temperatures. The wavelength distribution can be modified by allowing further multiple inelastic scattering (moderation) in a material at a different temperature to obtain a different equilibrium thermal distribution of neutrons. Cold sources are now present on many research reactors: for example, at the Institut Laue Langevin in Grenoble a 38 cm sphere of boiling liquid deuterium (at 25 K) is used to provide a distribution of neutrons that peaks at 0.6 nm on neutron guides that feed several different instruments, including the small-angle scattering instrument and spin-echo spectrometer that are described below. Other moderators are present to optimize the flux for instruments requiring other ranges of wavelength. Similar arrangements can be made around spallation sources: indeed, they are almost essential as the neutrons that emerge from a heavy metal target under bombardment from high-energy protons (or electrons) have themselves rather high energy or short wavelength.

In the last ten years, spallation sources have been built in several laboratories and have given rise to some interesting developments in instrumentation. Most of these sources operate in a pulsed mode (typical repetition rates in the range 10–50 Hz) and deliver a broad spectrum of wavelengths. It has proved advantageous to exploit this pulse structure by using instruments which measure the time-of-flight of each neutron to determine the momentum transfer, q. A wide range of q can be covered with a detector at a single angle, which can have distinct advantages in certain experiments. Measurement of the entire scattering pattern can be made simultaneously providing rapid measurements of entire spectra. The design of cells for complex sample environments such as high pressure is simplified. If detectors at a wide range of angles are provided then a very large range of momentum transfer can be measured simultaneously (for details of the principles of instrument design see reference 48).

High-flux reactors still provide a larger time-averaged neutron flux than accelerator sources of neutrons in the range of 'thermal' and particularly 'cold' wavelengths (0.1–3 nm). Most instruments on these sources [50] use a continuous beam that is monochromated to provide the defined incident neutron momentum.

4.3.2 Design of instruments

(a) Small-angle scattering instruments

Although most small-angle scattering experiments are still performed on reactor sources with instruments based on 'pin-hole' collimation as described by Abend *et al.* [51] and constructed by Ibel [52] and others, there are some significant improvements that have been made over the

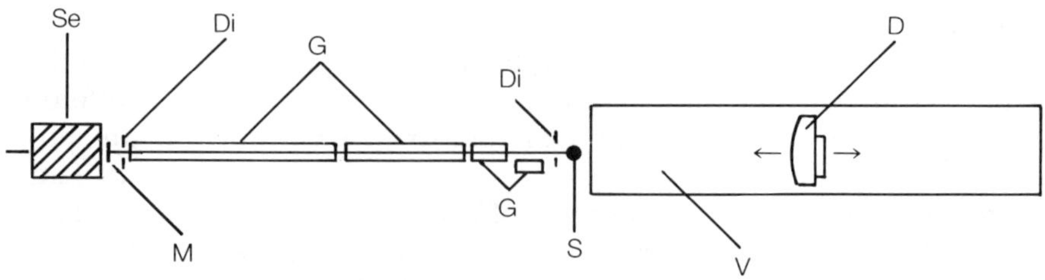

Fig. 4.2 A small-angle scattering spectrometer (D11 at the high-flux reactor of the ILL, Grenoble). Shown schematically are the mechanical velocity selector Se, an incident beam monitor M, removable neutron guides G and diaphragms Di to define the collimation and incident beam size. The detector D can move in the evacuated flight path V up to a maximum distance of 35 m from the sample S.

years as well as some novel designs. For example, a proposal to use focusing mirrors has been made which would have advantages in improving the flux [53]. A schematic diagram of a conventional instrument such as D11 at the Institut Laue Langevin in Grenoble is shown in Fig. 4.2. The basic elements include a helical slot velocity selector, neutron guides to bring the effective source closer to the sample when low resolution is acceptable, defining diaphragms for the beam and a two-dimensional position-sensitive detector [54, 55] situated in an evacuated tube. The principle is unchanged from the design of Ibel [53] but the flexibility and ease of operation is now greatly improved with automatic motion of the detector within the evacuated flight path for the scattered neutrons, motorized control of the collimation and the provision of automatic sample changers. Such improvements, although not materially affecting the resolution and flux, have led to improved science. For example, the ability to rapidly adapt the instrument to follow changes in samples has given rise to an increased interest in kinetic experiments. Several instruments of this type are now available at various research reactors [56, 57, 58]. Other instruments of this geometry that exploit the time structure of a white beam at spallation sources have also been described in the literature [59, 60].

Improvements in instrument performance have been matched with better data analysis. Although perhaps often not of great importance, the understanding of instrumental resolution has improved [61, 62, 63, 64]. Determination of absolute values for the size of molecules can be made within 1 or 2% if appropriate allowance is made for poly-dispersity in the sample and instrumental resolution. Relative values of the radius of gyration of a polymer can be measured to better than 0.5% in experiments that measure the change within a single sample, provided that data is collected to high statistical accuracy and due regard to statistical weights and errors are made in the curve-fitting procedure [65]. In some experiments, such as those concerned with elucidating the internal structure of

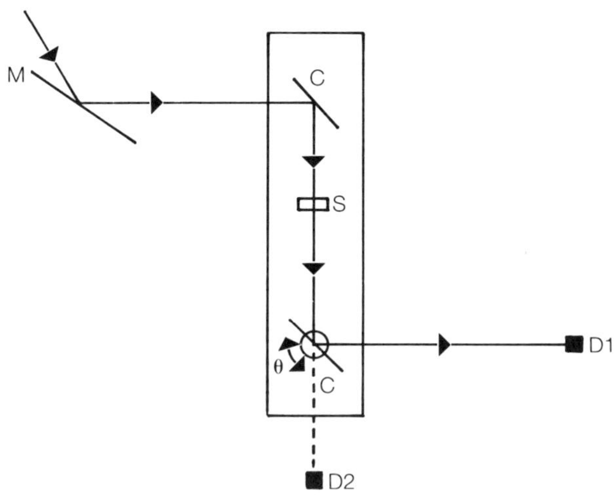

Fig. 4.3 Diagram of a double-crystal diffractometer. The incident beam is deflected from the premonochromator M on to the first of a pair of crystals C mounted on an optical bench and set to 'Bragg' reflect the incident beam on to the detector D1. The sample is placed between the two crystals and the broadening of the beam by small-angle scattering is measured by making a rocking curve (angular scan) of the second crystal. The second detector D2 provides a signal that is complementary to D1 and can be used to check for instabilities.

polymer lattices [66] (which are discussed further later) the effects of resolution are very significant in obtaining good fits of model functions to the data. Modern mathematical methods such as maximum entropy analysis have also been applied to SANS data [67, 68] but not as yet to systems concerning polymers. Other corrections such as those arising from time-varying background on pulsed sources have been described [69].

An entirely different design of instrument has attracted interest in recent years, double-crystal diffractometers (known to the X-ray community as Bonse–Hart cameras [70, 71]) are capable of much higher angular resolution and can measure to much lower momentum transfer. Values of 0.0001 nm^{-1} are readily accessible. The principle is indicated in Fig. 4.3. Two perfect crystals are each set at the Bragg condition for an incident beam. The angular broadening of the beam is measured as a rocking curve of the second crystal. When a sample that scatters at small angles is placed between the two crystals, the angular width of the beam effectively increases. Measurements are always made as a comparison of the empty container and the container with the sample. It is either necessary to deconvolute the measured sample scattering from that of the empty container or within fitting procedures convolute the instrumental resolution with model curves. Instruments of this design have been built in several laboratories and can be used to study materials such as composites, blends or colloidal dispersions for which the contrast or transparency available to neutrons is important and light scattering or other techniques are inappropriate. The requirement for good angular resolution can in part be offset by increases in sample size (the resolution is independent of sample area in these instruments) but the contrast must be fairly high to obtain useful data at very low scattering vectors. Some results from studies of spinodal decomposition in polymer blends [72] and the structure of latex particles [73] from such diffractometers have appeared in the literature.

(b) Reflection instruments

A major field of research which has developed since 1980 is the study of specular reflection of neutrons. Instruments dedicated to this type of measurement have now been built at a number of neutron-scattering centres. The simplest instrument uses different wavelengths at a fixed angle of incidence in order to measure the reflectivity at different momentum transfers. An example of such an instrument is CRISP on the pulsed spallation source ISIS at the Rutherford Appleton Laboratory [74]. This is shown in Fig. 4.4. In the simplest operation a single detector is aligned at an angle of reflection equal to that of the incident beam. The range of wavelengths from the hydrogen moderator that may be used is from 0.05 to 1.3 nm. The wavelength of each neutron is determined by time-of-flight measurements. Instruments at other laboratories include similar fixed-geometry instruments at both spallation [75, 76] and reactor sources as well as 'fixed-wavelength' instruments on which angular scans are made [77]. The first type of instrument has the advantage that alignment for a fixed geometry is straightforward and that the resolution is effectively constant over the entire range of measurement. The second

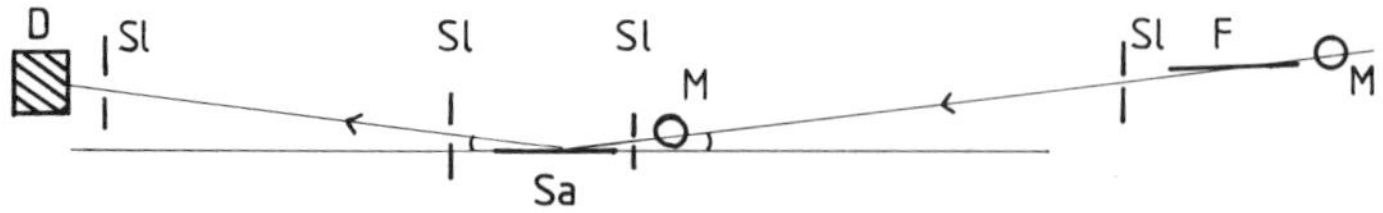

Fig. 4.4 Schematic diagram of a reflectometer, based on CRISP at ISIS, RAL. The incident beam, which is inclined to the horizontal at an angle of 1.5 degrees, strikes the sample Sa and is specularly reflected on to the detector D. Slits Sl provide primary and secondary collimation. Monitors M record the incident beam and a mirror F is inserted to remove the undesirably slow, long wavelength tail of each neutron pulse. All measurements are made at fixed angle and time-of-flight is used to determine the wavelength.

class of instrument may have advantages in flexibility of operation, for example, in the ability to scan a few points rapidly for kinetic studies.

(c) Some other neutron-scattering instruments

The instruments designed for experiments such as inelastic and quasi-elastic scattering as well as diffraction or wide-angle diffuse scattering are numerous and can only be described briefly. If energy transfer is to be measured, the wavelength of both the incident and scattered neutrons must be determined. The techniques for analysing energy in the scattered beam are broadly identical to the methods used in constructing monochromators. Thus if an incident monochromatic beam is pulsed, the time-of-flight of the neutron from the sample to the detector is a measure of the velocity or energy of the scattered beam. Many instruments exploit this principle to determine the spectrum of scattered energies.

An analyser crystal set at the Bragg condition for a given wavelength can, if placed in the secondary flight path, be used to determine the scattered intensity at different energy transfers. It will be obvious that both this technique, which requires a different setting of the analyser for each point in the spectrum of transferred energy, and the previous method of creating a pulsed or chopped beam (the duration of the pulse limiting the energy resolution) will result in much longer measuring times than those required for simple elastic or total scattering spectra.

The principles of most of these types of neutron-scattering instrument are now well known as are the choices for optimizing neutron flux on the sample and resolution. The modifications required for use on the new generation of pulsed sources are described in the book of Windsor [48]. Although improvements have been made in many instruments, mostly through technological advances such as those in monochromator production or detector design there are few novel aspects of interest to polymer scientists that cannot be found readily in the literature. An exception to this rule is found in extensive measurements made with spin-echo spectrometers since the end of the 1970s. This technique permits quasi-elastic scattering measurements with an energy resolution on the scale of nanoelectron volts over a wide range of momentum transfer.

The normal problem with very high-resolution spectroscopy is that a highly monochromatic beam is required and the incident flux on the sample therefore decreases as the resolution improves. The novel feature of spin-echo spectroscopy is to use the precessing spin on each neutron as an indicator of the change in energy occurring on scattering and this allows measurements to be made with a comparatively broad range of incident wavelengths (energies). The suggestion of Mezei [78] was that an instrument with an incident polarized beam and symmetrically disposed magnetic fields along the beam axis before and after the sample should be

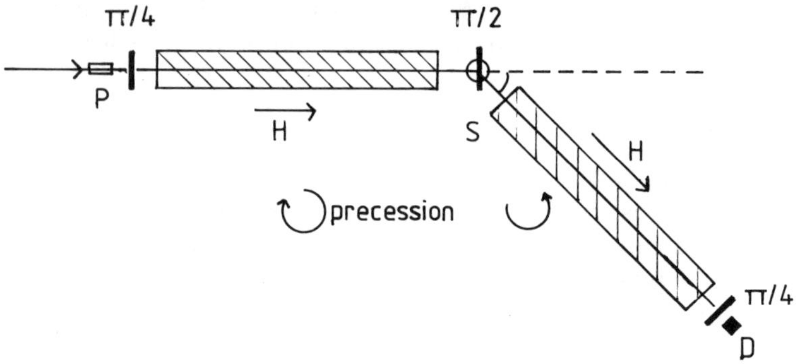

Fig. 4.5 Diagram of a neutron spin-echo spectrometer. The incident beam passes through a polarizer P to give a single incident spin. This is aligned perpendicular to the magnetic field H and beam direction with the $\pi/4$ coil (rotated through 90 degrees). The spin precesses in the field. At the sample S, the direction of the spin is inverted with the $\pi/2$ coil. If the scattering process is not perfectly elastic, the spin will precess to a different extent in the reverse direction during the time of passage through the second field H of equal amplitude. The difference in polarization is thus a measure of the energy transfer to the neutron. This can be analysed with a further coil, polarizer and detector D.

constructed. If the direction of the spin is inverted precisely at the position of the sample, the opposite precession after the sample will exactly compensate that before the sample for all wavelengths. If there is energy transfer on scattering, the velocity in the region of the second magnetic field will be different and a shift in the average direction of polarization will be observed. This is shown schematically in Fig. 4.5.

In practice the design of such a spectrometer is complex as a high degree of uniformity of the precession fields is required and correction for stray fields (including that of the Earth) must be made, particularly in the area around the sample, which should ideally be subject only to the field of a perfect 'spin-flipper'. However, such instruments can be realized [79] and have made important contributions to several fields of which studies of motion in polymer melts and solutions are perhaps the most significant. A review of the principles and application of this technique has been provided by Nicholson [80].

4.3.3 Problems of contrast and labelling

It has been indicated that one of the advantages of neutrons as a tool for the investigation of polymers is the possibility of using isotopic contrast to highlight individual molecules or parts of molecules. It is appropriate to question the assumption that physical and chemical properties are unaltered on isotopic substitution. Although the simple theories [81] would indicate that no differences would be found, there is ample evidence for immiscibility of deuterated and protonated species under certain circumstances and changes in some physical properties.

Bates *et al.* have studied high-molecular-weight polymers and shown that they will separate into two phases [82, 83]. The behaviour can be explained by the Flory–Huggins (see, e.g. reference 84) model for polymer interactions, which is usually applied to blends of chemically distinct molecules. It should be emphasized that the effects only become significant for molecular weights of the order of one or a few million. The explanation of this behaviour is not entirely clear. The effect of the different mass on molecular mobility and zero-point motion can be calculated but many more complex effects may play a role. For example, it has been observed that the melting point of alkanes and polyethylene varies with deuteration but calculations have not been in agreement with the observations [85].

A few general remarks on the effect of isotopic substitution can be made. In crystalline or semi-crystalline polymers, many properties are determined by kinetic processes which will depend on molecular mass. Even the small change in the mass of a deuterated and protonated hydrocarbon can significantly alter the rate of crystallization and thus lead to effective separation of deuterated and protonated species. Even bigger effects than the variation of mobility with the square root of monomer mass can be observed when the hydrogen atom is involved in a rate-determining chemical process. For this reason the synthesis of polymers from deuterated monomers can often give rise to significant differences in the molecular weight, poly-dispersity, microstructure and tacticity of the products to those obtained from a preparation under identical conditions with protonated monomer. Some care must be taken in distinguishing these effects from those that are truly due to a simple isotopic difference in interactions between polymers. At present, it is only the cases of polystyrene and polybutadiene that have provided clear indication of this last type of behaviour giving rise to a large effect. It is, however, now well characterized and has even been exploited for a study of a model polymer blend [86]. If hydrogen bonding is significant in any system, this will be modified by replacement with deuterium to produce very large changes in properties. A good example of this is the phase behaviour of polyethylene oxide in water: the theta conditions can be altered by many degrees and this change in interaction might be expected to play a role in other circumstances.

Some of the other problems associated with the preparation of deuterated polymers can be avoided by modification of a single parent polymer to provide two species with different contrasts. Several studies have been made on blends of polybutadienes that have been saturated with hydrogen and deuterium to give polymers essentially identical to polyethylenes [23, 65, 87, 88, 89] in the manner described by Crist *et al.* [90] with a palladium catalyst. The chemistry of the saturation reaction is such that no change in the polymer molecular weight occurs. If the same parent polymer is used the isotopically different products display no chemical differences.

Before leaving the topic of labels for scattering experiments, one new development can be mentioned. As described earlier, the neutron will interact with a nuclear spin. If hydrogen nuclei in a deuterated matrix are polarized, the contrast can be enhanced by an order of magnitude. Some successful test measurements have been made to exploit this advantage [91, 92] but there remain considerable technical difficulties to the widespread application of this technique. The nuclei must be cooled to temperatures of less than 0.5 K with a helium dilution refrigerator. Although there will be limits in application, the technique has interesting potential for elastic scattering studies on materials for which only small samples are available. At present this has attracted most interest in the study of biological polymers.

4.4 APPLICATIONS OF SANS

4.4.1 Introduction and general aspects

Small-angle neutron scattering has achieved popularity because it can provide direct information about the size of molecules in the solid state. In the case of a simple homopolymer, isotopic substitution such as the replacement of hydrogen with deuterium can be used to contrast individual molecules in solid solution. The data interpretation to extract a radius of gyration is usually straightforward and adequate statistics to determine this quantity to a few per cent can often be obtained in a few minutes. Most experiments now extend the analysis to obtain absolute values for $I(\mathbf{q}=0)$ and thus the molecular weight as well as extracting other quantities such as molecular interaction parameters or details of polymer conformation at distance scales less than the radius of gyration. Another development has been to look at more complex systems: there is a great deal of interest in determining molecular strain in drawn samples and morphology in composite materials.

The limits to which it is appropriate to continue detailed analysis of SANS data are now fairly well understood. For example, careful measurement of the incoherent background is necessary as this may be many times larger than the coherent signal of interest at large momentum transfer. Maconnachie [93] has pointed out that there is a large variation of this background with temperature (several per cent). It is therefore desirable to make measurements of background sample scattering under conditions identical to those of other samples. This can be understood in

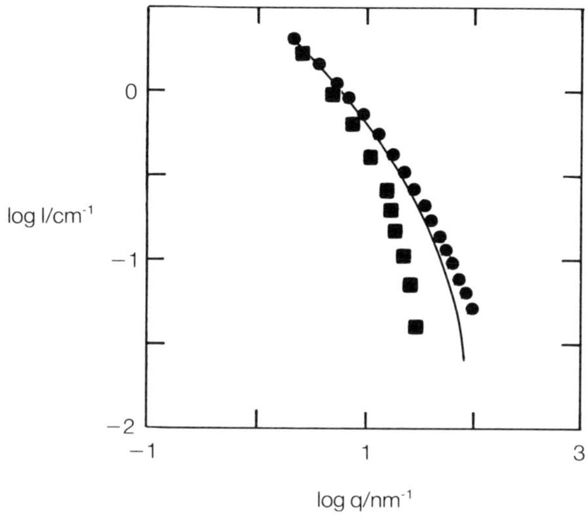

Fig. 4.6 Effect of different equations for correcting SANS for background scattering. The sample is a polyvinyl acetate gel and approximate background measurements were made with samples of different transmission T. The plot shows the correct allowance for incoherent scattering (line) (varying as $1-T$) and the effect of assuming that the background varies as T (for example coherent scattering) for samples with a slightly incorrect transmission (circles $T=0.83$ and squares $T=0.88$). The sample transmission was 0.82.

terms of multiple inelastic scattering [94]. Geissler *et al.* [95, 96] have demonstrated the importance of correctly subtracting the different components of the background. As the coherent and incoherent components of the background will vary in different ways with sample thickness, composition and transmission, this again can be a source of large errors. At large momentum transfer (greater than 1 nm^{-1}) this can significantly alter the shape of the scattering curve and indeed the interpretation of the results. The results of using a background spectrum that differs in transmission from that of the sample by as little as 1 per cent may have a large effect if the data is not scaled correctly. An example of this difficulty is shown in Fig. 4.6. The application of the correct formula gave results that were substantially independent of the transmission of the material used as a background [96]. For these reasons a considerable degree of caution must be exercised in accepting some of the analysis that has been presented of SANS data at large momentum transfer. These problems remain difficult to treat in contrast to the more straightforward effects of instrument resolution which were described previously [61, 62, 63].

4.4.2 Molecular conformation in crystals and glasses

The first major application of neutrons to polymer science in the early 1970s was to determine the size and conformation of macromolecules in the bulk of glasses and crystals. Experiments by various groups looked at amorphous poly(methyl methacrylate) and polystyrene [97, 98, 99, 100, 101]. Both were found to have a size similar to the molecules dissolved in theta solvents. These were the first direct verifications of Flory's theory of screening of excluded volume interactions. Attention turned very rapidly to some other problems such as the conformation of polymers in the crystalline and semi-crystalline state. The first experiments by Ballard *et al.* [102] on polyethylene have been followed by many other groups working on various systems such as isotactic polystyrene [103], polyethylene oxide [104, 105, 106], saturated polybutadienes [87, 88, 89, 90] as well as a great deal of further work on polyethylenes [107, 108, 109, 110, 111].

This study of crystalline materials has proved much more difficult. For a variety of reasons it has proved difficult to draw definitive conclusions of a general nature. In part this is due to the complexity of the problem: crystallization under different conditions from solution or melt can cause large differences in resulting morphology because the process is kinetically controlled and a true equilibrium is often not achieved. This does not make the study any less interesting, but can cause difficulty in comparing and generalizing results. It has been recognized that in the case of melt-crystallized polymers a major feature of interest is the chain folding and particularly the degree of adjacent re-entry in lamellae [107, 108] and the number of ties between lamellae [104]. Measurements to resolve these questions obviously extend beyond simple, small-angle scattering measurements of a single molecular dimension and will require data at larger momentum transfer to give information about distance scales within single molecules and lamellae. The usual approach has been to test data measured over as wide a range of momentum transfer as experimentally feasible against model calculations [111]. This has proved difficult as the signal is small at large momentum transfer, the background is high, and measurements must often be made at large angles where angular-dependent corrections for sample geometry, multiple scattering and non-isotropic background must be considered. Although some papers have reported such data [104, 109] it is not clear that the analysis either from the point of view of data reduction or model fitting can be considered as providing conclusive answers [114]. The behaviour in melt [113] and solution-crystallized [107, 112, 115] polyethylene and polyethylene oxide [104, 106] can be markedly different. In a slightly different approach, comparison of the diffraction pattern for a sample of deuterated polyethylene and that for blends of deuterated and protonated polyethylene [116, 117] has been made but no evidence for adjacent re-entry was found in a melt-crystallized sample.

4.4.3 Studies of kinetic processes and diffusion

A major development in recent years has been the use of small-angle scattering to study diffusive processes in polymers such as crystallization, phase separation in blends and co-polymers as well as diffusion in homo-polymers. The principles of such measurements have been discussed by Summerfield and Ullman [118]. Some of the distinctive advantages of SANS have been discussed in a review by Rennie [119]. The distance scales observed in SANS experiments are well matched to the overall dimensions of polymer molecules and it is on this scale that diffusive motion is often of greatest practical importance in controlling morphology, mixing or healing of cracks and crazes. The diffusion coefficients of polymer molecules are often as low as 10^{-20} $m^2 s^{-1}$. In these circumstances it is useful to resolve motion on as fine a scale as is appropriate to the processes involved, in order to avoid unnecessarily lengthy annealing periods that would be required by macroscopic analytical measurements of diffusion.

Several experiments have now been made to measure diffusion processes. These include measurements of the increase of scattering with time as alternately stacked thin sheets of deuterated and protonated polymer (polyethylene) inter-diffuse to form a molecular solution [120, 121]. A significant limit in such experiments has been the ability to produce accurate initial distributions of differently labelled polymers on the microscopic scale appropriate to scattering experiments. This problem has been elegantly circumvented in the work of Hellmann *et al.* [122, 123] who have prepared blends of deuterated and protonated polycarbonates in which one component contains weak links that break on heating. The initial distribution is a homogeneous blend of a polymer of high molecular weight. On heating, the resultant low-molecular-weight polymers will diffuse apart in a process that can be followed readily by SANS. This process is shown schematically in Fig. 4.7. This type of measurement has permitted diffusion coefficients as low as 1×10^{-16} $m^2 s^{-1}$ to be followed. Another series of related experiments has concerned inter-diffusion of latex particles during film formation [124, 125].

Several aspects of polymer chemistry have been the subject of SANS investigations. Those experiments concerned primarily with polymer lattices are left for discussion later. Of the remaining topics, perhaps the most frequently encountered is that of trans-esterification. Many groups have been surprised in studies of blends of hydrogenated and deuterated polyesters to find molecular weights from the SANS data that are much lower than expected from other measurements. This has then been the

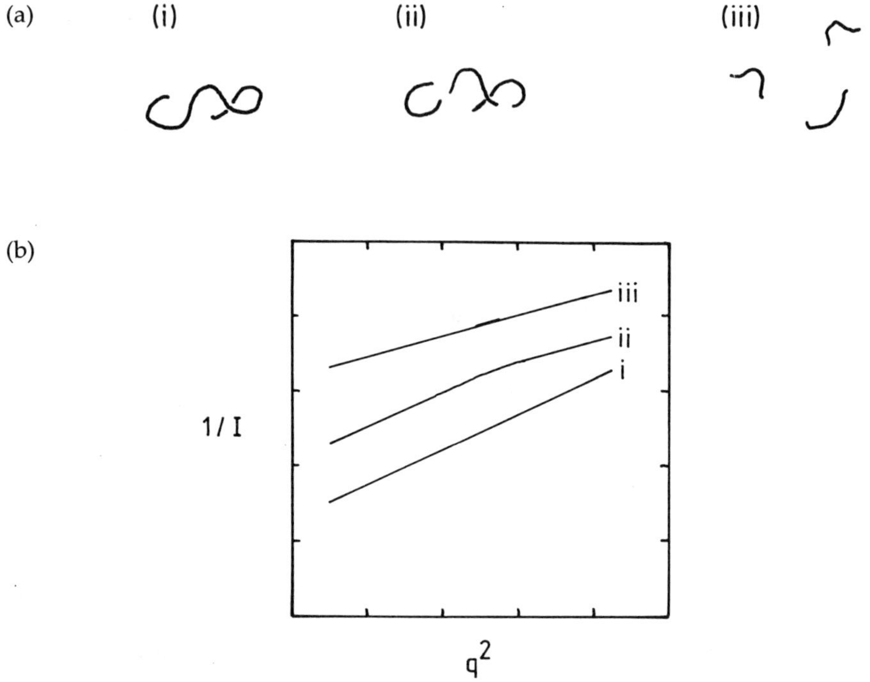

Fig. 4.7 (a) Schematic diagram of chain scission and diffusion process. (b) Typical change in SANS after chain scission during a diffusion process. This is shown in the 'Zimm' representation of $1/I$ against q^2. The apparent molecular weight and radius of gyration decrease with time (slope decreases and intercept with the vertical axis increases).

subject of several more detailed studies [126, 127, 128] in which the kinetics of trans-esterification are followed by SANS. Some attempts have been made at the theoretical analysis of this process [129] which has many similarities to the diffusion experiments mentioned in the previous paragraph, but it is clear that more work remains to be done in this area.

Although under favourable circumstances reasonable SANS data can be collected on polymer samples in one or a few minutes, this may still impose an unacceptably long lower time limit on studies of dynamic processes. If it is possible to reproduce a process in a cyclic manner, then this limit may be circumvented. The method, which had previously been applied to X-ray and infra-red studies [130, 131], involves phasing the data acquisition with the dynamic process cycle and adding equivalent time slices from successive cycles until sufficiently good counting statistics have been obtained. This method has been described by Rennie and Oberthür [132] for the investigation of deformation of elastomers. It can, however, be applied to many different systems and several experiments on phase separation of co-polymers in concentrated solution have been made in this manner [133], by repeating temperature jumps. Other experiments are in progress. The limit in time resolution in such measurements depends on the wavelength resolution and sample-to-detector distance [132] but is typically better than 1 ms. Several studies of mechanical deformation in polymers have been made with the aim of elucidating mechanisms of flow and could be considered as kinetic measurements, but they are left for discussion after the studies of equilibrium deformation.

4.4.4 Blends and co-polymers

In recent years the use of neutrons for the study of blends and co-polymers has grown to match their widespread practical application. Although it might be thought that neutrons would have no particular advantage in the study of systems which contain species that are chemically different, the small electron density difference between most organic materials gives considerable advantage to isotopic labelling techniques. Further, the careful choice of contrast can be used to determine the conformation of individual molecules within phase-separated regions of blends. The determination of the conformation of molecules in mixtures is one of the most direct ways of determining the interaction (or chi) parameter. The number of different systems that have been investigated by SANS is now very large and it is possible only to give a brief summary of the main lines of investigation.

The primary use of SANS in the study of polymer blends has followed directly from the early work of Kirste in determining the dimensions of polymer molecules in mixtures of silicone polymers that were chemically identical but differed in molecular weight [134] and in blends of chemically different polymers (PMMA and polystyrene-acrylonitrile co-polymers) [135]. It has also been useful to determine the limits of miscibility by SANS, as this is one of the few techniques that probes effectively on the scale of molecular dimensions. Systems that may appear to be uniform under optical microscopes have often proved to be phase separated on careful study at higher spatial resolution. For this reason neutron scattering has been a favoured tool for the investigation of kinetics of separation and in particular the early stages of, for example, spinodal decomposition.

These studies have now been extended to a wide variety of different systems which include polystyrene/polyvinyl methyl ether [72, 136, 137, 138, 139, 140, 141, 142] and blends of polystyrene with many other polymers such as polyphenylene oxide [143, 144, 145], polymethylmethacrylate [146], and polybutadiene/styrene co-polymers [147] and deuterated polystyrene [86]. Other systems that have received considerable attention are polymethylmethacrylate with solution-chlorinated polyethylene [148, 149, 150] and polybutadiene/polyisoprene blends [151, 152]. The origin of many of the early investigations was in tests of models of phase behaviour and the physics associated with separation kinetics. It is perhaps for this reason that much of the work was conducted with the system polystyrene/polyvinyl methyl ether for which there is a large domain of compatibility with a negative chi parameter. Unfortunately, some of these studies suffer from a lack of information about the polymer samples: tacticity, for example, can make a large difference in the phase diagram of this mixture. As might be expected, the use of polymers as engineering materials has prompted study of some specific systems, particularly those with desirable mechanical properties. Polycarbonates, for example, have featured in several studies [153, 154]. The chemical complexity of systems chosen for investigation is now rapidly increasing: for example a recent study has reported results for a blend of a co-polymer with ionomers [155].

The scattering from block co-polymers has many similarities to that of polymer blends. A range of homogeneous and phase-separated structures is seen depending on the chemical compatibility of the components. The theoretical basis for the interpretation of scattering data from homogeneous phases of di-block co-polymers in terms of the random-phase approximation was presented by Leibler [25]. A more complete description of the scattering from multi-block co-polymers is found in the work of Benoit and Hadziianou [26]. The most significant feature of the scattering from diblock co-polymers or their blends with a homo-polymer is the large peak observed in the small-angle scattering pattern. This is shown in Fig. 4.8. This characteristic behaviour is associated with a 'correlation hole' or the depletion of one monomeric species in a region around a polymer block of that species which is caused by the higher local concentration of the second type of monomer attached as a block to that chain. The position of the peak is largely determined by the block size and the width by the interaction parameter (or degree of separation) of the two components. Early work on these materials has been reviewed by Richards [27]. A large number of co-polymers and their blends with homo-polymers have been investigated [156, 157, 158, 159, 160], largely confirming the theoretical models.

Co-polymers are frequently encountered as surfactants or compatibilizers and several SANS studies have been made of polymer micelles [161, 162, 163, 164, 165], adsorbed layers at the surface of colloidal particles [166, 167, 168] and their application as 'compatibilizers' in polymer blends [169]. There are also reports on the structure of phase-separated systems and the molecular conformation and interphase structure in such systems [170, 171, 172, 173, 174, 175].

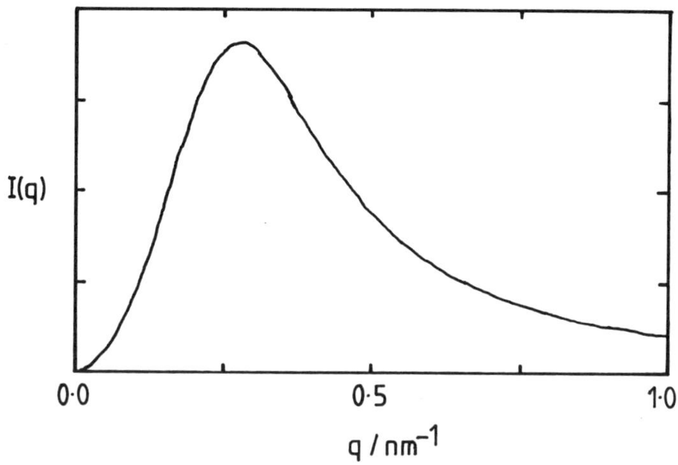

Fig. 4.8 Scattering from a di-block co-polymer. Model calculation for a symmetric polymer of total radius 7 nm. The pronounced peak is caused by the so-called correlation hole. Details of these scattering laws can be found in [25] and [26]. Fits to such data are valuable in determining the interaction between the two blocks.

4.4.5 Molecular alignment under applied stress

One distinctive advantage of neutrons in comparison to other forms of radiation is their penetrating power and the consequent wide choice of materials available as sample containers. It is possible to construct elaborate apparatus to provide conditions around samples that are typical of realistic service conditions. Studies of the effect of stress on the molecular organization within polymeric materials can be made readily. Simple tensile strain rigs have been used in the neutron beam as well as pre-stressed samples. More recently dynamic strain devices [132] have been constructed. Shear cells [176] in various geometries are also used but as yet have been applied primarily to colloidal dispersions [177] and polymer solutions [178] rather than bulk polymers. There is, however, interesting potential for experiments to investigate the mechanisms of obtaining orientation in solid samples by fibre spinning or melt flow.

SANS, as a technique that provides direct measurement of macromolecular dimensions, has been used in a variety of studies of deformation and elongation of polymers. The materials that undergo considerable molecular deformation are those in which there is a high degree of mobility. In consequence many studies of deformation in elastomers have been made, as well as investigations of hot-stretched samples of glassy or partially crystalline polymers which may be subsequently quenched.

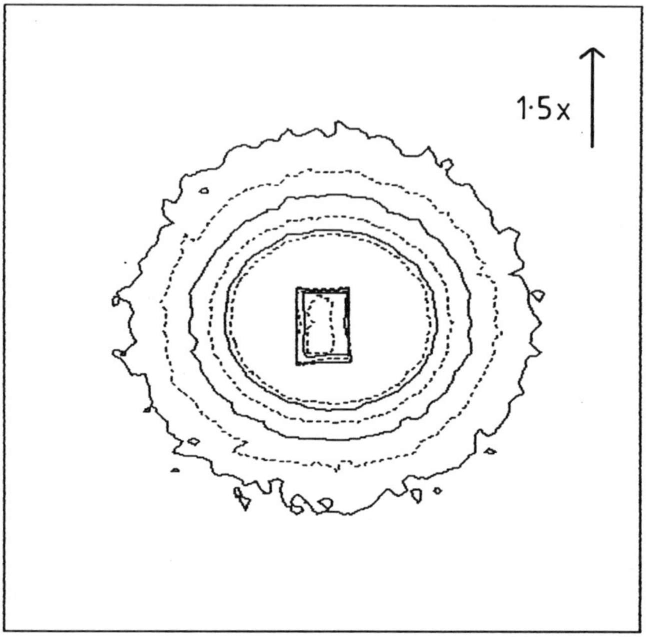

Fig. 4.9 Anisotropic scattering from stretched PDMS. This sample of tetrafunctional end-linked PDMS was stretched by 50% in the direction of the arrow. The sample contains a fraction of deuterated chains linked in the network and the small-angle scattering was measured with neutrons of wavelength 1 nm. The lines in the plot link points on the detector of equal scattering intensity. The elongation of the radius of gyration of these links in the direction of strain is seen by the anisotropy of the scattering pattern. The reciprocal relationship of scattering vector and distance means that the scattering pattern is compressed in the direction of elongation which is much less than affine in this material. In the middle, the effect of the rectangular beam-stop shielding the detector from the direct beam can be seen.

Most SANS experiments are conducted on instruments equipped with two-dimensional position-sensitive detectors. This provides an immediate indication of the anisotropy in the sample. An example of such data is shown in Fig. 4.9. Much data has been presented in the simple form of analysis by sector, emphasizing the differences observed between directions parallel and perpendicular to the applied stress field. While this may be a useful simplification, it is wasteful of the information obtained in other directions. Mildner [179] has emphasized the advantages of treating the full two-dimensional map of scattering data. This will not only give greater statistical accuracy but will also allow better tests to be made of the model for the mechanism of deformation and the consequent angular variation of, for example, molecular alignment.

The most straightforward experiments have been made on model elastomers with the aim of determining the molecular deformation for a given microscope strain. A series of experiments by Beltzung *et al.* [180] on regular silicone networks have provided evidence for the 'phantom' network model in which the cross-link points deform affinely but there is free motion of the interconnecting chains. Lightly cross-linked polystyrene networks have also been used in a variety of studies [181]. Boué has performed detailed measurements on these materials to determine not just the equilibrium structures but also the transient conformations of molecules during deformation [182, 183]. These experiments were made on rapidly quenched samples that had been hot-stretched and have been somewhat confusingly described as 'real-time' experiments by the authors. More detailed investigations of the orientation in polystyrene have been continued by Muller [184, 185] and others [186, 187]. Some interesting anisotropic scattering patterns have been observed in some of the recent experiments and have been designated 'butterfly' patterns from their similarity to the shape of these insects [188]. These appear to arise from strongly anisotropic molecular relaxation times in the directions parallel and perpendicular to deformation.

Experiments using the technique of cyclic deformation in the neutron beam have been reported. These have used the cyclic data acquisition mentioned above [132] for studies of model silicone networks [189].

There is less theoretical foundation but a great deal of practical interest in the studies of deformation of semi-crystalline polymers such as polyethylene and polypropylene. Although a part of the initial impetus for this work was to determine the interconnection of crystalline lammellae and the degree of regular chain folding [190, 191], detailed studies of the drawing and necking process in polyethylene [192, 193, 194, 195] and molecular relaxation [196] are now available.

Cold strain of solid, glassy polymers has presented more difficulties as the onset of voiding or crazing can render the interpretation of the scattering data difficult. The reversible, elastic strains in such samples are often small and better suited to diffraction experiments performed with X-rays. These difficulties notwithstanding, studies have been made on cold-drawn polymethyl methacrylate [197] and polyethylene terephthalate fibres [198].

4.4.6 Latex morphology and film formation

An interesting class of polymeric materials are those that originate as lattices or colloidal particles dispersed in a fluid. They are widely used as paints but also find other applications in situations such as those where synthesis as an emulsion has particular advantages, or where preparation of end products from a fine sintering powder is favoured. Several experiments have exploited neutrons in the study of such systems. The morphology of the particles may be important and several studies of heterogeneity in co-polymerized particles have been reported [73, 199, 200, 201, 202]. More directly relevant to the final applications, such as the use in paints, have been studies of coalescence of particles during drying. Hahn *et al.* have measured the inter-diffusion of polymer molecules between particles during the drying process [124, 125]. Similar experiments have now been conducted on a variety of other preparations [202, 203] by other groups. The organization of the particles and their interfaces have been studied by Cabane *et al.* [204] and Crowley *et al.* [205]. This last paper has concentrated on studying the scattering that arises from the interface structures. Another application of SANS has been to study the mechanisms of emulsion polymerization and the origin of possible heterogeneities [206, 207, 208].

4.5 WIDE-ANGLE SCATTERING AND DIFFRACTION

4.5.1 Diffraction and crystal structures

The major advantage of neutrons when compared to X-rays or electrons is the ability to locate the atoms of low atomic number in the structure. The general aspects of diffraction from polymers will not be reiterated here but may be found in standard texts [31, 32]. The large incoherent background that arises from hydrogenous samples does present a problem and it is often essential to make measurements on deuterated compounds. A further problem in the application of diffraction methods to polymers is the extreme difficulty in growing single crystals of an adequate size (typically several cubic millimetres). Some single-crystal diffraction studies are made on biological polymers which may crystallize more readily than synthetic polymers. Indeed a considerable effort has been made in low-resolution crystallography on proteins [209]. For these materials the lack of beam damage to the sample in neutron experiments can be of overriding importance. Experiments with synthetic polymers have been largely restricted to powder diffraction or, occasionally, fibre diffraction. It is the subject of diffraction from 'powders' that is of most importance in respect to synthetic polymers.

It should be emphasized that a 'powder' means a sample that has a sufficiently large number of randomly oriented crystals that, whatever the orientation of the sample, the Bragg condition for diffraction can be met for all crystallographic planes. The form of the sample is not necessarily a powder; it can also be a solid polycrystalline sample such as a sheet or film. Any preferred orientation of crystallites in the sample will cause a change in the relative intensity of the peaks. This can be exploited to determine the orientation in a sample but may pose problems if

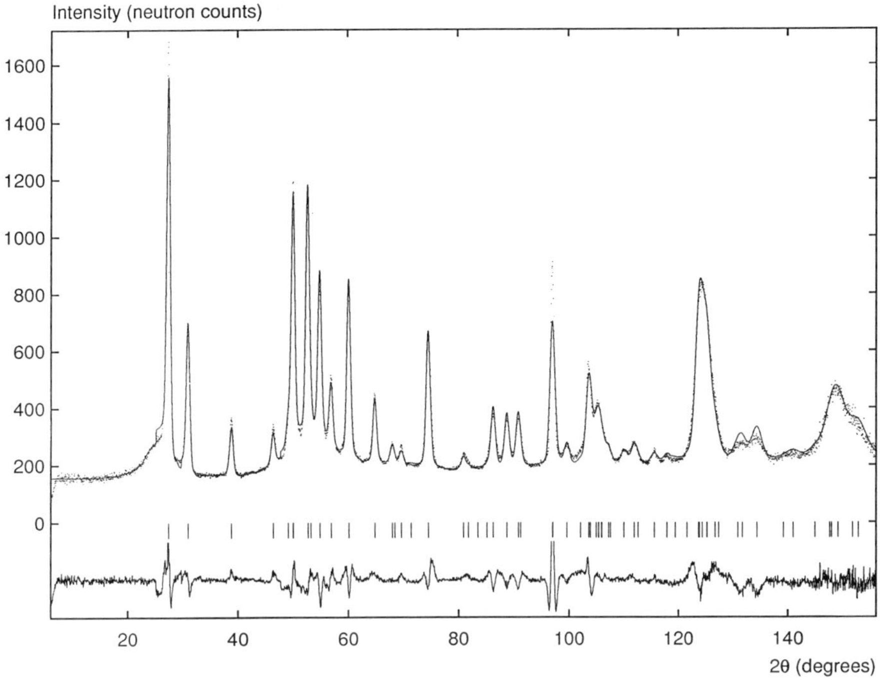

Fig. 4.10 Diffraction pattern of deuterated polyethylene measured with a neutron wavelength of 0.1911 nm at 1.5 K. The continuous line shows the fit that results from Rietveld refinement of the structural parameters [210].

measurements of peak intensity are required for refinement of crystal structures. An example which shows that good diffraction data can be obtained from a translucent semi-crystalline film is the diffraction pattern from a deuterated polyethylene sample [210] shown in Fig. 4.10.

Such data as that shown in Fig. 4.10 are essentially similar to those obtained by Avitabile *et al.* [211] on other samples of polyethylene. Although the basic structure has been known for many years from X-ray diffraction [212], the details of the positions of the deuterium and the thermal motion within the crystal can be obtained by refinement of the model structure with this data according to the methods of Rietveld [34]. As yet, rather few polymers have been studied in this way and no *ab initio* structure determinations of polymers have been made by neutron powder diffraction. The large unit cells and difficulties in obtaining highly crystalline samples are the main reasons for this paucity of study when compared to other classes of materials.

Other applications of neutron powder diffraction have exploited the contrast available by isotopic substitution to study the structure of doped polymer crystals [213, 214, 215] and complexes with small molecules [216]. This has particular application in the investigation of semiconducting polymers such as polyacetylenes [213, 215]. Kinetic studies of the doping of polyparaphenylene with AsF_5 have been reported [214]. The use of diffraction data from mixed crystals to test models of the folding of polymer chains was mentioned previously [116, 117].

4.5.2 Diffuse scattering

In many polymers there is no long-range crystalline order but the study of molecular packing and local order is still of interest. In such materials the wide-angle scattering will show diffuse, broad peaks rather than the sharp Bragg diffraction from crystals. In some respects this can be considered as a continuation of the diffuse scattering from isolated objects that is measured in SANS experiments to larger momentum transfer. However, there are distinct features which cause us to describe the experiments separately. Structure, as observed in scattering experiments, may arise from either intra-molecular or inter-molecular correlations. The analysis of such data is similar to that from other amorphous solids or liquids. The particular advantage of neutrons is that different contrasts can be used in samples of a given chemical composition. This will enable measurements of the different partial structure factors within a complex system [35]. As yet there have been rather few studies in detail of local structure in amorphous polymers but some recent work shows the potential of this technique.

Studies will be mentioned which typify work in this field: they are concerned with ionomers. These materials, which contain a few ionizable groups, may consist of several phases. A characteristic feature of the scattering pattern that is observed with X-rays or neutrons is a peak observed at scattering vectors of the order of $1\,nm^{-1}$ (known as an 'ionomer peak') which has been associated with clusters of the ionic groups and the spacing of such clusters. Attempts to model this structure are usually hampered by lack of information about the chemical composition of different phases: materials of interest are commonly copolymers that may partially crystallize, contain a fraction of ionized groups and have a large uptake of solvent (water). There has been some limited success in describing the scattering of these materials with models physically consistent with the known bulk composition and density [217, 218, 219, 220]. Recently Gabrys *et al.* [221, 222] have undertaken careful work with polarized neutrons which permits subtraction of the incoherent background to much higher precision than previous work and so provides a clearer indication of the pair correlations in such samples. They have shown that correlations in ionomers are similar to those found in polymers without the charged species [223]. This approach has also been applied to uncharged polymers [222]. In a slightly different approach, Lee *et al.* [224] have used contrast variation together with measurements over a very wide range of momentum transfer to deduce partial structure factors that can be Fourier transformed to give radial distribution functions. This wide range of measurements has avoided the usual problem of spurious structure that is created on transforming data sets that are poorly terminated (for example, with a step between a background level and zero) at small or large momentum transfer.

Total scattering measurements with diffractometers can also prove useful in experiments on motion in glasses or crystals as correlated motion can give rise to diffuse peaks. Some experiments on fluoroalkanes [225] have exploited this technique.

4.6 INELASTIC AND QUASI-ELASTIC SCATTERING

Direct study of the dynamics of solid polymers by means of neutron scattering may be conveniently divided into three distinct categories. Collective vibrational modes such as phonons can be measured in crystalline polymers. The complex molecular architecture of many polymers provides a rich variety of local motion of side groups and the polymer chain which can be observed as diffusive motion in crystalline or amorphous materials. Finally elastomers can display the modes of diffusive motion associated with polymer melts provided that the number of flexible units between cross-links is large.

The first category (of phonons in solids) includes some of the earliest applications of neutron scattering to polymers [226], dating back to the 1960s. The usual approach of solid-state physicists is to measure the dispersion relation with a triple-axis spectrometer. On such an instrument the monochromator and analyser crystals are scanned to provide known momentum and energy transfer while the third axis is used to select different crystallographic directions in a single crystal sample. While this type of study is of great value in determining the force constants within a crystal lattice, it is rarely applicable to polymers because of the inability to grow large single crystals of the order of 1 cm^3. It should also be noted that the large incoherent background from hydrogen in most organic polymers will prevent collection of useful data unless fully deuterated samples can be prepared. It is perhaps for this reason that, although results have been reported for polyoxymethylene [227] and polyethylene [228], work has often concentrated on perfluoro-olefins [226, 229, 230]. In some recent work the inability to grow single crystals has been compensated by the careful preparation of mats of biaxially orientated crystals [230]. A large degree of diffusional motion of the polymer chain may be present in some crystals, as seen in the rotator phases of long alkanes which have been the subject of considerable study [231, 232].

The major interest in dynamic properties of amorphous solid polymers has been in the changes that occur near the glass transition. This process is associated with the onset of diffusive motion over large distances. Recent models have been based on mode coupling theories and inelastic neutron-scattering experiments have proved useful in testing them. The wide range of momentum and energy transfer (corresponding to distances and frequencies) available with neutron experiments has proved of particular advantage. Detailed studies have been made of polybutadiene [233, 234, 235, 236] using a wide range of neutron spectroscopic techniques, ranging from the traditional time-of-flight spectrometers to neutron spin-echo experiments. Other experiments have been concerned with purely local motion, such as that of side groups in polystyrenes [237] and polymethyl methacrylates [238, 239].

The motion that can occur in cross-linked networks or rubbers has been investigated. The use of isotopic labelling of particular parts of molecules which are then linked to form networks has meant that the motion of cross-link points can be separated from that of the rest of the network [240, 241, 242]. This type of information is of particular value in testing theoretical models [243]. The motion of flexible chains in lightly linked elastomers is substantially similar to that in entangled polymer melts but the entropic forces are modified by the effective length of the flexible freely moving units. Studies have been made, using spin-echo techniques, [244] on monodisperse polydimethyl siloxane end-linked into networks. This last study also included the effects of deformation on the motion in elastomers.

4.7 REFLECTION

Although comparatively new as a technique, neutron reflection has attracted considerable attention from polymer scientists. There are two distinct fields of application: the study of thin solid films and their interfaces and secondly the study of adsorption of polymers from solution. An extensive review of both neutron and X-ray reflectivity measurements on polymers has been made recently by Russell [245], to which the interested reader is referred for a fuller discussion of most of the topics mentioned in this section. Some of the practical aspects of neutron reflection are described in the review of Penfold and Thomas [41]. At this stage, it is difficult to provide a brief review covering the full potential applications of this technique in the study of solid polymers. Comparatively few papers have appeared in the literature and many obvious developments of this technique have not yet been applied.

The following paragraphs give some indication of the subjects that have been published in the last two to three years.

Most of the work on polymer films has concentrated on the properties of polymer blends and co-polymers. At the surface an additional contribution to the thermodynamic description of the blend in the form of surface free energy must be considered. This will in general be different for chemically different polymers and may dominate over bulk properties in thin films. Indeed, interesting observations of the preference of one polymer or one block of a co-polymer for the free surface have been made with neutron reflection. Russell *et al.* have now performed detailed measurements on systems containing styrene/methyl methacrylate block co-polymers and mixtures of the homo-polymers with and without the co-polymer [246, 247, 248]. If layers are prepared on a silicon substrate, the polar acrylic group tends to prefer proximity to the silica formed on top of the silicon. The styrene blocks will be found at the air/film interface. It is possible to perturb this structure from the usual equilibrium spacing found in phase-separated co-polymers by constraining the overall film thickness [248, 249]. Interfaces of these two homo-polymers have also been studied [250]. The properties of block co-polymers as compatibilizers for blends have been explored: they have been found to broaden the interfaces [246, 247].

Other systems that have been studied are interfaces of polymethyl methacrylate with solution-chlorinated polyethylene [251, 252] and mixtures of hydrogenous and deuterated polystyrene [253, 254] and polystyrene with poly(bromostyrene) [255]. This technique has provided much new, detailed information about surfaces. For example, evidence for an unusually high density of polymethyl methacrylate at the surface of a spun film has been presented [256].

The concentration profile of the interface between polymers in the experiments mentioned above has been studied in some detail. The evolution with time can provide data on the rate of inter-diffusion as well as the compatibility of two components. Several experiments have been aimed at studying the mutual inter-diffusion of polymers [251, 257, 258]. In some respects this is similar to the work that can be done with small-angle neutron scattering [118, 119, 120, 121], as the contrast and length scales that can be investigated are similar. The need to produce and characterize only one initial interface is an advantage over the stacking of many thin films that is required by some SANS experiments [120, 121]. The major disadvantage is the additional complication of surface free energy which may cause significant differences to properties in the bulk of large samples.

Another field that has attracted considerable attention is the structure of adsorbed layers of polymers at solid/solution [259, 260, 261] and vapour/solution [262, 263, 264] interfaces. The first of these topics is important for the study of polymers as stabilizers of dispersions and other systems such as composite materials in which interfacial properties of polymers are important. It extends the range of measurements that have been made with SANS of polymers [265, 266] and co-polymers [267] adsorbed on colloidal particles, for which there are often difficulties in characterizing the substrate. The study of polymers at air/liquid interfaces is relevant to the preparation of Langmuir–Blodgett films and although these have been studied by X-ray reflectivity, there has as yet been little work with neutrons, although several groups are now interested in these problems.

4.8 CONCLUSIONS AND OUTLOOK

Neutron scattering has developed enormously as an analytical tool for polymer scientists in recent years and the trend is likely to continue. The two major drawbacks are the requirement for large facilities as a neutron source, and the need for many experiments to prepare isotopically labelled (usually deuterated) samples. The evidence of the large amount of literature in this field suggests that these are outweighed by the detailed information on molecular structure and dynamics that can be obtained for many samples. In many experiments the data analysis is straightforward, even to the extent that direct comparison of data from different samples or single samples in different directions may provide useful information. It is to be expected that the use of neutrons as a routine tool for the determination of molecular size will continue to increase. The technique of small-angle scattering is likely to develop in several ways. Extensions to larger distance scales will become more common as the full potential of double-crystal diffractometers is developed. This will have particular advantage in the study of composite materials. As the details of the technique are better understood it becomes possible to envisage careful experiments that will study differences between samples that are much smaller (less than 1%) [65].

Several new fields of application of neutrons are emerging. For example, experiments on materials under time varying applied fields (mechanical stress, electrical or magnetic fields) are clearly developing rapidly for the study of both equilibrium structure and dynamic properties. The use of reflectivity measurements is another field of much promise: for example, measurements of diffuse scattering in this geometry will provide additional information on lateral heterogeneity in surfaces [268, 269]. This field is largely unexplored. One can speculate that the application of dynamic polarization techniques [91, 92], currently being tested in small-angle scattering measurements, may prove of more advantage in the study of polymer films in reflection geometry.

It has been necessary to be selective in the material presented in this review and it has not been possible to mention all of the many different materials that have been the subject of neutron studies. The choice of references has been made with a view to providing an introduction to the literature through citation of recent work.

REFERENCES

1. Lovesey, S.W. (1984) *Theory of Neutron Scattering from Condensed Matter*, Vols 1 and 2, Oxford University Press, Oxford.
2. Squires, G.L. (1978) *Introduction to the Theory of Thermal Neutron Scattering*, Cambridge University Press, Cambridge.
3. Feigin, L.A. and Svergun, D.I. (1987) *Structure Analysis by Small-Angle X-ray and Neutron Scattering*, Plenum Press, New York.
4. Sadler, D.M. (1989) Neutron scattering from solid polymers, in *Comprehensive Polymer Science*, Vol. 1, Pergamon Press, Oxford pp. 731–63.
5. Maconnachie, A. and Richards, R.W. (1978) *Polymer*, **19**, 739–62.
6. Wignall, G.D. (1987) Neutron scattering, in *Encyclopedia of Polymer Science and Engineering*, 2nd edn, Vol. 10, J. Wiley & Sons, pp. 112–84.
7. Born, M. (1966) *Atomic Physics*, Blackie, Edinburgh.
8. Sears, V.F. (1984) *Thermal-Neutron Scattering Lengths and Cross Sections for Condensed-Matter Research*, Atomic Energy of Canada Limited Report AECL 8490, Chalk River.
9. Guinier, A. and Fournet, G. (1955) *Small-Angle Scattering of X-Rays*, J. Wiley & Sons, New York.
10. Glatter, O. and Kratky, O. (eds) (1982) *Small-Angle X-Ray Scattering*, Academic Press, London.
11. Debye, P. (1947) *Journal of Physical and Colloid Chemistry*, **51**, 18–32.
12. Jacrot, B. and Zaccai, G. (1981) *Biopolymers*, **20**, 2413–26.
13. Zimm, B.H. (1948) *Journal of Chemical Physics*, **16**, 1093–9 and 1099–116.
14. Huglin, M.B. (ed) (1972) *Light Scattering from Polymer Solutions*, Academic Press, London.
15. Benoit, H. and Benmouna, M. (1984) *Polymer*, **25**, 1059–67.
16. Benoit, H. and Benmouna, M. (1991) *Th. Vilgis Compte Rendue de L'Academie des Sciences*, **313**, Serie **II**, 869–72.
17. Wignall, G.D., Hendricks, R.W., Koehler, W.C. *et al.* (1981) *Polymer*, **22**, 886–9.
18. Sakurai, S., Hasegawa, H., Hashimoto, T. and Han, C.C. (1990) *Polymer*, **31**, 99–102.
19. Benoit, H. (1953) *Journal of Polymer Science*, **11**, 507–10.
20. Huber, K., Bantle, S., Burchard, W. and Fetters, L.J. (1986) *Macromolecules*, **19**, 1404–11.
21. Huber, K., Burchard, W., Bantle, S. and Fetters, L.J. (1987) *Polymer*, **28**, 1990–6.
22. Richter, D., Stuhn, B., Ewen, B. and Nerger, D. (1987) *Physical Review Letters*, **58**, 2462–7.
23. Boothroyd, A.T., Squires, G.L., Fetters, L.J. *et al.* (1989) *Macromolecules*, **22**, 3130–7.
24. Horton, J.C., Squires, G.L., Boothroyd, A.T. *et al.* (1989) *Macromolecules*, **22**, 681–6.
25. Leibler, L. (1990) *Macromolecules*, **13**, 1602–17.
26. Benoit, H. and Hadziioannou, G. (1988) *Macromolecules*, **21**, 1449–64.

27. Richards, R.W. (1985) *Advances in Polymer Science*, **71**, 1–39.
28. Benoit, H. and Doty, P. (1953) *Journal of Physical Chemistry*, **57**, 958–63.
29. Ullman, R. (1985) *Journal of Polymer Science: Polymer Physics Edition*, **23**, 1477–84.
30. Schelten, J. and Schmatz, W. (1980) *Journal of Applied Crystallography*, **13**, 385–90.
31. Kakudo, M. and Kasai, N. (1972) *X-Ray Diffraction by Polymers*, Kodansha/Elsevier, Tokyo.
32. Hukins, D.W.L. (1981) *X-Ray Diffraction by Disordered and Ordered Systems*, Pergamon Press, Oxford.
33. Bacon, G.E. (1971) *Neutron Diffraction*, 3rd edition, Clarendon Press, Oxford.
34. Rietveld, H.M. (1969) *Journal of Applied Crystallography*, **2**, 65–71.
35. Enderby, J.E. and Neilson, G.W. (1979) in *Water, a Comprehensive Treatise*, F. Franks (ed.) Vol. **6**. Plenum, London p. 1.
36. Born, M. and Wolf, E. (1975) *Principles of Optics*, 5th edn, Pergamon Press, Oxford.
37. Abelés, F. (1950) *Annales de Physique*, **5**, 596–640 and 706–82.
38. Lekner, J. (1987) *Theory of Reflection*, Martinus Nijhoff Publishers, Dordrecht.
39. Heavens, O.S. (1955) *Optical Properties of Thin Solid Films*, Butterworths, London.
40. Sears, V.F. (1989) *An Introduction to the Theory of Neutron Optical Phenomena and their Applications*, Oxford University Press, Oxford.
41. Penfold, J. and Thomas, R.K. (1990) *Journal of Physics: Condensed Matter*, **2**, 1369–412.
42. Crowley, T.L., Lee, E.M., Simister, E.A. *et al.* (1990) *Colloids and Surfaces*, **52**, 85–106.
43. Simister, E.A., Lee, E.M., Thomas, R.K. and Penfold, J. (1992) *Macromolecular Reports*, **A29**, 155.
44. Bee, M. (1989) *Quasi-Elastic Neutron Scattering*, Adam Hilger, Bristol.
45. Rouse, P.E. (1953) *Journal of Chemical Physics*, **21**, 1272–80.
46. Zimm, B.H. (1956) *Journal of Chemical Physics*, **24**, 269–78.
47. Doi, M. and Edwards, S.F. (1986) *The Theory of Polymer Dynamics*, Clarendon Press, Oxford.
48. Windsor, C.G. (1981) *Pulsed Neutron Scattering*, Taylor and Francis, London.
49. Carpenter, J.M., Lander, G.H. and Windsor, C.G. (1984) *Review of Scientific Instruments*, **55**, 1019–43.
50. Blank, H. and Maier, B. (eds) (1988) *Guide to Neutron Research Facilities at the ILL*, Institut Laue Langevin, Grenoble.
51. Abend, K., Schmatz, W., Schelten, J. and Muller, K.D. (1970) *Nuclear Instruments and Methods*, **83**, 111–4.
52. Ibel, K. (1976) *Journal of Applied Crystallography*, **9**, 296–309.
53. Alefeld, B., Schwahn, D. and Springer, T. (1989) *Nuclear Instruments and Methods*, **A274**, 210–6.
54. Allemand, R., Bourdel, J., Roudaut, E. *et al.* (1975) *Nuclear Instruments and Methods*, **126**, 29–42.
55. Convert, P. and Forsyth, J.B. (eds) (1987) *Position Sensitive Detection of Thermal Neutrons*, Academic Press, London.
56. Child, H.R. and Spooner, S. (1980) *Journal of Applied Crystallography*, **13**, 259–64.
57. Koehler, W.C. (1984) *Physica*, **137B**, 320–9.

58. Kjems, J.K., Bauer, R., Christensen, P. *et al.* (1985) in *Neutron Scattering in the Nineties*, IAEA Vienna, pp. 495–9.
59. Ishikawa, Y., Furusaka, M. and Niimura, N. (1986) *Journal of Applied Crystallography*, **19**, 229–42.
60. Hjelm, R.P. (1988) *Journal of Applied Crystallography*, **20**, 273–9.
61. Mildner, D.F.R. (1986) *Journal of Applied Crystallography*, **19**, 275–6.
62. Ramakrishnan, V. (1985) *Journal of Applied Crystallography*, **18**, 42–6.
63. Pedersen, J.S., Posselt, D. and Mortensen, K. (1990) *Journal of Applied Crystallography*, **23**, 321–33.
64. Wignall, G.D., Christen, D.K. and Ramakrishnan, V. (1988) *Journal of Applied Crystallography*, **21**, 438–51.
65. Boothroyd, A.T., Rennie, A.R. and Boothroyd, C.B. (1991) *Europhysics Letters*, **15**, 715–9.
66. Ottewill, R.H. (1990) Small angle neutron scattering on polymer colloids, in *Scientific Methods for the Study of Polymer Colloids and their Applications*, Candau, F. and Ottewill, R.H. (eds). Kluwer Academic Publishers, Dordrecht.
67. Potton, J.A., Daniell, G.J. and Rainford, B.D. (1986) *Institute of Physics Conference Series*, **81**, 81–6.
68. Potton, J.A., Daniell, G.J. and Rainford, B.D. (1988) *Journal of Applied Crystallography*, **21**, 663–8.
69. Epperson, J.E., Carpenter, J.M., Thiyagarajan, P. and Heuser, B. (1990) *Nuclear Instruments and Methods in Physics Research*, **A289**, 30–4.
70. Schwahn, D., Miksovsky, A., Rauch, H. *et al.* (1985) *Nuclear Instruments and Methods in Physics Research*, **A239**, 229–34.
71. Mikula, P., Lukas, P. and Eichorn, F. (1988) *Journal of Applied Crystallography*, **21**, 33–7.
72. Schwahn, D., Springer, T., Yee-Madeira, H. and Mortensen, K. (1988) *Springer Proceedings in Physics*, **29**, 296–300.
73. Hergeth, W-D., Bittrich, H-J., Eichorn, F. *et al.* (1989) *Polymer*, **30**, 1913–7.
74. Penfold, J., Ward, R.C. and Williams, W.G. (1987) *Journal of Physics E: Scientific Instruments*, **20**, 1411–7.
75. Felcher, G.P., Felici, R., Kampwirth, R.T. and Gray, K.E. (1985) *Journal of Applied Physics*, **57**, 3789–94.
76. Felcher, G.P., Hilleke, R.O., Crawford, R.K. *et al.* (1987) *Review of Scientific Instruments*, **58**, 609–19.
77. Stamm, M., Reiter, G. and Huttenbach, S. (1989) *Physica*, **B156/157**, 564–6.
78. Hayter, J.B. (1980) Theory of neutron spin-echo spectrometry, in *Neutron Spin Echo* (ed. F. Mezei) Lecture Notes in Physics, Vol. 128, Springer–Verlag, Berlin pp. 53–65.
79. Dagleish, P.A., Hayter, J.B. and Mezei, F. (1980) The IN11 neutron spin echo spectrometer, in *Neutron Spin Echo* (ed. F. Mezei) Lecture Notes in Physics, Vol. 128, Springer–Verlag, Berlin pp. 66–71.
80. Nicholson, L.K. (1981) *Contemporary Physics*, **22**, 451–75.
81. Landau, L. and Lifshitz, E. (1968) *Course of Theoretical Physics: Vol. 5 Statistical Physics* 2nd edn, Pergamon Press, Oxford.
82. Bates, F.S., Wignall, G.D. and Koehler, W.C. (1985) *Physical Review Letters*, **55**, 2425–8.
83. Bates, F.S. and Wignall, G.D. (1986) *Macromolecules*, **19**, 932–4.
84. de Gennes, P.G. (1979) *Scaling Concepts in Polymer Physics*, Cornell University Press, Ithaca.
85. Buckingham, A.D. and Hentschel, H.G.E. (1980) *Journal of Polymer Science*, Polymer Physics Edition 18, 853–61.

86. Schwahn, D., Hahn, K., Streib, J. and Springer, T. (1990) *Journal of Chemical Physics*, **93**, 8383–91.
87. Tanzer, J.D., Bartels, C.R., Crist, B. and Graessley, W.W. (1984) *Macromolecules*, **17**, 2708–14.
88. Crist, B. and Wignall, G.D. (1988) *Journal of Applied Crystallography*, **21**, 701–6.
89. Crist, B., Tanzer, J.D. and Finerman, T.M. (1989) *Journal of Polymer Science B: Polymer Physics*, **27**, 875–8.
90. Tanzer, J.D. and Crist, B. (1985) *Macromolecules*, **18**, 1291–4.
91. Knop, W., Meerwink, W., Schink, H.-J. *et al.* (1989) *Journal of Applied Crystallography*, **22**, 352–62.
92. Knop, W., Hirai, M., Meerwink, W. *et al.* (1991) *Physica*, **B174**, 275–90.
93. Maconnachie, A. (1984) *Polymer*, **25**, 1068–72.
94. Ghosh, R.E. and Rennie, A.R. (1990) *Institute of Physics Conference Series*, **107**, 233–44.
95. Horkay, F., Hecht, A-M., Mallam, S. *et al.* (1991) *Macromolecules*, **24**, 2896–902.
96. Geissler, E. and Hecht, A.M. (1991) Polymer Solutions and Gels: Incoherent Background Subtraction, unpublished proceedings of ILL Workshop on Analysis of SANS and NR Data, March 1991.
97. Kirste, R.G., Kruse, W.A. and Schelten, J. (1973) *Die Makromolekulare Chemie*, **162**, 299–305.
98. Kirste, R.G., Kruse, W.A. and Ibel, K. (1975) *Polymer*, **16**, 120–4.
99. Benoit, H., Decker, D., Higgins, J.S. *et al.* (1973) *Nature–Physical Sciences*, **245**, 13–5.
100. Cotton, J.P., Decker, D., Benoit, H. *et al.* (1974) *Macromolecules*, **7**, 863–72.
101. Ballard, D.G.H., Schelten, J. and Wignall, G.D. (1973) *European Polymer Journal*, **9**, 965–9.
102. Schelten, J., Wignall, G.D., Ballard, D.G.H. *et al.* (1976) *Polymer*, **17**, 751–7.
103. Guenet, J.M., Picot, C. and Benoit, H. (1979) *Faraday Discussions of the Chemical Society*, **68**, 251–62.
104. Fischer, E.W., Hahn, K., Kugler, J. *et al.* (1984) *Journal of Polymer Science: Polymer Physics Edition*, **22**, 1491–513.
105. Fischer, E.W. (1978) *Pure and Applied Chemistry*, **50**, 1319–41.
106. Fischer, E.W. (1985) *Polymer Journal*, **17**, 307–20.
107. Sadler, D.M. and Keller, A. (1977) *Macromolecules*, **10**, 1128–40.
108. Yoon, D.Y. and Flory, P.J. (1976) *Polymer*, **18**, 509–13.
109. Stamm, M., Fischer, E.W., Dettenmaier, M. and Convert, P. (1979) *Faraday Discussions of the Chemical Society*, **68**, 263–78.
110. Yoon, D.Y. and Flory, P.J. (1979) *Faraday Discussions of the Chemical Society*, **68**, 288–96.
111. Guttman, C.M., Hoffman, J.D. and DiMarzio, E.A. (1979) *Faraday Discussions of the Chemical Society*, **68**, 297–309.
112. Spells, S.J. and Sadler, D.M. (1984) *Polymer*, **25**, 739–48.
113. Sadler, D.M. and Spells, S.J. (1984) *Polymer*, **25**, 1219–26.
114. Guest, M.W. and Mansfield, M.L. (1991) *Macromolecules*, **24**, 276–81.
115. Sadler, D.M. and Spells, S.J. (1989) *Macromolecules*, **22**, 3941–8.
116. Stamm, M. (1982) *Journal of Polymer Science: Polymer Physics Edition*, **20**, 235–44.
117. Wignall, G.D., Mandelkern, L., Edwards, C. and Glotin, M. (1982) *Journal of Polymer Science: Polymer Physics Edition*, **20**, 245–58.
118. Summerfield, G.C. and Ullman, R. (1987) *Macromolecules*, **20**, 401–4 and (1988) *Macromolecules*, **21**, 2643–4.

119. Bartels, C.R., Crist, B. and Graessley, W.W. (1984) *Macromolecules*, **17**, 2702–8.
120. Rennie, A.R. (1988) *Springer Proceedings in Physics*, **29**, 72–85.
121. Stamm, M. (1983) *American Chemical Society Polymer Preprints*, **24**, 380.
122. Meurer, E., Kuhn, K.J., Hellmann, G.P. and Rennie, A.R. (1990) *Journal of Polymer Science: Part B: Polymer Physics*, **28**, 893–905.
123. Hellmann, E.H., Hellmann, G.P. and Rennie, A.R. (1991) *Macromolecules*, **24**, 3821–7.
124. Hahn, K., Ley, G. and Shuller, H. and Oberthür, R.C. (1986) *Colloid and Polymer Science*, **264**, 1092–6.
125. Hahn, K., Ley, G. and Oberthür, R.C. (1988) *Colloid and Polymer Science*, **266**, 631–9.
126. McAlea, K.P., Schultz, J.P., Gardner, K.H. and Wignall, G.D. (1986) *Polymer*, **27**, 1581–4.
127. Kugler, J., Gilmer, J.W., Wiswe, D. *et al.* (1987) *Macromolecules*, **20**, 1116–9.
128. MacDonald, W.A., McLenaghan, A.D.W., McLean, G. *et al.* (1991) *Macromolecules*, **24**, 6164–7.
129. Benoit, H.C., Fischer, E.W. and Zachmann, H.G. (1989) *Polymer*, **30**, 379–85.
130. Kawai, H., Itoh, T., Keedy, D.A. and Stein, R.S. (1964) *Journal of Polymer Science*, **B2**, 1075–8.
131. Burchell, D.J., Lasch, J.E., Farris, R.J. and Hsu, S.L. (1982) *Polymer*, **23**, 965–8.
132. Rennie, A.R. and Oberthür, R.C. (1984) *Revue de Physique Appliquée*, **19**, 765–8.
133. Connell, J.G., Richards, R.W. and Rennie, A.R. (1991) *Polymer* **32**, 2033–42.
134. Kirste, R.G. and Lehnen, B.R. (1976) *Die Makromolekulare Chemie*, **177**, 1137–43.
135. Kruse, W.A., Kirste, R.G., Haas, J. *et al.* (1976) *Die Makromolekulare Chemie*, **177**, 1145–60.
136. Herkt-Maetzky, Ch. and Schelten, J. (1983) *Physical Review Letters*, **51**, 896–9.
137. Jelenic, J., Kirste, R.G., Oberthür, R.C. *et al.* (1984) *Die Makromolekulare Chemie*, **185**, 129.
138. Brereton, M.G., Fischer, W.W., Herkt-Maetzky, Ch. and Mortensen, K. (1987) *Journal of Chemical Physics*, **87**, 6144–9.
139. Schwahn, D., Mortensen, K., Springer, T. *et al.* (1987) *Journal of Chemical Physics*, **87**, 6078–87.
140. Han, C.C., Bauer, B.J., Clark, J.C. *et al.* (1988) *Polymer*, **29**, 2002–14.
141. Bauer, B.J., Briber, R.M. and Han, C.C. (1989) *Macromolecules*, **22**, 940–48.
142. Dudowicz, J. and Freed, K.F. (1991) *Macromolecules*, **24**, 5112–23.
143. Kambour, R.P., Bopp, R.C., Maconnachie, A. and MacKnight, W.J. (1980) *Polymer*, **21**, 133–4.
144. Wignall, G.D., Child, H.R. and Li-Aravena, F. (1980) *Polymer*, **21**, 131–2.
145. Maconnachie, A., Kambour, R.P. and Bopp, R.C. (1984) *Polymer*, **25**, 357–64.
146. Russell, T.P., Hjelm, R.P. and Seeger, P.A. (1990) *Macromolecules*, **23**, 890–3.
147. Sakurai, S., Izumitani, T., Hasegawa, H. *et al.* (1991) *Macromolecules*, **24**, 4844–51.
148. Tomlins, P.E. and Higgins, J.S. (1985) *Polymer*, **26**, 1554–60.
149. Tomlins, P.E. and Higgins, J.S. (1988) *Macromolecules*, **21**, 425–32.
150. Higgins, J.S., Fruitwala, H., Tomlins, P.E. and Clark, J.N. (1988) *Springer Proceedings in Physics*, **29**, 274–84.
151. Hasegawa, H., Sakurai, S., Takenaka, M. *et al.* (1991) *Macromolecules*, **24**, 1813–9.

152. Sakurai, S., Jinnai, H., Hasegawa, H. *et al.* (1991) *Macromolecules*, **24**, 4839–43.
153. Hellmann, E.H., Hellmann, G.P. and Rennie, A.R. (1991) *Colloid and Polymer Science*, **269**, 343–52.
154. Yu, D., Hellmann, E.H. and Hellmann, G.P. (1991) *Die Makromolekulare Chemie*, **192**, 2749–66.
155. Clark, J.N., Higgins, J.S. and Peiffer, D.G. (1992) *Polymer Engineering and Science*, **32**, 49–56.
156. Hadziioannou, G., Picot, C., Skoulios, A. *et al.* (1982) *Macromolecules*, **15**, 263–7.
157. Selb, J., Marie, P., Rameau, A. *et al.* (1983) *Polymer Bulletin*, **10**, 441–51.
158. Bates, F.S. (1985) *Macromolecules*, **18**, 525–8.
159. Bates, F.S. and Hartney, M.A. (1985) *Macromolecules*, **18**, 2478–86.
160. Marie, P., Selb, J., Rameau, A. and Gallot, Y. (1988) *Die Makromolekulare Chemie: Macromolecular Symposia*, **16**, 301–12.
161. Marie, P., Duplessix, R., Gallot, Y. and Picot, C. (1979) *Macromolecules*, **12**, 1180–6.
162. Higgins, J.S., Blake, S., Tomlins, P.E. *et al.* (1988) *Polymer*, **29**, 1968–78.
163. Schouten, M., Dorrepaal, J., Stassen, W.J.M. *et al.* (1989) *Polymer*, **30**, 2038–46.
164. Wanka, G., Hoffmann, H. and Ulbricht, W. (1990) *Colloid and Polymer Science*, **268**, 101–17.
165. Cogan, K.A., Gast, A.P. and Capel, M. (1991) *Macromolecules*, **24**, 6512–20.
166. Higgins, J.S., Tomlins, P.E., Dawkins, J.V. *et al.* (1988) *Polymer Communications*, **29**, 122–5.
167. Cosgrove, T., Finch, N., Vincent, B. and Webster, J. (1988) *Colloids and Surfaces*, **31**, 33–46.
168. Cosgrove, T., Finch, N. and Webster, J. (1990) *Colloids and Surfaces*, **45**, 377–89.
169. Jung, W.G. and Fischer, E.W. (1988) *Die Makromolekulare Chemie: Macromolecular Symposium*, **16**, 281–300.
170. Bates, F.S., Berney, C.V., Cohen, R.E. and Wignall, G.D (1982) *Polymer*, **24**, 519–24.
171. Richards, R.W. and Thomason, J.L. (1983) *Macromolecules*, **16**, 982–92.
172. Richards, R.W. and Thomason, J.L. (1983) *Polymer*, **24**, 1089–96.
173. Miller, J.A., Pruckmayr, G., Epperson, J.E. and Cooper, S.L. (1985) *Polymer*, **26**, 1915–20.
174. Matsushita, Y., Mori, K., Saguchi, R. *et al.* (1990) *Macromolecules*, **23**, 4387–91.
175. Sakurai, S., Izumitani, T., Hasegawa, H. (1991) *Macromolecules*, **24**, 4844–51.
176. Lindner, P. and Oberthür, R.C. (1984) *Revue de Physique Appliquée*, **19**, 759–63.
177. Ottewill, R.H. and Rennie, A.R. (1990) *International Journal of Multiphase Flow*, **16**, 681–90.
178. Lindner, P. and Oberthür, R.C. (1988) *Colloid and Polymer Science*, **263**, 443–53 and (1988), **266**, 886–97.
179. Mildner, D.F.R. (1983) *Macromolecules*, **16**, 1760–3.
180. Beltzung, M., Picot, C. and Herz, J. (1984) *Macromolecules*, **17**, 663–9.
181. Clough, S.B., Maconnachie, A. and Allen, G. (1980) *Macromolecules*, **13**, 774–5.
182. Boué, F., Nierlich, M., Jannink, G. and Ball, R.C. (1982) *Journal de Physique Lettres*, **43**, L585–91.
183. Boué, F. (1987) *Advances in Polymer Science*, **82**, 47–101.

184. Muller, R., Picot, C., Zang, Y.H. and Froehlich, D. (1990) *Macromolecules*, **23**, 2577–82.
185. Muller, R. (1988) Comportement rhéologique et mesures d'orientation molé culaire de polymères á l'état fondu dans un écoulement elongationnel simple. Thesis, Universite Louis Pasteur, Strasbourg.
186. Lefebvre, J.M., Escaig, B. and Picot, C. (1982) *Polymer*, **23**, 1751–4.
187. Schroeder, J.R., Hammouda, B., Bubeck, R.A. and Chang, J.W. (1991) *Journal of Polymer Science: Polymer Physics Edition*, **29**, 1607–12.
188. Boué, F., Bastide, J., Buzier, M. *et al.* (1991) *Colloid and Polymer Science*, **269**, 195–216.
189. Oeser, R., Picot, C. and Herz, J. (1988) *Springer Proceedings in Physics*, **29**, 104–11.
190. Ballard, D.G.H., Burgess, A.N., Crowley, T.L. *et al.* (1979) *Faraday Discussions of the Chemical Society*, **68**, 279–87.
191. Ballard, D.G.H., Cheshire, P., Janke, E. *et al.* (1982) *Polymer*, **23**, 1875–83.
192. Sadler, D.M. and Barham, P.J. (1983) *Journal of Polymer Science: Polymer Physics Edition*, **21**, 309–17.
193. Sadler, D.M. and Barham, P.J. (1990) *Polymer*, **31**, 36–42.
194. Sadler, D.M. and Barham, P.J. (1990) *Polymer*, **31**, 43–5.
195. Sadler, D.M. and Barham, P.J. (1990) *Polymer*, **31**, 46–50.
196. Crist, B., Tanzer, J.D. and Finerman, T.M. *Journal of Polymer Science B: Polymer Physics*, **27**, 875–8.
197. Dettenmaier, M., Maconnachie, A., Higgins, J.S. *et al.* (1986) *Macromolecules*, **19**, 773–8.
198. Gilmer, J.W., Wiswe, D., Zachmann, H.G. *et al.* (1986) *Polymer*, **27**, 1391–4.
199. Wai, M.P., Gelman, R.A., Fatica, M.G. *et al.* (1987) *Polymer*, **28**, 918–22.
200. Fisher, L.W., Melpolder, S.M., O'Reilly, J.M. *et al.* (1988) *Journal of Colloid and Interface Science*, **123**, 24–35.
201. Bootle, G.A., Lye, J.E. and Ottewill, R.H. (1990) *Die Makromolekulare Chemie, Macromolecular Symposia*, **35/36**, 291–305.
202. Linne, M.A., Klein, A., Miller, G.A. and Sperling, L.H. (1988) *Journal of Macromolecular Science Physics*, **B27**, 217–31.
203. Yoo, J.N., Sperling, L.H., Glinka, C.J. and Klein, A. (1991) *Macromolecules*, **24**, 2868–76.
204. Joanicot, M., Wong, K., Maquet, J. *et al.* (1990) *Progress in Colloid and Polymer Science*, **81**, 175–83.
205. Sanderson, A.R., Crowley, T.L., Morrison, J.D. *et al.* (1993) *Langmuir* (in press).
206. Feeney, P.J., Geissler, E., Gilbert, R.G. and Napper, D.H. (1988) *Journal of Colloid and Interface Science*, **121**, 508–13.
207. Yang, S-I., Klein, A., Sperling, L.H. and Casassa, E.F. (1990) *Macromolecules*, **23**, 4582–90.
208. Mills, M., Gilbert, R.G., Napper, D.H. *et al.* (Manuscript in preparation).
209. Finch, J.T., Lewit-Bentley, A., Bentley, G.A. *et al.* (1980) *Philosophical Transactions of the Royal Society of London*, **B290**, 635–8.
210. Cockcroft, J.K. and Rennie, A.R. (unpublished).
211. Avitabile, G., Napolitano, R., Pirozzi, B. *et al.* (1975) *Journal of Polymer Science: Polymer Letters Edition*, **13**, 351–5.
212. Bunn, C.W. (1939) *Transactions of the Faraday Society*, **35**, 482–91.
213. Riekel, C., Haesslin, H.W., Menke, K. and Roth, S. (1982) *Journal of Chemical Physics*, **77**, 4254–5.
214. Haesslin, H.W. and Riekel, C. (1982) *Synthetic Metals*, **5**, 37–50.

215. Riekel, C., Haesslin, H.W., Menke, K. and Roth, S. (1984/5) *Synthetic Metals*, **10**, 31–42.
216. Point, J.J., Damman, P. and Guenet, J.M. (1991) *Polymer Communications*, **32**, 477–80.
217. Roche, E.J., Stein, R.S. and MacKnight, W.J. (1980) *Journal of Polymer Science: Polymer Physics Edition*, **18**, 1035–45.
218. Roche, E.J., Pineri, M., Duplessix, R. and Levelut, A.M. (1981) *Journal of Polymer Science: Polymer Physics Edition*, **19**, 1–11.
219. Aldebert, P., Dreyfus, B. and Pineri, M. (1986) *Macromolecules*, **19**, 2651–3.
220. Dreyfus, B., Gebel, G., Aldebert, P. *et al.* (1990) *Journal de Physique France*, **51**, 1341–54.
221. Gabrys, B., Higgins, J.S. and Scharpf, O. (1986) *Journal of the Chemical Society Faraday Transactions I*, **82**, 1923–7.
222. Gabrys, B., Higgins, J.S. and Scharpf, O. (1986) *Journal of the Chemical Society Faraday Transactions I*, **82**, 1929–34.
223. Scharpf, O. and Gabrys, B. (1990) *ILL Technical Report 90SC26T*, Institut Laue Langevin, Grenoble.
224. Lee, E.M., Thomas, R.K., Burgess, A.N. *et al.* (1992) *Macromolecules*, **25**, 3106.
225. Albrecht, T., Elben, H., Jaeger, R. *et al.* (1991) *Journal of Chemical Physics*, **95**, 2807–16.
226. LaGarde, V., Prask, H. and Trevino, S. (1969) *Discussions of the Faraday Society*, **48**, 15–8.
227. Anderson, M.R., Harryman, M.B.M., Steinman, D.K. *et al.* (1982) *Polymer*, **23**, 569–77.
228. Jobic, H. (1982) *Journal of Chemical Physics*, **76**, 2693–6.
229. Fanconi, B. and Rabolt, J.F. (1985) *Journal of Polymer Science: Polymer Physics Edition*, **23**, 1201–15.
230. Albrecht, T., Jaeger, R., Petry, W. *et al.* (1991) *Journal of Chemical Physics*, **95**, 2817–22.
231. Ewen, B. and Richter, D. (1978) *Journal of Chemical Physics*, **69**, 2954.
232. Guillaume, F., Doucet, J., Sourisseau, C. and Dianoux, A.J. (1988) *Springer Proceedings in Physics*, **29**, 54–8.
233. Richter, D., Frick, B. and Farago, B. (1988) *Physical Review Letters*, **61**, 2465–8.
234. Frick, B., Farago, B. and Richter, D. (1990) *Physical Review Letters*, **64**, 2921–4.
235. Richter, D., Zorn, R., Frick, B. and Farago, B. (1991) *Berichte der Bunsen-Gesellschaft*, **95**, 1111–8.
236. Richter, D., Zorn, R., Frick, B. *et al.* (1992) *Physical Review Letters*, **68**, 71–4.
237. Spells, S.J., Shepherd, I.W. and Wright, C.J. (1977) *Polymer*, **18**, 905–12.
238. Gabrys, B., Higgins, J.S., Ma, K. and Roots, J.E. (1984) *Macromolecules*, **17**, 560–6.
239. Gabrys, B., Higgins, J.S. and Young, D.A. (1985) *Polymer*, **26**, 355–63.
240. Higgins, J.S., Ma, K. and Hall, R.H. (1981) *Journal of Physics C*, **14**, 4995–5001.
241. Oeser, R., Ewen, B., Richter, D. and Farago, B. (1988) *Physical Review Letters*, **60**, 1041–4.
242. Richter, D., Farago, B., Ewen, B. and Oeser, R. (1989) *Physica*, **B 156** and **157**, 426–9.
243. Vilgis, T.A. and Boué, F. (1988) *Journal of Polymer Science B: Polymer Physics*, **26**, 2291–302.

244. Oeser, R., Boué, F., Bastide, J. *et al.* (1988) *Springer Proceedings in Physics*, **29**, 208–13.
245. Russell, T.P. (1990) *Materials Science Reports*, **5**, 171–271.
246. Russell, T.P., Anastasiadis, S.H., Menelle, A. *et al.* (1991) *Macromolecules*, **24**, 1575–82.
247. Russell, T.P., Menelle, A., Hamilton, W.A. *et al.* (1991) *Macromolecules*, **24**, 5721–4.
248. Russell, T.P., Menelle, A., Anastasiadis, S.H. *et al.* (1991) *Macromolecules*, **24**, 6263–9.
249. Menelle, A., Russell, T.P., Anastasiadis, S.H. *et al.* (1992) *Physical Review Letters*, **68**, 67–70.
250. Fernandez, M. L., Higgins, J.S., Penfold, J. *et al.* (1988) *Polymer*, **29**, 1923–8.
251. Fernandez, M.L., Higgins, J.S., Penfold, J. and Shackleton, C.S. (1991) *Journal of the Chemical Society Faraday Transactions*, **57**, 2055–61.
252. Fernandez, M.L., Higgins, J.S., Penfold, J. *et al.* (1990) *Polymer*, **31**, 2146–51.
253. Russell, T.P., Karim, A., Mansour, A. and Felcher, G.P. (1988) *Macromolecules*, **21**, 1890–3.
254. Jones, R.A.L., Norton, L.J., Kramer, E.J. *et al.* (1990) *Europhysics Letters*, **12**, 41–5.
255. Zhao, W., Zhao, X., Rafailovich, M.H. *et al.* (1991) *Physica*, **B173**, 43–6.
256. Fernandez, M.L., Higgins, J.S., Penfold, J. and Shackleton, C.S. (1990) *Polymer Communications*, **31**, 124–7.
257. Felcher, G.P., Karim, A. and Russell, T.P. (1991) *Journal of Non-Crystalline Solids*, **131–3**, 703–8.
258. Stamm, M., Huttenbach, S., Reiter, G. and Springer, T. (1991) *Europhysics Letters*, **14**, 451–6.
259. Lee, E.M., Thomas, R.K. and Rennie, A.R. (1990) *Europhysics Letters*, **13**, 135–41.
260. Satija, S.K., Majkrzak, C.F., Russell, T.P. *et al.* (1990) *Macromolecules*, **23**, 3860–4.
261. Cosgrove, T., Heath, T.G., Phipps, J.S. and Richardson, R.M. (1991) *Macromolecules*, **24**, 94–8.
262. Sun, X., Bouchard, E., Lapp, A. *et al.* (1988) *Europhysics Letters*, **6**, 207.
263. Rennie, A.R., Crawford, R.J., Lee, E.M. *et al.* (1989) *Macromolecules*, **22**, 3466–75.
264. Dai, L., White, J.W., Kerr, J. *et al.* (1989) *Synthetic Metals*, **28**, D69–D79.
265. Auvray, L. and Cotton, J.P. (1987) *Macromolecules*, **20**, 202–7.
266. Cosgrove, T., Heath, T.G., Ryan, K. and Crowley, T.L. (1987) *Macromolecules*, **20**, 2879–82.
267. Cosgrove, T., Finch, N., Vincent, B. and Webster, J. (1988) *Colloids and Surfaces*, **31**, 33–46.
268. Sinha, S.K., Sirota, E.B., Garoff, S. and Stanley, H.B. (1988) *Physical Review*, **B38**, 2297–311.
269. Garoff, S., Sirota, E.B., Sinha, S.K. and Stanley, H.B. (1989) *Journal of Chemical Physics*, **90**, 7505–15.

5

Mixed-crystal infra-red spectroscopy

S.J. Spells

5.1 INTRODUCTION

A wide range of polymers, obtained by crystallization from either the melt or the solution state, are now known to form lamellar crystals. After the initial identification of single crystals in polyethylene crystallized from dilute solution (e.g. reference 1), it was recognized that to accommodate a particular molecule within the crystal required that the molecule should fold back on itself and re-enter the crystal. Since then, the nature of the 'fold surface' has continually been a subject of research, conjecture and debate. Controversy has been prolonged by the lack of experimental techniques capable of providing direct information. At the same time, the central importance of this feature, with its relevance to crystallization mechanisms, to crystal morphology and to physical properties has never been doubted. This chapter is concerned with a technique developed over recent years to provide detailed information on chain conformation in the case of polyethylene.

It is necessary to question the extent to which polyethylene can be regarded as a 'typical' semi-crystalline polymer: the structure is characteristic of a highly flexible molecule, without specific inter- or intra-molecular interactions. In this sense, polyethylene can be regarded as a model system, and features in other polymers such as increased molecular rigidity or the presence of hydrogen bonding can be expected to produce structural differences.

The formation of single crystals, rather than multi-layered aggregates, requires the crystals to be grown isothermally at low concentrations (typically 0.05% w/v). The types of polyethylene crystals produced in this way have been characterized in some detail, although it is only recently that the variety of habits obtainable has been documented [2]. At the

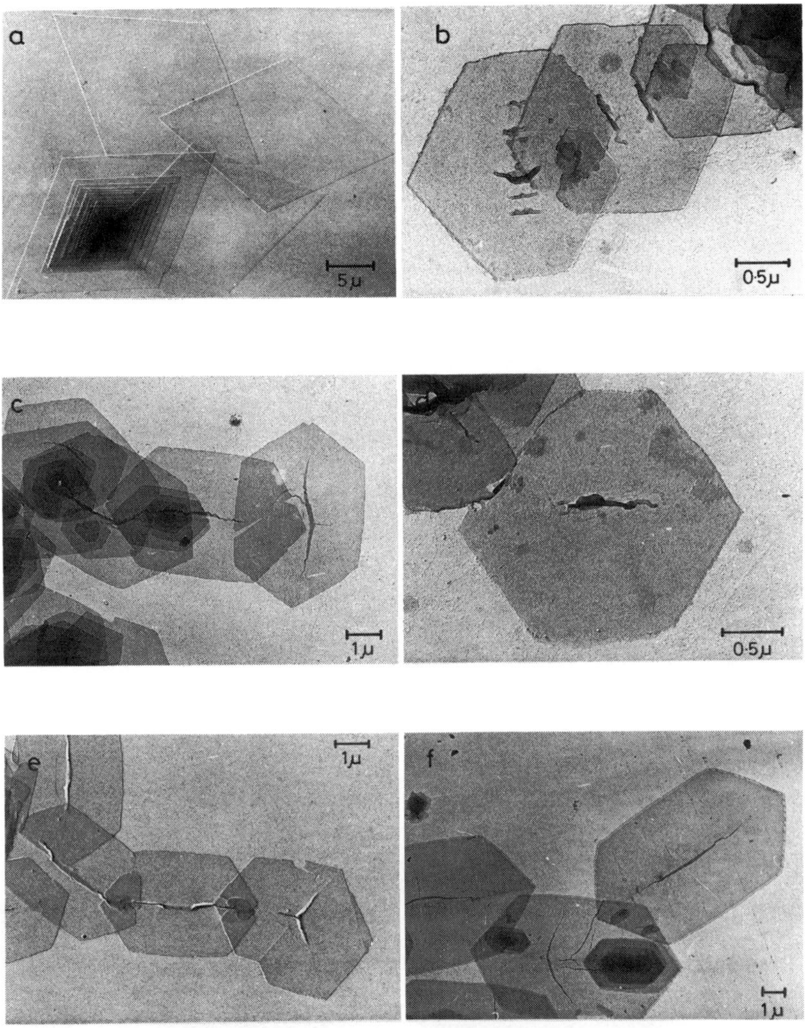

Fig. 5.1 Electron micrographs showing single crystals of polyethylene grown from 0.05% solution in the solvents, and at the crystallization temperatures, shown. (a) Xylene, $T_c < 70°$ C; (b) xylene, $T_c = 86.0°$ C; (c) octane, $T_c = 93.0°$ C; (d) xylene, $T_c = 94.0°$ C; (e) hexadecane, $T_c = 96.5°$ C; (f) dodecane, $T_c = 98.8°$ C; (g) hexadecane, $T_c = 103.1°$ C; (h) hexatriacontane, $T_c = 105.6°$ C; (i) ethyl ester, $T_c = 107.0°$ C; (j) tetradecanol, $T_c = 107.8°$ C; (k) ethyl ester, $T_c = 111.8°$ C; (l) tetradecanol, $T_c = 111.8°$ C; (m) hexatriacontane, $T_c = 112.8°$ C; (n) hexatriacontane, $T_c = 115.0°$ C. (Reproduced from [2] by permission of Chapman and Hall Ltd.) Cont. on next page.

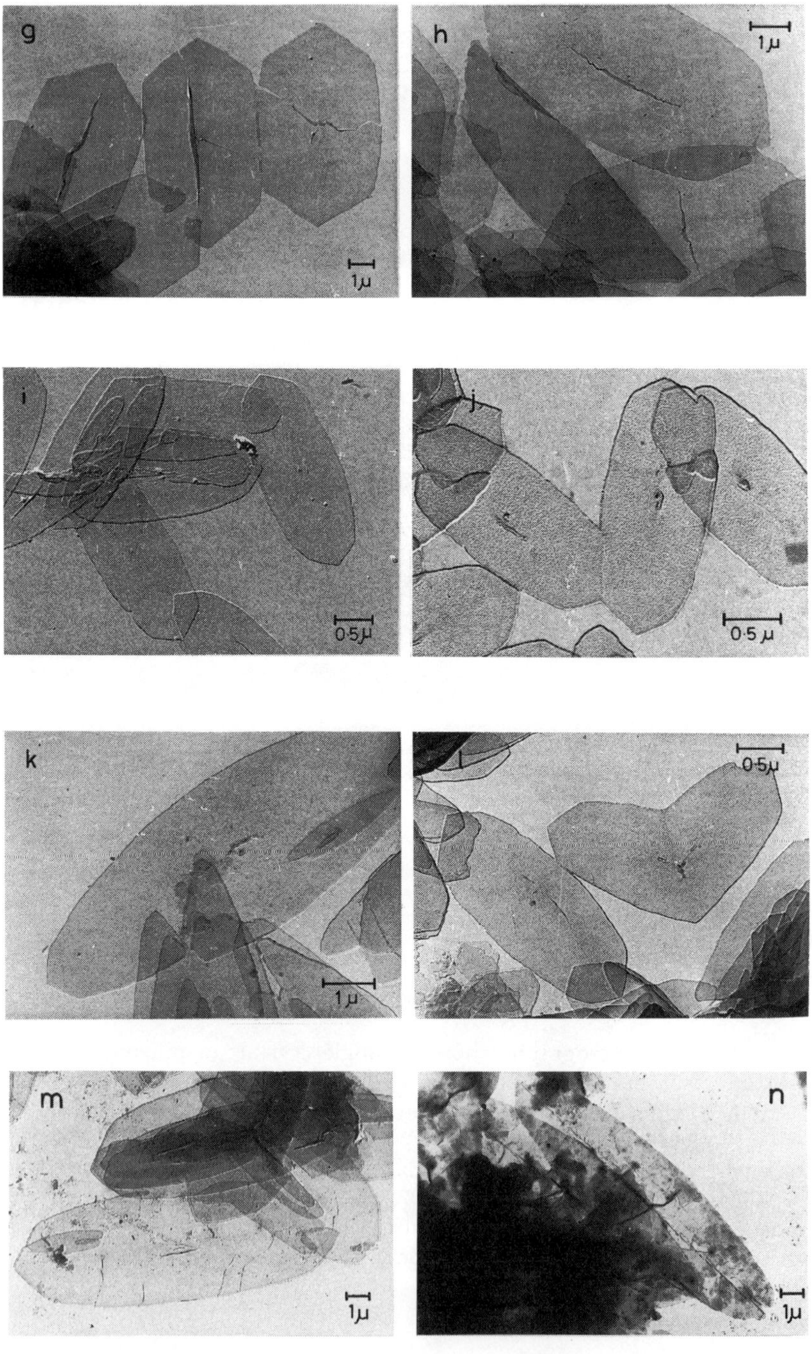

Fig. 5.1 cont.

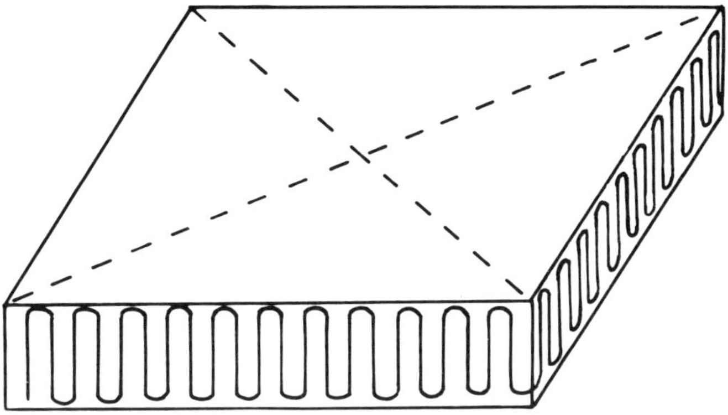

Fig. 5.2 Schematic view of a chain-folded lamella, with dashed lines representing sector boundaries.

lowest crystallization temperatures, corresponding to the highest supercoolings, the crystals are lozenge-shaped and bounded by {110} faces [Fig. 5.1(a)]. This crystal morphology is the subject of the majority of the present chapter. As the crystallization temperature is increased (or, alternatively, the polymer concentration is increased), the lozenges become truncated with {100} faces [Fig.1(b) to Fig.1(e)]. Eventually, curvature appears in faces which are nominally {100}, the curvature becoming more pronounced at higher temperatures [Fig.1(f) to Fig.1(j)].

A schematic diagram of a lozenge-shaped chain-folded lamella is shown in Fig. 5.2. The diagram is schematic in two main respects: first the folding is taken to be as regular as possible, with all folds connecting neighbouring crystal 'stems' (the molecular segments which traverse the crystal) and with all the folds of the same type. Second, the crystal stems themselves are represented as straight lines, whereas in fact they have a planar zigzag conformation. The distance between top and bottom extremities of a chain-folded molecule lying in the crystal (the 'fold length') effectively corresponds to the crystal thickness. For the solution-grown crystals described here, the crystal thickness can be taken as not significantly different from the long period, as measured by small-angle X-ray scattering. Crystals grown from xylene solution at 70°C show a typical long period of about 10.6 nm. Variation of crystallization temperature, normally achieved through different solvents, produces long periods as high as 23 nm [3].

Solution-grown crystals of polyethylene have high crystallinity, typically around 80% on the evidence of DSC data and it is convenient to characterize the crystals from a given preparation by experimentally determined values of the long period and the crystallinity.

Crystals of polyethylene formed from the cooled melt, either by quenching or by isothermal crystallization, are not considered in great detail here. The reason lies in the inherent limitations to the experimental technique described, rather than in any lack of interest in melt crystallization. Nevertheless, some progress has been made with the technique and this serves to indicate some of the structural differences between a polymer crystallized from solution and from the melt. Crystallization of polyethylene from the melt usually leads to a spherulitic morphology, the individual spherulites consisting of lamellar chain-folded crystals radiating from a central nucleus. A wide range of long periods can be obtained from melt crystallization over a range of crystallization temperatures. Typically, values above 20 nm are obtained, and values for crystallinity are significantly lower than for solution crystallization.

5.2 THE PROBLEM: CHAIN CONFORMATION

Detailed knowledge of the chain conformation within chain-folded polymer crystals provides the opportunity to answer some of the fundamental questions of polymer science. Among the unique features of polymers is the ability to form these chain-folded crystals, leading to the question of why these molecules fold. Underlying this question is another – what is the crystallization mechanism responsible?

Considerable debate still centres around this subject. The widely accepted kinetic theory [4] involves the nucleation and growth of chain-folded 'ribbons' along the crystal growth face. As new features of polymer crystallization behaviour have emerged (including the observation of polyethylene crystals with curved faces), the basic theory has been considerably modified and at the same time new approaches have emerged, including models based on an entropic barrier to crystal growth (e.g. reference 5). Important evidence of the manner in which a polymer has crystallized exists in the path of an individual molecule within the crystal. Determination of the chain conformation therefore provides vital information which is related to the crystallization mechanism. The wide range of crystal habits obtainable in polyethylene has already been noted (see Fig. 5.1). Figure 5.3 demonstrates the basic relationship between crystallographic faces and the crystal lattice itself, but does not provide any direct clue to the origin of the different habits. Again, the key to this information is likely to be found in the differences in chain conformation which occur.

The lamellar morphology and high crystallinity of polyethylene suggest that there is intimate contact between crystals, after both solution- and melt-crystallization. More specifically, the close proximity of fold surfaces from neighbouring lamellae, and also the possibility of tie molecules between them, indicates considerable interactions between lamellae. To what extent are the physical properties influenced by the

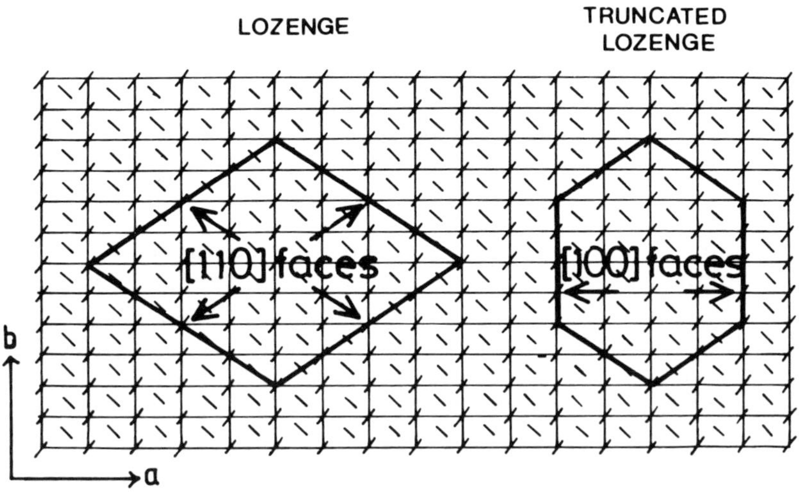

Fig. 5.3 Typical morphologies of polyethylene single crystals and their relationship to the unit cell. (Reproduced from [2] by permission of Chapman and Hall Ltd.)

nature of the fold surface? The **presence** of crystals within a polymer is known to have a considerable effect on mechanical properties, but the effect of more subtle differences, such as might occur at the fold surface, has not been studied *per se*. Clearly, there is considerable potential here for further study.

The complete specification of the conformation of an individual molecular path within a polymer crystal essentially involves two components: the crystal stems and the chain folds (leaving aside the possibility of tie molecules linking nearby lamellae). If the location of stems from one molecule within the crystal is known, then a range of possible fold types can be visualized, all equally capable of linking the same set of stems [6]. Conversely, if the position and conformation is known for every fold arising from one molecule, the corresponding crystal stem sites can be determined unambiguously.

Attempts to predict or model the possible fold conformations in polyethylene crystals have been made over many years (e.g. reference 7). Generally, results have been dependent to a large extent on the contributions to the potential energy included in any particular model (e.g. reference 8).

Electron microscopy provides experimental evidence of the nature of the fold surface. The observation of distinct sectors in lozenge-shaped polyethylene crystals provides strong evidence for folds which occur preferentially along specific crystallographic directions. However, this gives no information on the regularity of folding within the preferred direction, although this direction is demonstrated to be {110} [9].

More recently, Wittman and Lotz have developed the method of fold surface decoration of polyethylene crystals [10]. Vapours of crystallizable polymers, such as polyethylene, are used to deposit elongated lamellae which show a tendency to orient with respect to the fold direction. The geometry of the rod-like crystals which form the sub-structure of the lines on polyethylene single crystals was investigated in detail, with the conclusion that the direction of surface loops and folds is directly responsible for the decoration patterns. In both lozenge-shaped polyethylene crystals and those with rounded faces, decorating lines are observed to lie perpendicular to the (local) growth faces and clearly reveal the different crystal sectors with their differing fold directions. A spherulitic polyethylene film, prepared by slow cooling, showed a preferential orientation of decoration normal to the lateral faces of the lamellae involved. This provides strong evidence against the possibility of random folding occurring in any of the polyethylene systems considered.

Small-angle X-ray diffraction measurements also have a significant bearing on the question of fold surface structure. Various methods have been used, including study of the variation of diffracted intensity with diffraction order, analysis of the shape of diffraction lines and a comparison between the long period from X-ray diffraction and the chain length as determined by the Raman-active longitudinal acoustic mode. The results generally seem to favour a narrow region over which crystallographic register is lost, probably less than 2–3 nm. [6].

Evidently, the early techniques applied to the chain conformation in polyethylene crystals achieved limited success. Clear indications emerged of a preferred fold direction and of the restricted thickness of the fold region, but no detailed chain conformational model had yet been formed. Nevertheless, certain limiting cases for the fold arrangement were formulated, and these will provide useful reference points.

The most regular form of chain folding would give rise to a sequence of adjacent crystal stems from the same molecule. Here we have adjacent re-entry with folding, as indicated by the observed crystal sectors, preferentially along {110} planes. At the opposite extreme, if chain folding incorporates a high degree of randomness, both in the directions and the lengths of folds, we arrive at a model with randomly re-entrant chains.

The relationship between fold regularity and the transition region from crystalline to amorphous material has been widely discussed. Frank has shown [11] that for a 'switchboard model' (one with random re-entry of chains), an 'overcrowding' occurs in the transition region with a density estimated to be about three-times greater than that of the crystal lattice, a result incompatible with the observed **reduction** of density on reducing the overall crystallinity. Two solutions were proposed: the first involved a high proportion of adjacent folding, while the second envisaged lamellae with crystal stems oblique to their surfaces, although the latter required an improbably large degree of obliquity.

Having seen the limited progress, both experimental and theoretical, in determining the chain conformation, we now consider the potential of techniques based on isotopic substitution and in particular mixed-crystal infra-red spectroscopy.

5.3 ISOTOPIC SUBSTITUTION

In order to 'see' an individual molecule within a partially crystalline polymer, conventional characterization techniques are generally inadequate. A labelling technique becomes necessary, with the aim of modifying specific molecular properties while simultaneously leaving all others unchanged. Deuterium substitution in polymers has long been used in vibrational spectroscopy, primarily as an aid to band assignments. The basis of this method is the shift in vibrational frequency which results from a change in mass of one (or more) of the vibrating atoms. The frequency shifts are large: for a simple C–X stretching vibration (X = H or D), the frequency is expected to change by a factor of $\sqrt{2}$, while the factor is observed to be similar for other vibrational modes. A clear separation between spectral features arising from polyethylene (PEH) and perdeuterated polyethylene (PED) is therefore usually obtained.

Vibrational data and assignments for polyethylene have been reported extensively over the years [12]. In order to account for factor group splittings and lattice modes observed in crystalline polyethylene, a realistic model should include inter-molecular interactions. Tasumi and Shimanouchi used a force field including lattice interactions [13], with only the four shortest H . . . H separations considered. An extension [14] to include dipole–dipole interactions showed these to have a small but significant influence on factor group splittings. Difficulty in optimizing the calculations simultaneously for both lattice frequencies and the splittings of intra-molecular vibrations led to two alternative sets of force constants being quoted. This work concerned both PEH and PED, and formed the foundation for the method of mixed-crystal infra-red (IR) spectroscopy.

Tasumi and Krimm calculated the vibrational frequencies for solid solutions of PEH and PED with a variety of arrangements of 'guest' stems within a 'host' lattice [15]. They showed that differences in the calculated splittings resulting from inter-stem interactions might provide a test for various models of chain folding (and hence crystal stem arrangements) in polyethylene single crystals and their approach is summarized later.

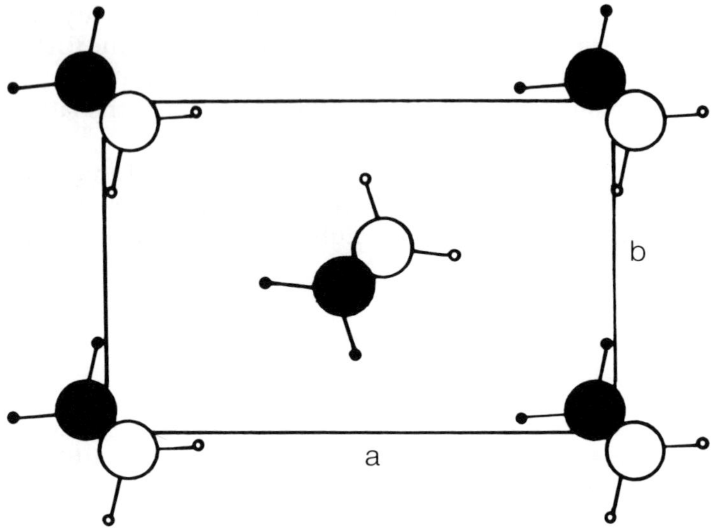

Fig. 5.4 Cross-section of the unit cell for the polyethylene orthorhombic crystal structure.

Polyethylene usually crystallizes with an orthorhombic structure, the unit cell of which is shown in Fig. 5.4. The unit cell contains two chains which are not equivalent in the crystallographic sense, so that some vibrations appear as doublets in the IR spectrum. The two components correspond to vibrations in neighbouring stems which are either in-phase or out-of-phase with each other, giving rise to the difference in frequency. The CH_2 rocking (ν_r) vibration in PEH appears as a doublet at 720/730 cm^{-1} and the CH_2 bending (ν_b) vibration as a doublet at 1463/1473 cm^{-1}. Tasumi and Krimm calculated vibrational frequencies for various mixtures of PEH and PED, using the two sets of force constants mentioned above for the lattice frequencies and the methylene group frequencies [15]. The most significant of the models considered are shown in Fig. 5.5. The results obtained for rocking and bending vibrations, through the H . . . H and D . . . D interactions, were dependent on three factors: first, there are no significant interactions between neighbouring PEH and PED stems; second, the extent of H . . . H (or D . . . D) interactions was assumed to be limited to the nearest neighbour stems. This assumption was later tested experimentally and found to be valid (see section 5.7.1). The third factor was the absence of interactions leading to doublet formation in all crystallographic directions other than <110>. Three possible cases can be considered [16] for systems with a small proportion of PED crystallized together with a majority of PEH.

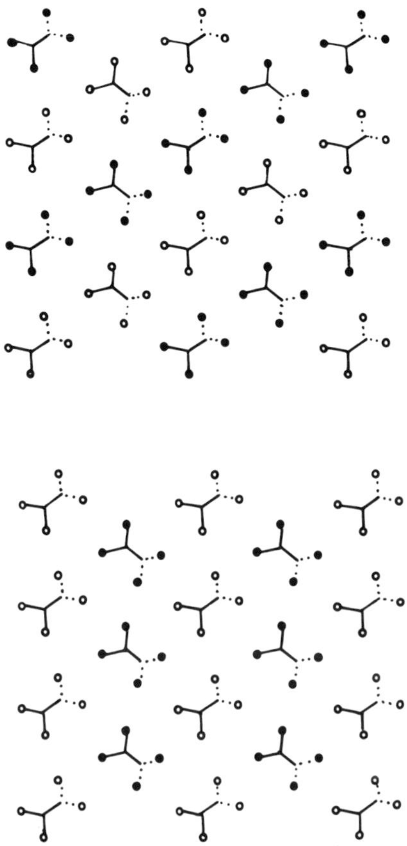

Fig. 5.5 Cross-section of the crystal lattice containing PEH stems (open circles) and PED stems (shaded circles). (a) Chain stems in (110) planes; (b) chain stems in (200) planes. (After [15] with permission of John Wiley and Sons, Inc., copyright © 1968.)

1. A random distribution of PED stems within the PEH lattice. Most PED stems will be isolated with respect to other PED stems. In the absence of nearest neighbour PED stems in <110> directions, the ν_b and ν_r modes will appear now as singlets. The small number of PED stems with <110> nearest neighbour PED stems will give rise to small splittings – smaller than in pure PED, because of the smaller average number of <110> interactions involved.
2. A distribution of PED stems along the (110) crystal planes. This situation is shown in Fig. 5.5(a) (but note that a repetition of this structure would produce a mixture with 50% PED and 50% PEH). The ν_r and ν_b modes are now expected to be doublets. However, the

splittings will be smaller than in pure PED, since each PED stem in Fig. 5.5(a) has two PED nearest neighbours, whereas in pure PED each PED stem has four such neighbours.
3. A distribution of PED stems along the (200) crystal planes. Figure 5.5(b) shows this model, although again a regular repetition would imply 50% PED. The spectrum would be expected to show ν_r and ν_b singlets, because neighbouring stems in the (200) plane are equivalent.

Calculations for models 2 and 3 above provided vibrational frequencies corresponding to room temperature [15], and established the basis for an experimental method capable of distinguishing between structures such as those listed above.

Since the analysis described here is concerned with the arrangement of crystal stems, it is equally applicable to *n*-alkane mixtures. Clearly, short-chain alkanes are not capable of chain folding, so that a co-crystallized isotopic mixture would be expected to show a random arrangement of normal and deuterated crystal stems. Mixtures of this type hence provide useful 'model' systems to compare with polyethylene mixed crystals. A comparison of an alkane mixture and a polyethylene mixture, with the same overall concentration of labelled stems, would show a larger CD_2 bending splitting in the case of the polymer system undergoing regular adjacent (110) folding, for example. Measurements on model *n*-alkane systems are discussed in section 5.7.1.

The development of the mixed-crystal IR technique occurred simultaneously with the development of neutron scattering to provide similar information on the chain conformation in polymer systems generally. Neutron-scattering techniques are reviewed in Chapter 4, and in relation to this work on polyethylene the relevant studies have been discussed elsewhere [17]. The two techniques are inevitably often discussed together, and indeed they were used in combination to maximize the information obtainable. For this reason, it will be necessary to consider results from neutron scattering and their influence on the development of molecular models. The types of neutron-scattering experiment applied to polyethylene crystals will therefore be outlined here.

1. Small-angle (SANS). $q < 0.01$ nm^{-1} (q is the scattering wave vector). This is the region over which the radius of gyration (R_g) can be determined. The zero-angle extrapolated intensity, I_0, provides a check on the molecular weight of the polymer. Any significant isotopic fractionation (see section 5.4) can be inferred from any disagreement between the value obtained from I_0 and that found by an independent technique such as GPC.
2. Intermediate-angle (IANS). $0.04 > q > 0.01$ nm^{-1}. Analytical expressions are available here for certain geometries of scatterer [18],

including sheet-like scattering objects and thin rods, although they are not free from ambiguity.

3. Wide-angle (WANS). $q > 0.04\ nm^{-1}$. This region clearly includes crystallographic peaks, but only the scattering behaviour up to the first crystallographic peak is relevant here. Use can be made of the two-dimensional nature of the crystal stem projections for an oriented single-crystal mat [19]. In this way, a correlation function for the labelled stems can be obtained for scattering data.

In combination, these three regions of scattering wave vector provide structural information over a range of distance scales. Nevertheless, some ambiguity can arise – different structures can give rise to the same scattering function. It is wise to consider neutron-scattering measurements in conjunction with other experimental information. In particular, the sensitivity of the mixed-crystal IR technique to local structure (through the limited range of inter-stem interactions) has proved to be invaluable in restricting the range of models which provide an adequate fit to neutron data.

Isotopic substitution is an essential requirement for both the IR and neutron-scattering techniques described here. Ideally the only influence of deuteration on the polyethylene molecule would be to change the mass of the repeat unit, and hence the vibrational frequencies, and to change the coherent scattering length to provide sufficient contrast for neutron-scattering experiments. Any effect on the crystallization behaviour and hence on the crystal stem arrangements is therefore undesirable. Unfortunately, this ideal is not realised in practice: the melting point of PED is lower than that of PEH by 4°C. The kinetics of crystallization are expected to be modified accordingly. In extreme cases, isotopic fractionation is expected to lead to a non-uniform distribution of isotopic species through the polymer crystals. Fractionation is discussed in the next section, alongside an analysis of early IR work on solution-crystallized polyethylene.

5.4 EARLY MIXED-CRYSTAL MEASUREMENTS AND ISOTOPIC FRACTIONATION

Early IR measurements demonstrated the usefulness of the mixed-crystal technique, although inappropriate sample preparations in some cases and the use of room temperature for sample measurements led to some erroneous conclusions with regard to the structures concerned. Unfortunately, some of these structural conclusions have since been widely propagated in the general polymer literature, and it is necessary to explain the shortcomings of this work before considering later results. The issue of isotopic fractionation has been widely studied in relation to both IR and neutron-scattering measurements and is also considered here.

The origins of the mixed-crystal IR technique lie in the work of Hornig and Hiebert on the low-temperature orthorhombic crystal phases of HCl and HBr [20, 21]. It was shown that the spectrum in the HCl stretching region for HCl is modified if the HCl is co-crystallized with DCl, with the doublet being replaced by a singlet [20]. The spectra of isotopic mixtures, consisting of zigzag hydrogen-bonded 'chains' of molecules, were obtained for a range of isotopic compositions. Taking into account the interruption of zigzag chains of one isotopic species by the other, the data were used to calculate the angle made between alternate molecules and the 'chain' axis.

The earliest mixed-crystal IR measurements on polyethylene crystals were made by Bank and Krimm on samples where neither isotopic species was a narrow-molecular-weight fraction [22]. It has already been noted that the presence or absence of vibrational doublet splittings from the 'guest' species in an isotopically mixed sample can be used to distinguish between certain simple conformational models. In addition, doublet splittings can be calculated for models such as those shown in Fig. 5.5. Spectra were compared for solution-cast films and isothermally solution-crystallized mats of polyethylene, along with melt-crystallized polyethylene and the alkane *n*-hexatriacontane. Solution casting and solution crystallization of polyethylene produced the same results, which were significantly different from those for *n*-alkane mixtures: for the polyethylene samples, minority species doublets persisted to the lowest concentrations used (about 9% molar volume), whereas the *n*-alkane mixtures showed minority species singlets at low concentrations. This was interpreted as evidence of {110} chain folding for both solution-crystallized and cast samples of polyethylene, with the preferential adjacency of labelled crystal stems giving rise to larger splittings than for the *n*-alkane mixtures. The latter, without the ability to form chain folds, are expected to provide a random arrangement of labelled stems, with no preference for particular crystallographic planes. For solution-crystallized polyethylene, the absence of any dominant singlet contribution to the minority species methylene bending vibration led to the conclusion that chain folds were predominantly connecting adjacent stems, with adjacent re-entry the main mode of folding. Thus far, the conclusions reached are not in dispute, having been essentially substantiated by later work.

Through melt crystallization of cast polyethylene films, the minority species doublets were found to transform into singlets, indicating a change in the arrangement of labelled stems. Splittings for the **host** species were found to be significantly higher than in alkane mixtures with the same concentration of labelled species, a difference later found to be dependent on the temperature of measurements. At this stage it led to the erroneous conclusion of adjacent {200} folding in melt-crystallized polyethylene.

The effect of the crystallization temperature on the IR splittings was investigated: samples crystallized isothermally from solution at 90°C were compared with films cast from solution at 55°C [23] (where non-isothermal crystallization would produce a range of morphologies). The larger doublet splittings observed for isothermal crystallization at 90°C were interpreted as evidence of isotopic segregation, through the different isotopic crystallization rates.

The question of isotopic fractionation is central to the validity of mixed-crystal measurements. Stehling *et al.* investigated both mixed alkane and mixed polyethylene systems [24] and suggested that the previous results of Bank and Krimm were dominated by isotopic fractionation effects. An equilibrium phase diagram for $C_{36}H_{74}/C_{36}D_{74}$ showed selective partitioning of the two isotopic species between phases, with enrichment of the crystalline phase by $C_{36}H_{74}$. Isotopic differences in the crystallization rate were demonstrated for mixed polymer systems, leading to large compositional differences between the crystals and solution as a function of time. In this case, a crystallization temperature of 89°C provided both slow and incomplete crystallization. It was suggested that the larger doublet splittings observed in polyethylene crystals than in *n*-alkane mixtures were explicable in terms of compositional heterogeneity, rather than a preferred fold plane.

Krimm and Ching refuted this suggestion [25], showing that, provided one isotopic alkane species was in large excess, the compositional range over the course of crystallization was narrow. The narrow IR CD_2 bending singlet observed in samples with low $C_{36}D_{74}$ concentration also indicated no significant departures from random stem organization (although there is later experimental evidence of isotopic fractionation in alkane mixtures – see section 5.7.1). It was pointed out that fractionation effects in polymer crystals may be minimized by using normal and deuterated polymers with similar melting temperatures, and also by using a lower crystallization temperature. Importantly, it was shown that compositional heterogeneity, in combination with random molecular re-entry into the crystal, was unable to account for the observed polyethylene doublet splittings. Some evidence of fractionation under certain conditions was later provided from IR data for melt-crystallized polyethylene [26].

The importance of isotopic fractionation was also realized at an early stage by those involved in neutron-scattering work. The technique was used to determine the experimental parameters which influence the extent of fractionation, and to quantify the effect. A large increase in the forward scattering, by comparison with predicted values, was observed for polyethylene quenched from the melt and was found to disappear in the melt state [27, 28]. This was attributed to clustering of the isotopic species. The cooling rate was shown to influence the degree of fractionation, the effect being minimized for rapid quenching [29]. The parameters determining the magnitude of fractionation were investigated in detail [30, 31] for narrow-molecular-weight fractions of PED as the 'guest' species. The dependence of fractionation on molecular weight (both 'guest' and 'host'), PED concentration and crystallization temperature and rate were studied. The results for solution-grown crystals are of particular relevance here, showing that for a crystallization temperature of 70°C from xylene solution the forward scattering is consistent with values for molecular weight determined independently by GPC. A crystallization temperature of 85°C, by contrast, generally produced an enhanced signal. The q^{-4} dependence of the 'excess' small-angle scattering allowed the use of Porod's law to obtain the quantity $\Delta\rho^2 A$ ($\Delta\rho$ = deuterium density difference across the boundaries of the structure and A = surface area per unit sample volume), characteristic of the structures obtained. Artificially fractionated samples, prepared by growing one isotopic species over previously grown crystals of the other, showed large values of $\Delta\rho^2 A$, as did samples slowly cooled from the melt state. Alternatively, small-angle data may be corrected by separating the contributions to scattering from density and concentration fluctuations [32]. Fractionation by molecular weight (through the variation of melting point with molecular weight) was recalled by Sadler and Keller [30], and used to develop a picture of crystals with a continuous change in isotopic concentration from the centre to the periphery. Experimental estimates of $\Delta\rho^2 A$ indicated a typical concentration variation across a crystal from about 1% to 4% PED for a sample with overall PED concentration of 1.4%. It is, therefore, appropriate to consider a small degree of isotopic enrichment, rather than the 'clustering' proposed by others. Experimental parameters affecting fractionation during melt crystallization were also investigated [31], with the similar conclusion that, to minimize fractionation, the molecular mobility and crystallization times should be minimized.

Buckingham and Hentschel [33] investigated the possibility of partial miscibility in isotopically mixed melts, by considering the Helmholtz free energy of mixing (ΔF_{mix}). It emerged that the upper consolute temperature, T_{con}, was proportional to the average degree of polymerization, decreasing with mismatching of the two isotopic species. T_{con} is critically dependent on $\Delta V/V$ (where V is the molar volume), which is known neither for alkanes nor for polyethylene. Nevertheless, use of an upper limit of $\Delta V/V$ for $C_{36}H_{74}/C_{36}D_{74}$ and PEH/PED systems gave values for T_{con} of 1.9 and 438 K respectively, while a lower limit for PEH/PED gave T_{con} = 4.1 K. Fractionation in the alkane melt therefore appears unlikely, whereas for polyethylene the likelihood depends on the parameters involved, including the molecular weight. The 'optimum concentration' for isotopic fractionation shifts from 50% for matched polymers to an excess of the smaller molecule where they are unmatched.

Isotopic effects are not unique to polyethylene and have been observed in several polymer systems (e.g. references 34 and 35). However, as far as polyethylene is concerned, the evidence outlined here suggests that under suitable crystallization conditions the effect can be reduced to negligible proportions. For solution crystallization, high crystallization rates are necessary, and these can be achieved, for example, at high supercoolings. For melt crystallization, it is only by rapid crystallization from the melt state that the degree of fractionation obtained is negligible. Furthermore, neutron scattering may be used to quantify fractionation effects, enabling us to eliminate unsuitable samples from consideration.

5.5 EXPERIMENTAL ADVANCES

The use of broad-molecular-weight samples of polyethylene and non-isothermal crystallization conditions gives rise to poorly defined morphologies, from which it is difficult to draw unambiguous conclusions about chain conformation. For this reason, in all but the earliest mixed-crystal measurements, narrow-molecular-weight fractions were used, at least for the minority species. Early measurements made use of conventional dispersive IR spectrometers, often without the capability of adding data from successive scans. Under these circumstances, the weak absorption from a small proportion of PED in the sample may produce a signal/noise ratio which prevents the resolution of doublet components [36]. This problem can be reduced by sample cooling.

Thermal contraction of the orthorhombic unit cell of polyethylene with decreasing temperature is well known (see, for example, reference 37), leading to increased interactions between crystal stems. For this reason, doublet splittings of both CH_2 rocking and bending vibrations increase as the temperature is reduced [38]. Band separation in the multiplet profiles obtained in mixed-crystal spectra can therefore be improved by sample cooling. The earliest reported measurements involving sample cooling showed substantial differences in 'guest' PED splittings, by comparison with room-temperature values [39]. The dependence of splittings on 'guest' concentration was also found to change, with splittings reaching a constant value for PED concentrations below about 10%. Perhaps a more important finding from low-temperature work was that differences in unit cell dimensions for polyethylene and *n*-alkanes at room temperature are, from the evidence of IR doublet splittings, eliminated on cooling with liquid nitrogen [40]. This implies that the earlier conclusion of {200} folding in melt-crystallized polyethylene, based on differences between 'host' splittings and those for *n*-alkane samples, was erroneous.

Despite the improvement in mixed-crystal IR data on cooling solution-grown polyethylene mats, the 'guest' PED CH_2 bending vibration generally still shows evidence of unresolved components. Curve fitting can provide some indication of the number and characteristics of the components, but the results are never free from ambiguity. Fourier self-deconvolution allows the spectral resolution of IR data to be improved, and gives a better estimate of the number of components present. In terms of modelling the molecular conformations and calculating IR band profiles, it will become evident later (see section 5.7) that a comparison between deconvoluted experimental data and calculated spectra provides a more stringent test for any molecular model.

Detailed accounts of the Fourier self-deconvolution method appear elsewhere [41, 42, 43]. The technique involves the deconvolution of an intrinsic line-shape function from the experimental spectrum. If the number of scans provides an adequate signal/noise ratio (typically at least 1000), the 'guest' PED CH_2 bending vibration is a suitable subject for deconvolution. The adjustable parameters available in this process can be demonstrated for a Lorentzian intrinsic line-shape function:

$$E_o(\bar{\nu}) = \frac{\sigma/\pi}{\sigma^2 + \bar{\nu}^2} \tag{5.1}$$

Here 2σ is the width of $E_o(\bar{\nu})$, the intrinsic line-shape function, at half-height. The ratio of the line widths before and after deconvolution is denoted as K. The parameters 2σ and K are commonly used as variables in commercial software based on this technique (e.g. Nicolet and Mattson systems).

For the purpose of determining the number of multiplet components and their frequencies, Fourier self-deconvolution has obvious advantages over curve fitting. The latter method can give, at best, only the minimum possible number of peaks present. By comparison with other methods of peak separation in common use, several features of self-deconvolution are relevant: first, the integrated intensity of a spectrum is unchanged on deconvolution, making the method suitable (at least in principle) for the measurement of peak areas. Secondly, for optimum conditions, deconvolution produces only one peak per spectral component, with no zero crossings, unlike methods of analysis based on derivatives.

The result of Fourier self-deconvolution of mixed-crystal IR data is illustrated in Fig. 5.6. Here, the original spectrum of the solution-crystallized sample shows three peaks, of which A and C show marked asymmetry. Deconvolution reveals several components in both A and C, some of which are still not completely resolved.

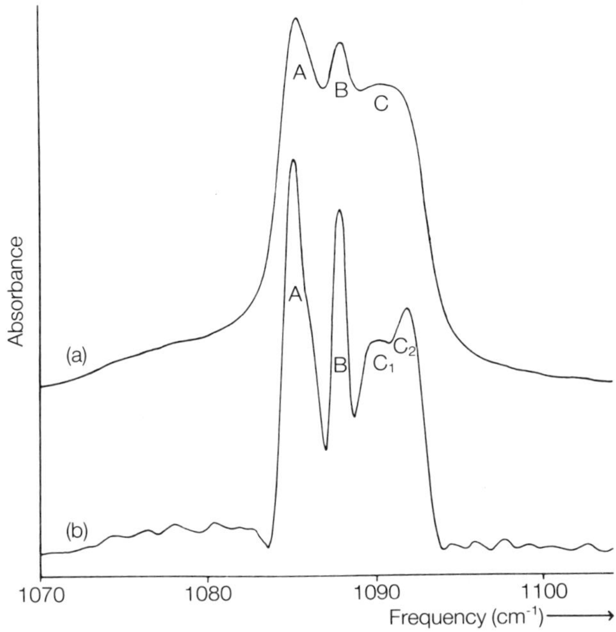

Fig. 5.6 (a) IR spectrum of a sample with 3% PED of $\bar{M}w$ 54000. Resolution $1\,cm^{-1}$. (b) Spectrum after Fourier self-deconvolution, using parameters $2\sigma = 1.7\,cm^{-1}$ and $K = 2.0$. (After [50] by permission of the publishers, Butterworth–Heinemann Ltd, © 1984.)

Curve fitting has also been used in the context of mixed-crystal IR measurements, largely by Krimm and co-workers. Data for cooled solution-crystallized mats were thus fitted to the minimum possible number of singlet and doublet components necessary for an adequate fit [40]. The procedure also involved the removal of the local background, the use of partially Lorentzian and partially Gaussian line-shapes and a least-squares-fitting method. The information obtainable in this way is clearly restricted, and the results suffer from ambiguity. Nevertheless, the raw data provide some indication of multiplet structure, and hence allow an initial estimate for the number of components.

5.6 THE THEORY OF MIXED-CRYSTAL SPECTROSCOPY

Possible structures for mixed polyethylene crystals, and their implications in terms of the presence or absence of CD_2 bending splittings, were discussed in section 5.4. The basic principle was that a labelled crystal stem must have labelled nearest neighbour(s) in {110} directions in order for a doublet splitting to be observed. The earliest mixed-crystal work made use of the simple presence of splittings, in the case of solution-crystallized samples, along with comparisons of splittings in samples prepared by different means. The absolute values of splittings were not utilized until calculations were made for specific regular arrangements of stems. It then became evident that these 'regular groups' were themselves inadequate to describe realistic chain conformations incorporating a certain amount of disorder. The calculations were therefore extended to take account of 'irregular groups'. These stages in the development of the modelling of IR spectra are now described, together with an alternative method for the calculation of band profiles in mixed *n*-alkanes.

5.6.1 Coupled oscillator model

Early mixed-crystal IR measurements for solution-grown PE crystals [22], which showed doublet splittings consistent with adjacent re-entry in {110} planes, suggest that the rows of labelled PE stems in effect form 'ribbons' or 'sheets' of labelled polymer. Symmetry arguments allow the calculation of splittings in some trivial cases: hence the splitting for an infinite isolated 'sheet' of labelled stems is expected to be half the value for pure PED, since the number of like nearest neighbour interactions for each PED stem is two for the single sheet and four for the pure polymer. Similarly, a pair of labelled non-equivalent PED stems surrounded by PEH nearest neighbours should have a splitting of a quarter of that for pure PED. There remains the need to determine splittings for finite sheets.

A complete normal vibrational analysis for a general mixed-crystal system presents a formidable problem. Cheam and Krimm used a full analysis in certain special cases [40], together with a coupled oscillator model to calculate splittings for other sheet-like arrangements of crystal stems. The full calculations followed earlier work in using H . . . H interactions [15], but also included C . . . C interactions. Refinement of force constants provided good agreement with low-temperature unit cell parameters. Values of 8.76 and 4.38 cm^{-1} were obtained for the CD_2 bending splitting, $\Delta\nu_{\infty\times\infty}$ and $\Delta\nu_{\infty}$, in pure PED and a single infinite (110) sheet of PED stems in a PEH matrix respectively. For an isolated pair of PED stems (although surrounding PEH stems were not included) a value for $\Delta\nu_0$ of 2.25 cm^{-1} was obtained. These results approximate to the expected relationship:

$$\Delta\nu_{\infty\times\infty} = 2\nu_{\infty} = 4\nu_{o} \tag{5.2}$$

Cheam and Krimm recognized that mixed-crystal systems with infinite sheets of labelled crystal stems provided a poor model for low-molecular-weight PED fractions [40], and developed the simple coupled oscillator theory in a semi-empirical manner. The method is based on that described by Gribov [44] and Zbinden [45], and relies on the simple relationships noted in equation 5.2 for limiting cases. In this way, splittings were calculated for particular regular two-dimensional arrays of labelled stems.

Following the nomenclature of Cheam and Krimm [40], we consider a row of M equivalent oscillators, each with an isolated (uncoupled) frequency of ν_0. This corresponds to a one-dimensional array of parallel dipoles, each representing a crystal stem. (In the context of the vibrational behaviour of a polymer chain of dipoles, the boundary conditions applied were those of a chain with fixed ends.) If the coupling force between adjacent stems is assumed to be proportional to the difference in amplitude between the two oscillators, then the frequencies of the normal modes are given by:

$$\nu_{M,s}^2 = \nu_0^2 + 4\nu'^2 \sin^2 \frac{(M-s)\pi}{2M} \tag{5.3}$$

for $s = 1, 2, \ldots M$. Here

$$\nu_0^2 = \frac{K}{4\pi^2 m} \tag{5.4}$$

and

$$\nu'^2 = \frac{K'}{4\pi^2 m} \tag{5.5}$$

where K is the internal force constant for each oscillator with reduced mass m and K' is the coupling force constant between adjacent oscillators. By reference to an infinite sequence of oscillators, equation 5.3 can be rewritten as

$$\nu_{M,s}^2 = \nu_{\infty,\infty}^2 + (\nu_{\infty,1}^2 - \nu_{\infty,\infty}^2) \sin^2 \frac{(M-s)\pi}{2M} \tag{5.6}$$

Only the $s = M$ mode has significant IR intensity, whereas the $s = 1$ mode is the only one with appreciable Raman intensity. To obtain the separation between these frequencies, we use the approximation that M is large, so that:

$$\nu_{\infty,1} + \nu_{\infty,\infty} = \nu_{M,1} + \nu_{M,M} \tag{5.7}$$

From equations 5.6 and 5.7,

$$\Delta\nu_M = \Delta\nu_\infty \sin^2 \frac{(M-1)\pi}{2M} \tag{5.8}$$

where $\Delta\nu_M = \nu_{M,1} - \nu_{M,M}$ and $\Delta\nu_\infty$ is the corresponding frequency separation for an infinite sequence. If every alternate oscillator in the linear array is now rotated about its long axis to reproduce the arrangement of crystal stems in a <110> direction, we obtain a sequence

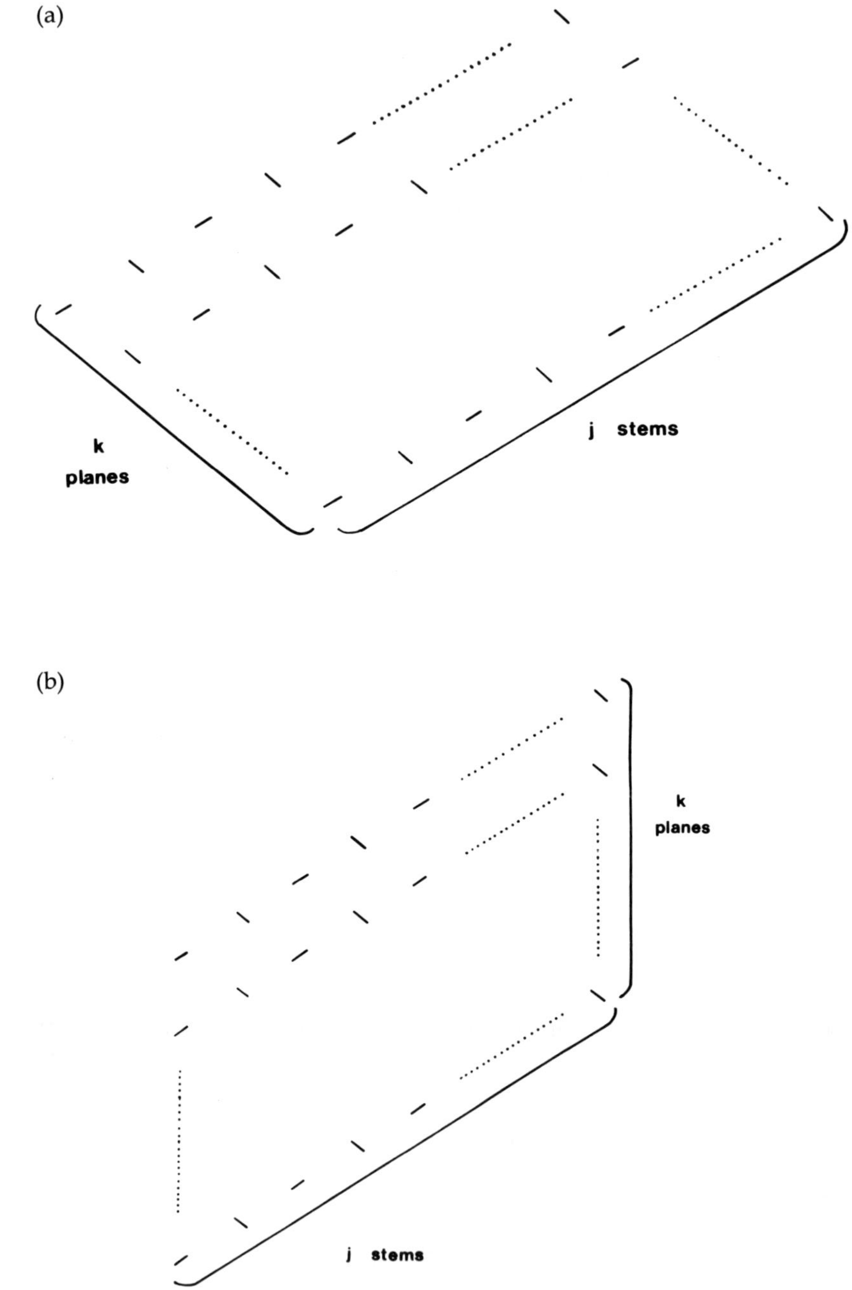

Fig. 5.7 Schematic views of (a) $\{(110)_j\,[110]_k\}$ and (b) $\{(110)_j\,[010]_k\}$ arrangements of crystal stems. (After [40] by permission of John Wiley and Sons, Inc., copyright © 1981.)

where neighbouring oscillators are non-equivalent and where the corresponding dipoles are no longer parallel [46]. Both limiting modes ($s = 1$ and $s = M$) are IR-active in this case.

Taking as the basic unit now a pair of (non-equivalent) crystal stems, the IR splitting for a row of N such pairs can be calculated in terms of the splitting for an isolated pair, $\Delta\nu_0$, by making use of the form of the assumed dispersion relation. Each frequency component for the isolated pair gives rise to an IR-active mode and a Raman-active mode for the sequence of N pairs, with separation given by equation 5.8. The symmetry of the dispersion relation then allows the overall IR splitting to be written as:

$$\Delta\nu_N = \Delta\nu_o + (\Delta\nu_\infty - \Delta\nu_o)\sin^2\frac{(N-1)\pi}{2N} \tag{5.9}$$

where $\Delta\nu_\infty$ now refers to an infinite sequence of stem pairs.

The analysis can now be extended to consider crystal stems arranged on adjacent (110) planes. Figure 5.7 shows two possible arrangements, with Fig. 5.7.(a) having j adjacent stems in the (110) plane. There are k adjacent planes, each with j stems and each plane starting in register along the [110] direction. The splitting for such an arrangement, denoted $\Delta\nu\{(110)_j\,[110]_k\}$, can be obtained by considering a single plane of j stems to be a single oscillator with a splitting $\Delta\nu\{(110)_j\,[110]_1\}$, which can be calculated using equation 5.9. This gives:

$$\Delta\nu\{(110)_j[110]_k\} = \Delta\nu\{(110)_j[110]_1\} + \Delta\nu\{(110)_j[110]_\infty\} - \Delta\nu\{(110)_j[110]_1\}\sin^2\frac{(k-1)\pi}{2k} \tag{5.10}$$

Experimental data for PED and simple symmetry relations enabled the determination of several terms in equation 5.10. Pure PED gave a CD_2 bending splitting of 10.2 cm^{-1} [40], corresponding to $\Delta\nu\{(110)_\infty\,[110]_\infty\}$. A single infinite (110) plane of stems, with half the number of nearest neighbour interactions per stem would give half this value, so that

$$\Delta\nu\{(110)_\infty[110]_1\} = 5.1\ \mathrm{cm}^{-1} \tag{5.11}$$

The splitting for an isolated pair of PED stems, $\Delta\nu_0$, was taken as the value from a normal vibration calculation (2.25 cm^{-1}), scaled to the experimental splitting obtained for PED (10.2 cm^{-1}) by comparison with the normal vibration calculation (8.76 cm^{-1}) [40] to give $\Delta\nu_0 = 2.62\ cm^{-1}$.

Semi-empirical calculations based on equation 5.10, with the parameters noted above, provided the splittings listed in Table 5.1. Analogous calculations for CH_2 bending vibrations are shown in Table 5.2. Although splittings for a single sheet of PED stems increase significantly only for a number of stems up to about eight, a group of stems with $j = k$ shows a splitting which increases significantly for a number of stems up

Table 5.1 Predicted splittings, in cm^{-1}, of CD_2 bending modes for $\{(110)_j[110]_k\}$ (reprinted from reference 40 by permission of John Wiley and Sons, Inc., copyright © 1981.)

	$\Delta\upsilon(cm^{-1})$								
j/k	*1*	*2*	*3*	*4*	*5*	*6*	*8*	*10*	∞
2	2.6	5.1	6.4	6.9	7.2	7.3	7.5	7.5	7.7
3	3.3	6.1	7.7	8.1	8.4	8.6	8.7	8.8	8.9
4	3.9	6.7	8.1	8.6	8.9	9.1	9.2	9.3	9.5
5	4.2	7.0	8.3	8.9	9.2	9.4	9.5	9.6	9.7
6	4.5	7.2	8.5	9.1	9.3	9.5	9.7	9.7	9.9
7	4.6	7.3	8.6	9.2	9.4	9.6	9.7	9.8	9.9
8	4.7	7.4	8.7	9.2	9.5	9.7	9.8	9.9	10.0
9	4.8	7.4	8.7	9.3	9.6	9.7	9.9	9.9	10.0
10	4.9	7.5	8.8	9.3	9.6	9.7	9.9	10.0	10.1
12	4.9	7.5	8.8	9.4	9.6	9.8	9.9	10.0	10.1
14	5.0	7.6	8.8	9.4	9.6	9.8	9.9	10.0	10.1
16	5.0	7.6	8.8	9.4	9.7	9.8	10.0	10.0	10.2
∞	5.1	7.7	8.9	9.5	9.7	9.9	10.0	10.1	10.2

Table 5.2 Predicted splittings, in cm^{-1}, of CH_2 bending modes for $\{(110)_j[110]_k\}$

	$\Delta\upsilon(cm^{-1})$								
j/k	*1*	*2*	*3*	*4*	*5*	*6*	*8*	*10*	∞
2	3.5	6.8	8.5	9.1	9.5	9.7	9.9	10.0	10.1
3	4.4	8.1	10.0	10.7	11.1	11.3	11.5	11.6	11.8
4	5.1	8.8	10.7	11.4	11.8	12.0	12.2	12.3	12.5
5	5.6	9.2	11.0	11.8	12.2	12.4	12.6	12.7	12.9
6	5.9	9.5	11.3	12.0	12.4	12.6	12.8	12.9	13.0
7	6.1	9.6	11.4	12.1	12.5	12.7	12.9	13.0	13.2
8	6.3	9.8	11.5	12.2	12.6	12.8	13.0	13.1	13.2
9	6.4	9.8	11.6	12.3	12.6	12.8	13.0	13.1	13.3
10	6.4	9.9	11.6	12.3	12.7	12.9	13.1	13.2	13.3
12	6.5	10.0	11.7	12.4	12.7	12.9	13.1	13.2	13.4
14	6.6	10.0	11.7	12.4	12.8	13.0	13.1	13.2	13.4
16	6.6	10.0	11.7	12.4	12.8	13.0	13.2	13.3	13.4
∞	6.8	10.1	11.8	12.5	12.9	13.0	13.2	13.3	13.5

to 100 or so. The mixed-crystal IR technique should, therefore, be particularly sensitive to systems containing groups of labelled stems within this range of size and shape.

Arrangements of stems like those shown in Fig. 5.7(b) gave splittings which were consistently smaller than those in Table 5.1, with differences becoming less significant for larger numbers of stems [40]. This is expected, since only stems near the group boundaries should be affected by the detailed arrangement.

Normal vibration calculations were used to test equation 5.10 for the case $j = \infty$ and $k = 2$, for both PEH and PED [40]. Both CD_2 rocking and bending doublets gave splittings in agreement with calculations from equation 5.10 to within $0.2\,cm^{-1}$, providing confidence in the method.

Cheam and Krimm used the results in Table 5.1 to analyse data for solution-crystallized PE samples (see section 5.7.2). Curve fitting was used to obtain the minimum possible number of doublet contributions to the spectrum in each case. Doublet splittings were then compared with values in Table 5.1 to give dimensions of the labelled groups. It became apparent from this and later work [47] that, although the results clearly supported a multiple sheet ('superfolded') structure for sufficiently high molecular weights, the group sizes assigned to observed doublet splittings were often significantly smaller than would be expected for a complete molecule. In addition, infra-red data in many cases showed a surprisingly large number of doublets, particularly after using enhancement techniques such as Fourier self-deconvolution [48]. This gave rise to ambiguities in assignment, with doublets possibly due to a variety of superfolded conformations. Taken together with earlier evidence of the dilution of labelled stems with other molecules [36], this points to a somewhat more irregular arrangement of stems than those shown in Fig. 5.7.

5.6.2 Statistical models

As an alternative to the model of regularly arranged superfolded sheets of equal dimensions (section 5.6.1), a statistical model was developed to describe less regular arrangements of labelled stems within single crystals [19, 48]. In particular, the model was intended to provide a detailed basis for comparison with both the complex multiplet structure of the CD_2 bending vibration and also the information obtained from neutron-scattering measurements. The model is dependent on four parameters:

a. the probability of adjacent folding along a (110) plane (p);
b. the proportion of lattice sites in the planes bounded by a labelled molecule which are occupied by that molecule (ϕ) [i.e. the fold plane is 'diluted' by other molecule(s)];
c. the number of labelled stems per superfolded sheet (n);
d. the number of superfolded sheets occupied by one labelled molecule (N).

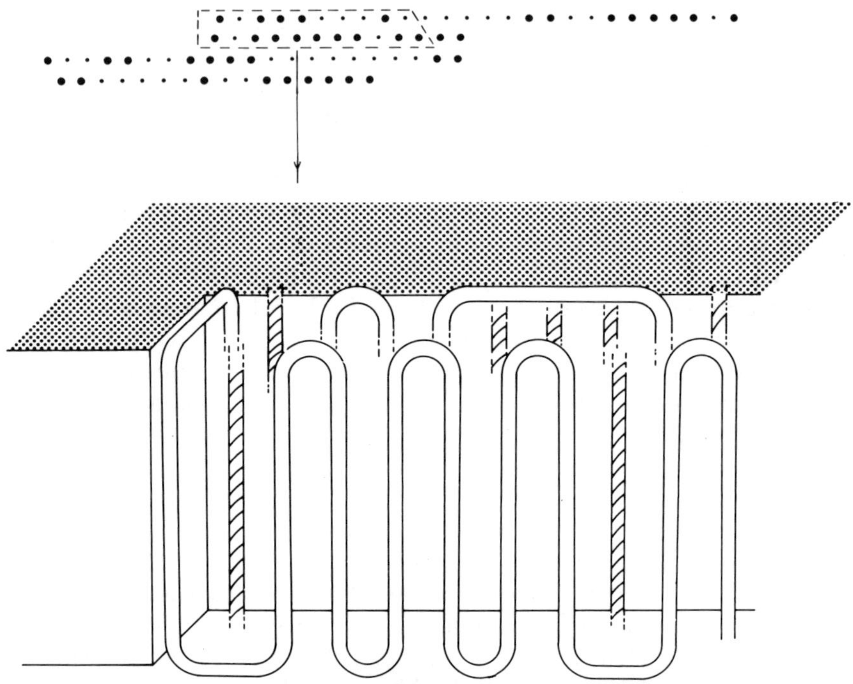

Fig. 5.8 Schematic representation of the molecular conformation of a labelled polyethylene chain within the lamella. The large spots in the crystal stem projection at the top correspond to the unshaded molecule in the perspective sketch. The [110] direction is horizontal. (Reprinted from [48] by permission of the publishers, Butterworth–Heinemann Ltd, © 1984.)

In the context of single-crystal measurements, experimental data from several sources are relevant to these parameters: neutron-scattering measurements showed the radius of gyration to be insensitive to molecular weight, suggesting (along with IR evidence quoted above) that superfolded sheets were present [49]. Second, evidence from neutron-scattering work [30] and from earlier IR spectroscopy [36] indicated a value for ϕ in the region of 0.5. Further, the use of mixed-crystal samples relies on the assumption that the two isotopic species behave identically, apart from the differences in mass and in neutron-scattering length for hydrogen and deuterium. In this case, the crystallization behaviour of the two polymers must be identical, with each polymer giving rise to the same statistical distribution of stems within the crystals. The probability for a 'guest' molecule to deposit a second stem adjacent to the first stem must, therefore, be the same as for a 'host' molecule to deposit a second stem adjacent to its first. This has the important implication that the overall dilution along the fold plane, ϕ, must be a ratio of integers (i.e. 1/2, 1/3, 1/4, etc.). Clearly, $\phi = 0.5$ satisfies this condition.

In the development of the model, two basic assumptions were made: (a) if a labelled molecule has deposited a crystal stem at one site on the growth face, then the probability of a second labelled stem being deposited at the adjacent site is enhanced; (b) the probability of occupation of the second site depends only on the type of chain occupying the first (adjacent) site and not on the occupation of more distant sites. If only two molecules are crystallizing simultaneously and the first site is occupied by a labelled stem, then the probability of an unlabelled stem occupying the second site is simply $(1-p)$.

In principle, the arrangement of labelled stems in one sheet of the superfolded conformation may influence the arrangement of labelled stems in the neighbouring sheet. It was shown [19] that such a correlation has only a minor effect on the predicted behaviour of mixed crystals. The transition from one sheet to the next, as required for superfolding, was assumed to occur at random, with the number of stems per sheet, n, allowed to vary randomly between fixed limits.

Implementation of this computer model allows 'molecules' to be generated for chosen sets of parameters p, ϕ, n and N. In general, this produces less regular arrangements of labelled stems than does the original model of Cheam and Krimm (see previous section). A typical conformation is shown in Fig. 5.8, where the three-dimensional sketch shows part of the labelled molecule as it has deposited stems on the growth face. It is now necessary to consider the contribution of such irregular stem arrangements to the CD_2 bending vibration.

5.6.3 Irregular groups

The general procedure adopted [48] was to take each of the molecular conformations generated (such as that in Fig. 5.8) and sort the labelled stems into **groups** of interacting stems. As will become evident, to a good approximation the inter-stem interactions responsible for doublet splittings can be regarded as extending only to (labelled) nearest neighbours. Such groups are effectively independent of others and make separate contributions to the IR band profile. Next, each group was transformed into an **equivalent group** with properties identical, insofar as is possible, to the original group. The purpose here was to simplify the geometry and hence the calculation of vibrational frequencies. A method of interpolation was then used to calculate the splitting of each equivalent group, by reference to the simple regular groups already considered by Cheam and Krimm [40]. To aid understanding, the simplified terminology used will now be described.

The analysis of a molecular conformation is initially in terms of **groups**

Fig. 5.9 Illustration of the notation used to describe equivalent groups of stems: (a) an (8,3) group; (b) an (8 × 3) group. (Reprinted from [48] by permission of the publishers, Butterworth–Heinemann Ltd, © 1984.)

of labelled stems, in which stems are connected by adjacent {110} folds (or, more precisely, by adjacent {110} interactions, since the method gives no **direct** information on the nature of the folds themselves). An **equivalent group** has a similar number of stems to the original group, and a similar distribution of {110} interactions between them. The arrangement of stems in an equivalent group is simpler than in the original group. Each sheet starts in register along a {110} plane which is not the principal fold plane. The sheets may have unequal numbers of labelled stems, but no unlabelled stems are allowed to interrupt these (unlike in the original groups). For example, Fig. 5.9(a) shows a conformation with an uncompleted second sheet. This is termed an (8,3) group, indicating eight stems in one sheet and three in the second, with the two sheets starting in register along a {110} plane.

By contrast, a group with all sheets 'completed' (i.e. having equal numbers of stems in each sheet) is called a **closed group** and is labelled ($X \times Y$). This indicates Y adjacent sheets, each with X adjacent stems, with the first stem of each sheet in register along a {110} plane. Figure 5.9(b) shows the example of an (8 × 3) group. A group which is not itself closed [e.g. Fig. 5.9(a)] nevertheless contains a closed component. Here, six of the labelled stems form a double sheet (3 × 2) array. This is called a **closed subgroup**.

As an illustration, Fig. 5.10 shows a computer-generated molecular conformation with $\phi = 0.5$ and $N = 4$. The structure has been divided into groups, some of which are simply isolated stems giving no doublet splittings. Isolated pairs (B), or longer, fully adjacent, single sheet conformations (C and F) have splittings which may be calculated using the model of Cheam and Krimm [40]. Group E is also, effectively, a single sheet. Groups A and D do not have such a simple form, and such cases require the formation of equivalent groups.

To characterize a group, we need to specify the number of stems, and the numbers of nearest neighbour {110} interactions for each of these. The number of sheets necessary for the equivalent group can then be determined, and the numbers of stems for each sheet can be calculated [48]. Overall, the distributions of nearest neighbour interactions in the original and equivalent groups are similar, but not identical. The

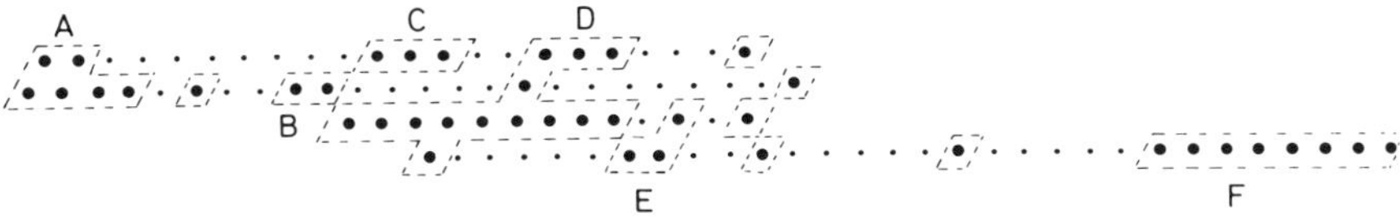

Fig. 5.10 Division of a computer-generated model, shown as a crystal stem projection, into groups of stems which are linked by {110} interactions. The [110] direction is horizontal. (Reprinted from [48] by permission of the publishers, Butterworth–Heinemann Ltd, © 1984.)

main difference lies in the number of stems with one nearest neighbour interaction, corresponding to those stems at the 'ends' of the original group. By comparison with the stems with larger numbers of interactions, these have only a minor effect on the overall doublet splitting.

The doublet splittings for equivalent groups must now be calculated. Direct calculation from the coupled oscillator model, with appropriate estimates for the unperturbed oscillator frequency and for the splitting for a single infinite sheet of stems, led to calculated splittings of 5.71 and 5.84 cm^{-1} for (3,2) and (5,2) groups, respectively [48]. This method is not readily applicable to the variety of large groups generated by the computer model, so a simpler model was sought.

The interpolation method adopted [48] makes use of the tabulated splitting for the largest closed subgroup within the equivalent group. Consider the equivalent group (x,y), where $x \geqslant y$. This is a double-sheet group with closed subgroup $(y \times 2)$. The group splitting, $\Delta\nu_{x,y}$, can be represented to a first approximation by $\Delta\nu_{y\times 2}$. A correction term can be added, corresponding to an increase in the size of a single-sheet group of $2y$ stems to $(x-y)$ stems:

$$\Delta\nu_{x,y} = \Delta\nu_{y\times 2} + (\Delta\nu_{(x+y)\times 1} - \Delta\nu_{(2y)\times 1}) \tag{5.12}$$

Extending this method to three sheets of stems in the equivalent group (x,y,z), where $x \geqslant y \geqslant z$, gives:

$$\Delta\nu_{x,y,z} = \Delta\nu_{z\times 3} + (\Delta\nu_{(y+z/2)\times 2} - \Delta\nu_{(3z/2)\times 2}) + (\Delta\nu_{(x+y+z)\times 1} - \Delta\nu_{(2y+z)\times 1}) \tag{5.13}$$

for $z \geqslant 2$.

By comparison with the direct method of calculation quoted above, this method gave splittings of 5.40 and 5.80 cm^{-1} for (3,2) and (5,2) groups respectively. These results are in agreement with direct calculations to well within 0.5 cm^{-1}, which is considered satisfactory when the range of experimentally observed splittings is taken into account.

This technique provides a method of calculating the doublet splittings expected for any statistical model of the stem arrangement. In the first application, a common (frequency) position for each low-frequency doublet component was assumed [48]. However, the use of Fourier self-deconvolution of mixed-crystal data showed that additional components could be detected in the multiplet structure, on both high- and low-frequency sides of the central singlet [50]. A more complete treatment of the doublet components was clearly necessary. The presence of a greater number of components than had previously been suspected was itself of interest, emphasizing the need for a statistical model.

Using a range of solution-crystallized mixed polyethylene samples with different molecular weights and subjected to a variety of heat annealing treatments, it was found that in many cases a high- and a low-frequency component could be assigned as a pair, with a corresponding doublet splitting [51]. It was generally necessary to have one doublet of significantly higher intensity than any others. The frequencies of both components could be accurately determined from self-deconvoluted spectra. The effect of annealing will be considered in more detail in section 5.7.2.(c), and it will simply be stated here that the larger splittings are usually progressively lost, allowing a range of splittings to be measured. Plotting both doublet component frequencies versus splitting for PEH matrix samples gave two straight lines. For PED matrix samples, both plots showed curvature, although an adequate fit was obtained using two straight lines in each case. This difference in behaviour with isotopes may be a result of interaction between IR- and Raman-active CH_2 bending vibrations. The frequency difference for CH_2 bending is small (approximately 10 cm^{-1}) by comparison with the corresponding CD_2 bending value (approx. 65 cm^{-1}) [52]. The calculated splittings for equivalent groups of labelled stems (equations 5.12 and 5.13) can be combined with these calibration results to obtain the positions of both high- and low-frequency components for the group.

For simplicity, the central singlet was not included in the earliest model calculations [48]. Fourier self-deconvolution showed no evidence of more than one singlet contribution [50], contrary to previous suggestions [40]. The central peak was always observed at the same frequency, and the smallest observed doublet splitting corresponded closely to that predicted for an isolated pair of labelled stems lying on a {110} plane. The singlet was therefore included in later calculations [51]. Contributions to this peak arise from labelled crystal stems which are isolated, having no labelled 110 nearest neighbours, and also from non-crystalline segments of labelled molecules. The first of these contributions can be quantified by inspection of the computer-generated molecular models. For the second, an independent value for sample crystallinity is required, typically from DSC data.

Having determined the frequencies of singlet and doublet contributions to the CX_2 bending vibration, it is necessary to consider the intensities expected. In the case of doublets, the intensities of both high-

and low-frequency components were taken to be proportional to the numbers of stems in the original groups generated by the model, with the same constant of proportionality for both components. Similarly, the number of isolated stems created was used to obtain the singlet intensity, with a contribution added from non-crystalline chains in proportion to their molar fraction. No allowance was made for the effects of the direction of dipole moment change during the bending vibration for different groups of stems.

Finally, some account must be taken of natural and instrumental line-broadening. A Lorentzian line-shape was used for each (singlet and doublet) contribution to the multiplet band profile, with a bandwidth equal to the experimental resolution used (normally 1 cm^{-1}). A summation of all doublet and singlet components can then be compared with self-deconvoluted data.

Further modifications to this model will be discussed under the headings for the appropriate sample types.

5.6.4 Lattice dynamical calculations and the electro-optical parameter method

A different approach to the calculation of CX_2 bending vibrational frequencies and intensities was developed by Fanconi *et al.* [52] for *n*-alkane mixed crystals. They used lattice dynamical methods to calculate frequencies and electro-optical parameters to derive intensities. The lattice dynamical calculations involved infinite lattices with a translational unit cell containing from two to 18 ethylene repeat units, depending on the arrangement of normal and deuterated crystal stems considered. Non-bonded interactions within a radius of 0.45 nm were included, with interactions beyond nearest neighbour ones neglected, as in other analyses. Twenty-four different unit cells were generated, each containing between one and four deuterated stems. For a specified overall labelled stem concentration, probabilities for each unit cell were calculated. Particular emphasis was placed on deuterated stem concentrations of up to 10%, since the restriction to four deuterated stems in a unit cell (i.e. a group of four stems) only accounts for a reasonable proportion of groups formed at these low deuterium concentrations. A comparison between the splittings calculated using coupled oscillator (see earlier) and lattice dynamical models for groups of up to four stems showed differences of less than 0.4 cm^{-1}.

Molecular geometry and lattice parameters were taken from previous data [53, 54], with modifications to an earlier vibrational force field [12] to fit spectroscopic data for deuterated molecules. For pure deuterated hexatriacontane, the calculated CD_2 bending splitting was 0.4 cm^{-1} lower than the observed low-temperature value.

The electro-optical parameter treatment of IR intensities [55] has become more widely used over recent years. This enabled the relative intensities of high- and low-frequency doublet components to be obtained. It was shown that the assumption used in the coupled oscillator model – namely that for a particular group the IR intensity is distributed between two modes (one doublet) only – is not strictly valid, and that a small proportion of the intensity can be distributed among the modes with intermediate frequencies.

5.7 EXPERIMENTAL RESULTS

5.7.1 Mixed *n*-alkane crystals

Mixed *n*-alkanes provide model systems, without the complication of chain folding to influence the deposition sites of labelled crystal stems. With the condition that isotopic fractionation is not significant, then labelled molecules are expected to occupy a random selection of lattice sites. 'Guest' doublet splittings should be markedly less than those in mixed polyethylene crystals, for the same overall molar concentration of labelled molecules. Hence, alkane samples enable the methods of calculation of group splittings to be verified. Additionally, phenomena such as diffusion of alkane chains in isotopic mixtures have been investigated with mixed-crystal techniques.

As the concentration of 'guest' species in an *n*-alkane isotopic mixture is increased, the 'guest' doublet splitting should increase as the average group size increases. At very low 'guest' concentrations, the band will appear as a singlet, since most labelled molecules are isolated from others. As the 'guest' concentration increases, doublets will appear with splittings increasing with concentration. At the same time the intensity ratio of singlet to doublet components is expected to decrease. Eventually, the group size effectively becomes infinite (analogous to the percolation problem), and a splitting characteristic of the pure 'guest' alkane is expected. Bank and Krimm [22] observed this behaviour in mixed *n*-hexatriacontane samples crystallized from benzene solutions, with splittings in reasonable agreement with calculations [14]. Systems with either normal or deuterated 'guest' species behaved in a similar way. The variation in CD_2 bending band-shape with concentration of deuterated species is illustrated in Fig. 5.11 for mixed *n*-dotriacontane samples.

Even without the benefits of FTIR and sample cooling, it was noted that for a doublet splitting to be resolvable requires a much higher concentration of labelled molecules in the case of *n*-alkane mixtures than for polyethylene mixtures [36]. The width of the band profile was compared for the two sets of samples, with the conclusion that the average labelled

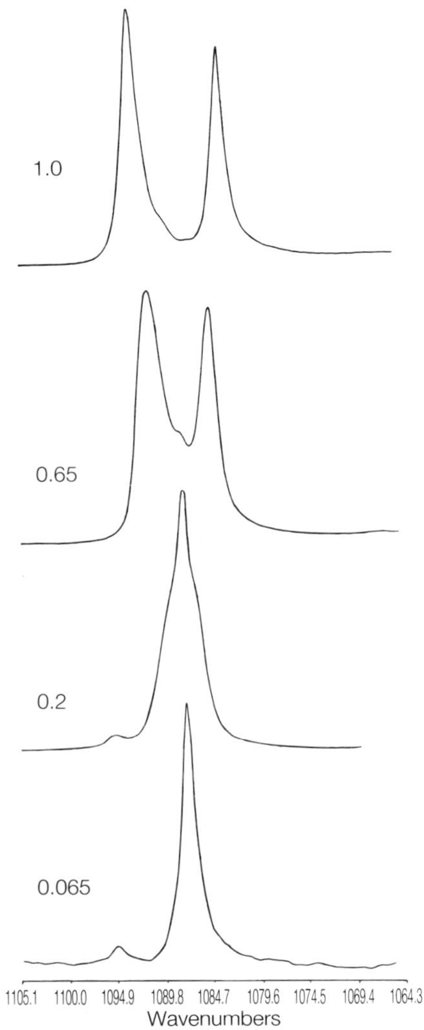

Fig. 5.11 IR CD_2 bending spectra for mixed crystals of dotriacontane. The proportion of d_{66}-dotriacontane is indicated in each case.

stem concentration in solution-grown polyethylene crystals, in the planes occupied by labelled molecules, was in the range 50–80% (i.e. $\phi = 0.5$ to 0.8). At liquid-nitrogen temperature, the unit cell dimensions of polyethylene and sufficiently long *n*-alkanes become very similar, allowing a more meaningful comparison between splittings. Using curve fitting, low-temperature CD_2 bending data for hexatriacontane samples (crystallized from benzene solutions) with 5 and 50% deuterated species were

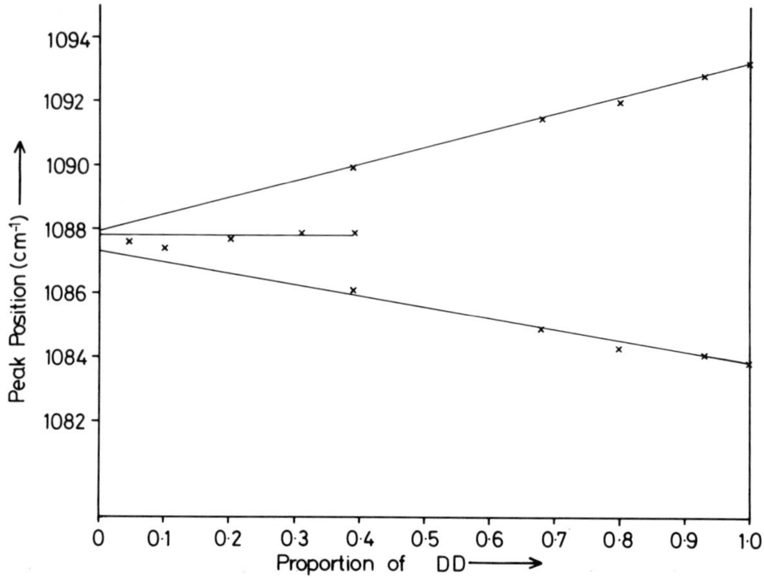

Fig. 5.12 IR peak frequency positions for mixed crystals of dotriacontane plotted versus concentration of the deuterated *n*-alkane (DD). The central singlet is only resolved at lowest DD concentrations. (Reprinted from [48] by permission of the publishers, Butterworth–Heinemann Ltd, © 1984.)

fitted to a singlet and a singlet plus one doublet, respectively [40]. A small (0.4 cm^{-1}) shift in singlet frequency was interpreted as indicating different environments of labelled molecules, namely isolated molecules in the first case and molecules with labelled (200) neighbours in the second. However, FTIR measurements show no systematic change in singlet frequency with 'guest' concentration (see Fig. 5.12). Furthermore, Fig. 5.12 shows both high- and low-frequency components with a linear dependence of frequency on concentration of deuterated molecules, for dotriacontane samples crystallized from benzene solutions.

The random arrangement of labelled molecules in mixed *n*-alkanes has been verified using WANS measurements [19]. A dotriacontane sample, crystallized from benzene solution with 10% deuterated molecules, produced a linear Kratky plot (with intensity modified to take into account the finite diameter of crystal stems). This behaviour is expected for a random array of labelled stems, and differs significantly from that observed for polyethylene samples. Fanconi *et al.* have since studied *n*-hexatriacontane mixtures in detail, using FTIR spectroscopy, cooled samples and Fourier self-deconvolution of spectra [52]. Calculations of IR spectra involved lattice dynamical and electro-optical parameter methods (see section 5.6.4). The experimentally observed CD_2 bending bandwidth was found to show considerable scatter as a function of concentration of labelled molecules. Configurational calculations were made using two different sets of assumptions: first, random placement of labelled molecules was considered and, second, isotopic fractionation was taken into account. Fractionation was found to have a small effect on the CD_2 bandwidths, and the effect of instrumental broadening was shown to be larger (albeit still very small). Calculated and observed spectra for isotopic mixtures showed good agreement.

Molecular migration of *n*-alkanes has also been studied by mixed-crystal spectroscopy [56]. It is interesting to note that the coupled oscillator model described in section 5.6.1 was also used to calculate the correlated domain size in a rotator phase in *n*-alkane solid solutions, on the basis of measurements of the CH_2 rocking mode [46].

5.7.2 Solution-crystallized polyethylene

It has already been noted (section 5.4) that early measurements primarily concerned crystal growth by cooling dilute solutions. It is clear that isothermal crystallization is necessary to obtain well-defined crystal morphologies, and that high supercoolings (typically $T_c = 70°C$, from xylene solution) are required to avoid excessive isotopic fractionation. These conditions formed the basis for extensive later studies, which continued alongside neutron-scattering work and the development of molecular models. The refinement of such models is discussed here.

(a) Crystallization at high supercoolings

In early room-temperature dispersive IR measurements, doublet splittings were not universally observed [36], so the CD_2 bending bandwidth was used to characterize samples. The bandwidth was found to increase generally with both increasing PED concentration and molecular weight. Even with such a crude measure of inter-stem interactions, differences between solution- and melt-crystallized blends were significant. Using a comparison of bandwidths for *n*-alkane and polyethylene isotopic mixtures, an estimate of 0.5 was obtained for the fraction of lattice sites occupied by a labelled molecule (ϕ). In the light of preliminary FTIR measurements, with sample cooling, this figure was adjusted to a range of 0.5–0.8 [36]. A similar conclusion was reached on the basis of discrepancies between intermediate-angle neutron-scattering intensities observed experimentally and calculated.

Cheam and Krimm used dispersive IR spectroscopy with sample cooling to study samples crystallized on cooling solutions to 55°C [40].

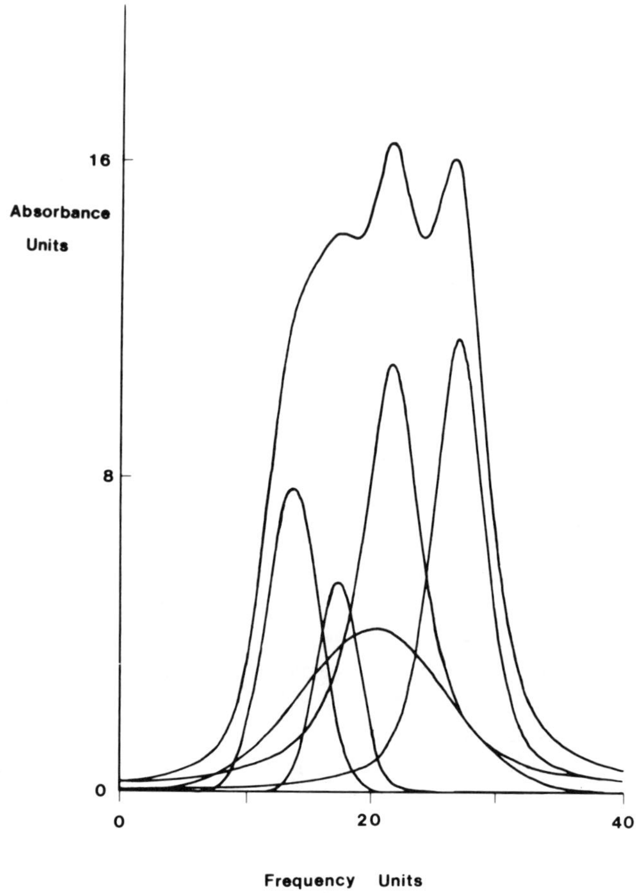

Fig. 5.13 IR spectrum for 40 PEH/1 PED mixed crystals in the CD_2 bending region for a PED fraction with $\bar{M}_w = 57100$. The top curve represents the sum of the individual components, and agrees well with experimental data. Frequency scale: $0.0 = 1100\,cm^{-1}$, $40.0 = 1080\,cm^{-1}$. (After [40] reprinted by permission of John Wiley and Sons, Inc., copyright © 1981.)

Data for narrow-molecular-weight fractions were analysed by a curve-fitting procedure, using crystalline and non-crystalline singlets plus the minimum number of doublets necessary to obtain a satisfactory fit. An example is shown in Fig. 5.13, where the broad central band is the non-crystalline singlet. The low-frequency components of each doublet were assumed to coincide in frequency. It was noted that at low temperatures, CD_2 bending splittings remained constant for PED concentrations up to 5%, indicating no significant isotopic fractionation. At higher PED concentrations, splittings were found to increase because of the enhanced probability of neighbouring PED molecules. A comparison between the central singlet frequency in *n*-alkanes and polyethylene showed a difference of 0.4 cm^{-1}, with the higher value of 1089.2 cm^{-1} for polyethylene attributed to adjacency along (200) planes and the lower frequency corresponding to isolated *n*-alkane molecules. Normal-mode calculations indicated the possibility of such a frequency shift [15], although the sign and magnitude depend critically on the force field used. In general, when taking into account the number of crystal stems corresponding to one molecule, the larger splittings could not be attributed to stems deposited along a single-fold plane. A multiple-sheet structure was therefore proposed, with superfolding between consecutive sheets, except for the lowest molecular weight used ($\bar{M}w = 8700$). Quantitative spectral interpretation was pursued using the coupled oscillator model with regular stem groups and superfolding.

This work was extended to include samples grown isothermally at 80°C, showing the characteristic lozenge-shaped single crystals [57]. Curve fitting was carried out for pure PEH spectra in the 1070–1110 cm^{-1} region, using three Gaussian components. For mixed-crystal spectra, a background of the same form was initially subtracted from spectra, before fitting the CD_2 bending components as previously. Separate low-frequency components were now assumed for the different doublets present. Results for samples isothermally crystallized at 80°C were compared with those for samples which crystallized on cooling to 55°C. Since the latter group would probably involve crystallization both above and below 80°C, it is difficult to ascertain trends with crystallization temperature from this work. Nevertheless, the two groups of samples showed different splittings. Samples with T_c=80°C again showed evidence of superfolding for all but the lower molecular weights.

Solution-crystallized polyethylene samples cooled to liquid-nitrogen temperature generally show a complex multiplet CD_2 bending structure (e.g. references 40 and 48], which is especially evident after using spectral enhancement techniques such as Fourier self-deconvolution [50]. The sizes of regular groups of labelled stems corresponding to the best resolved doublet components [40] clearly do not represent all the information obtainable from the IR spectrum. At best, these figures represent the most numerous arrangements of labelled stems within the polymer microstructure. The figures are often ambiguous in their interpretation and, where two (or more) doublets are obtained, there is no obvious explanation for labelled stems being contained in two (or more) groups of quite different dimensions. The use of curve fitting, and in particular of the constraint (necessary for the sensible use of curve fitting) of the minimum number of doublets needed for a satisfactory fit to data, has led to this over-simplification of the underlying structure.

A more realistic approach relies on features of molecular conformation which have been unambiguously established, chiefly a sheet-like structure with superfolding and the dilution of fold planes with unlabelled chain stems. Of these three features, a sheet-like conformation was determined from the angular dependence of IANS data [30], with a degree of adjacent {110} folding indicated by IR splittings [22]. Superfolding was first suggested as an explanation for the insensitivity of the radius of gyration (from SANS measurements) to molecular weight [49], and was confirmed from the analysis of IR doublet splittings, some of which, as seen earlier, were larger than could be achieved with a single-sheet structure [40]. Dilution of the fold planes was signalled by both the low IANS intensities and a 'local' deuterated stem concentration (ϕ) from IR spectroscopy of about 0.5. These details formed the framework for the development of molecular models [48].

Molecular conformations were generated repetitively, in accordance with several statistical parameters. The calculation of IR spectra from these conformations is detailed in section 5.6.3, using the coupled oscillator model extended to handle less uniform arrangements of stems. The computer-generated molecular conformations were also used to determine a correlation function for labelled stems, and hence to calculate the neutron-scattering intensity. Taking the figure quoted earlier of ϕ=0.5, Yoon and Flory showed that a model with a preference for occupation of alternate lattice sites (i.e. a minimum of adjacent re-entry) would reproduce experimental neutron-scattering data in the small- and intermediate-angular regions [58]. However, they also predicted a peak in intensity at a q value corresponding to twice the nearest neighbour separation, which was not observed experimentally [19]. Optimum agreement between neutron-scattering data and model calculations was obtained for a probability of adjacency (p) of 0.75. To deal with a range of sample molecular weights, the number of superfolded sheets was allowed to vary. The number of labelled stems per superfolded sheet was allowed to vary randomly within the range 10–20, with the mean value of 15 defining the number of sheets, N, required to represent a particular molecular weight. One computer-generated conformation is illustrated in Fig. 5.8, showing both the stem arrangement for the whole labelled

Table 5.3 Key parameters derived for polyethylene single crystals grown from dilute xylene solution at 70°C

Molecular weight per sheet	21000
Probability of adjacent folding along a (110) plane (p)	0.75
Probability of 'next-to-adjacent' folding along a (110) plane	0.07
Proportion of lattice sites occupied by the labelled molecule (ϕ)	0.5
Average number of labelled stems per superfolded sheet (n)	15

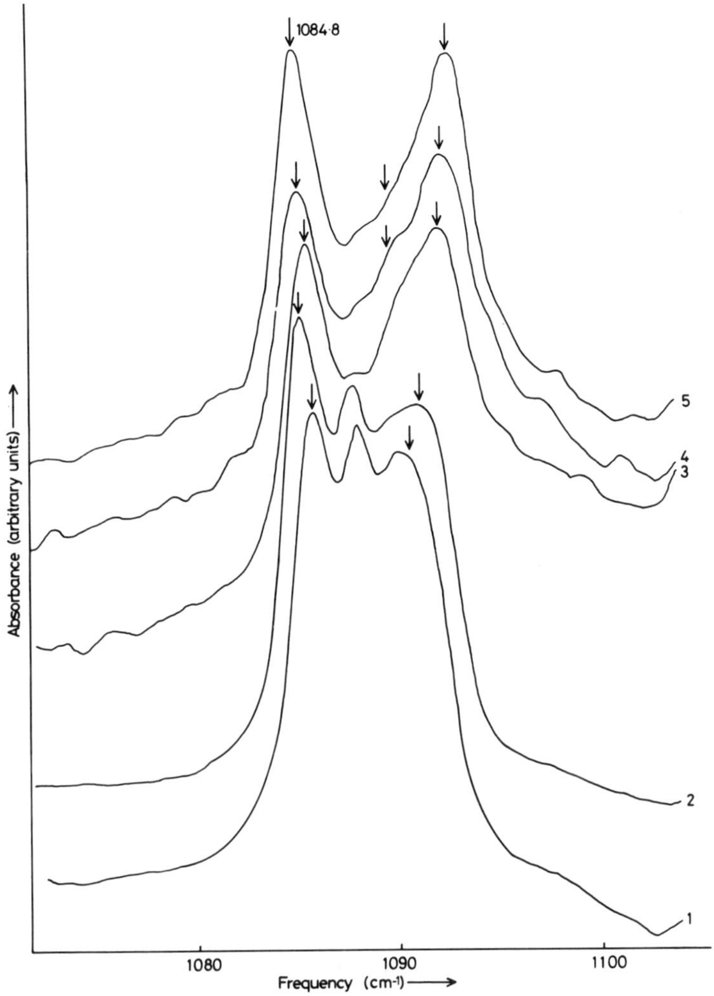

Fig. 5.14 IR spectra for mixed crystals with 0.5% PED of molecular weights ($\bar{M}_w$) (1) 46000; (2) 69000; (3) 155000; (4) 216000; (5) 386000. A reference PEH spectrum has been subtracted from all spectra. The positions of the doublet components obtained by curve fitting are arrowed. (Reprinted from [48] by permission of the publishers, Butterworth–Heinemann Ltd © 1984.)

molecule and a perspective sketch of part of the crystal containing a labelled molecule. The key parameters are listed in Table 5.3.

FTIR spectra were obtained for blends with 0.5% PED, crystallized from dilute xylene solution at 70°C [48]. Allowance was made for weak features in the region 1070–1100 cm^{-1} in the spectrum of pure PEH. A PEH reference spectrum was subtracted, using a variable scaling factor, from sample spectra until a broad band around 1080 cm^{-1} was removed. Data after this subtraction procedure are shown in Fig. 5.14. An increase in the maximum doublet splitting is evident with increasing molecular weight, while the central minimum in absorbance becomes more apparent.

Initial model calculations made use of a common low-frequency component for all doublets, and omitted the central singlet from consideration [48]. On increasing the number of sheets, N, in the model calculation, an increase in the outermost splitting was found, arising from a progressive increase in group sizes. The positions of components with smaller splittings changed minimally with N, although their intensities were clearly affected. As N increased beyond a figure of four, differences in calculated spectra became minor and beyond $N = 11$ no significant differences were observed. A distribution of labelled stem groups, obtained with the parameters listed in Table 5.3, therefore gives rise to a large number of doublets in the IR spectrum. One notable feature of the calculated spectra was the appearance of some discrete peaks, although not all components were resolved when using a Lorentzian broadening function of bandwidth 1 cm^{-1} (equal to the experimental resolution). One reason for this behaviour, rather than a continuous curve, is the presence of large intervals between the doublet splittings, particularly for the smaller splittings (e.g. $\Delta\nu_{3\times1} - \Delta\nu_{2\times1}$).

The use of Fourier self-deconvolution of sample spectra enables the separation of certain doublet components [50], providing a better test for molecular models than the raw data. This technique allowed the identification of pairs of components from any particular doublet, so that two vibrational frequencies could be assigned to any doublet splitting [51] (see section 5.6.3). The assumption of a common position for low-frequency components of all doublets then becomes unnecessary. Furthermore, estimates of the amorphous content of any sample, together with the proportion of isolated labelled stems derived from model calculations, allowed the singlet peak to be included in calculations.

Comparisons between experimental data, for a range of sample molecular weights, and the results of calculations based on the model shown in Fig. 5.8 (parameters from Table 5.3) are shown in Fig. 5.15. In all cases, the calculated curve corresponds to a set of computer-generated crystal stem models, each of which has been divided into groups of labelled stems. Each such group was then used to derive an equivalent group, whose doublet splitting was calculated. The calibration results [51] were used to obtain component frequencies and the number of labelled stems in the group was used to scale the intensities. The singlet intensity was obtained by adding the contributions from isolated crystal stems (from the computer model) and from non-crystalline material. From DSC measurements, the latter was assumed to constitute 20% of the material.

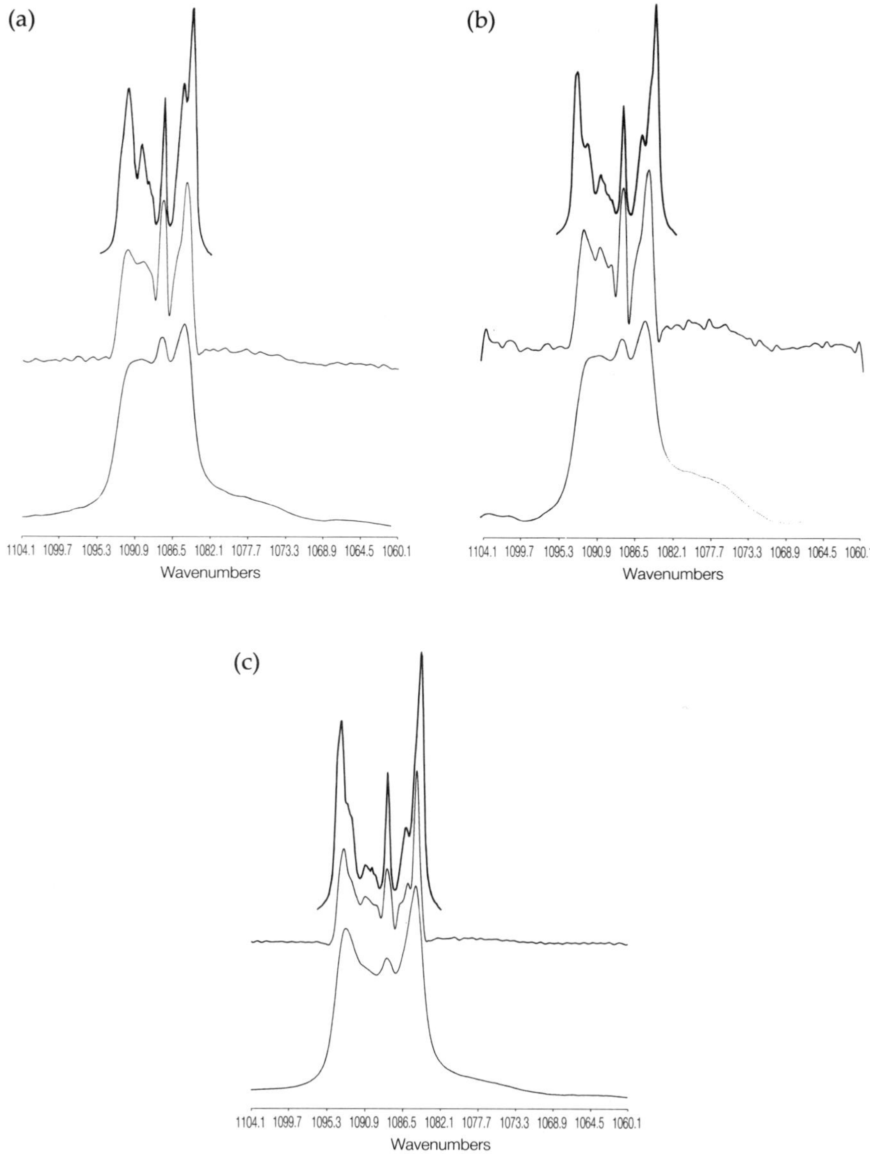

Fig. 5.15 Bottom: experimental data; middle: data after Fourier self-deconvolution; top: calculation based on the model of Fig. 8.8 for (a) 3% PED of $\bar{M}w = 54000$ and three-sheet model; (b) 1% PED of $\bar{M}_w = 155400$ and seven-sheet model; (c) 3% PED of $\bar{M}_w = 386000$ and 11-sheet model. (Reprinted from [51] by permission of the publishers, Butterworth–Heinemann Ltd, © 1985.)

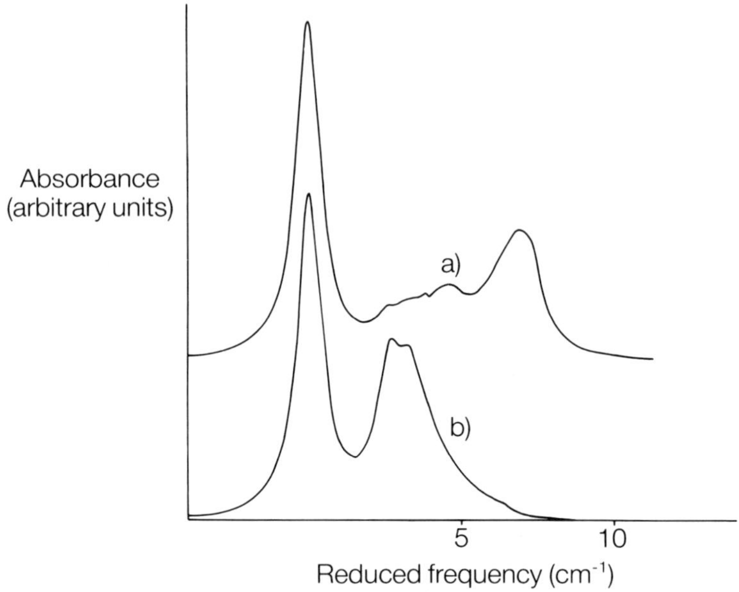

Fig. 5.16 Computer model calculations for a molecule occupying two sheets, with a dilution of 0.5. The singlet contribution has been omitted and a common low-frequency component assumed and shifted to zero frequency. A Lorentzian broadening function of bandwidth 1 cm^{-1} has been used. (a) Model with 0.75 as proportion of adjacent re-entry; (b) model with 0.19 as that proportion. (Reprinted from [50] by permission of the publishers, Butterworth–Heinemann Ltd, © 1984.)

Peak broadening involved a Lorentzian function with half-width 0.5 cm^{-1}.

In each case, the calculation shown is based on models with the (integer) number of sheets of crystal stems closest to $\bar{M}_w/21000$, where $\bar{M}_w$ is the appropriate figure obtained from GPC data. Generally, the quality of agreement between deconvoluted experimental data and the calculated curves is good. The outermost splitting is usually in agreement to within 1 cm^{-1}. The positions of other peaks are also predicted well, but their intensities are generally somewhat higher than experimental values. The singlet intensity is in good agreement, especially at higher molecular weights. The second moment of the absorbance was used as a further criterion by which to judge the molecular model [51], reinforcing the conclusion reached here, namely that the molecular model with parameters listed in Table 5.3 provides a good description of the chain conformation for crystals obtained at 70°C from dilute xylene solution.

Only one set of model parameters has been considered so far. As an example of a different model, we now examine a structure proposed on

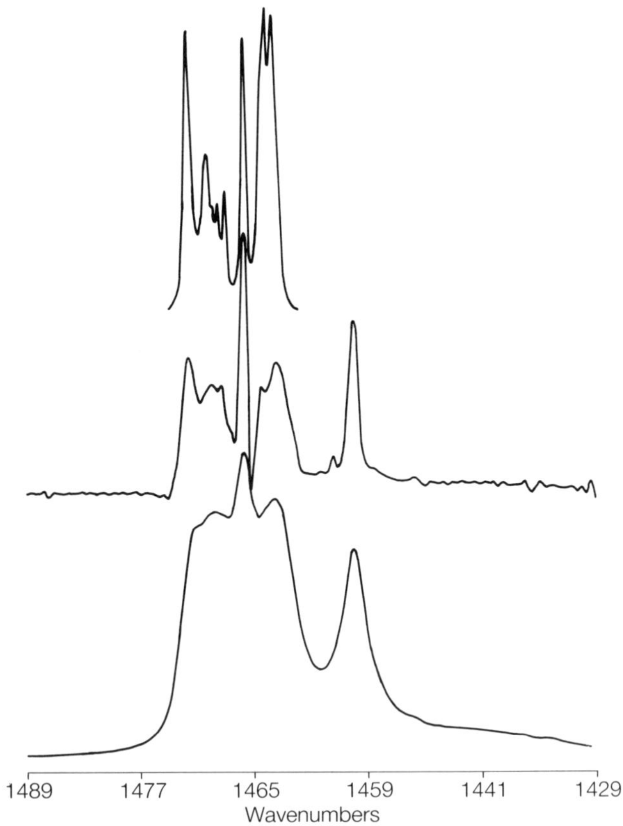

Fig. 5.17 Bottom: experimental spectrum for 3% PEH of $\bar{M}_w = 28600$ in PED matrix; middle: deconvoluted spectrum; top: calculated spectrum for two-sheet model. (Reprinted from [51] by permission of the publishers, Butterworth–Heinemann Ltd, © 1985.)

the basis of a lattice theory of interfacial chain packing [59]. The thickness of the lamellar surface layer in this case influences both the proportion of adjacent re-entry (p) and the dilution (ϕ). The value of p is predicted to be less than 20%, so that only this parameter need be altered among those listed in Table 5.3. Typical calculations are shown in Fig. 5.16, by comparison with the model previously discussed. The outermost splittings in the two cases are 3.2 and 7.0 cm^{-1} respectively, showing that the spectrum is very sensitive to the value of p. The proportion of isolated stems, giving rise to the singlet peak, is also considerably increased for the system with less adjacency, although this contribution is omitted in the figure. Clearly, the mixed-crystal IR technique strongly favours a model with a high proportion of adjacent re-entry.

Similar measurements have also been made using mats with a deuterated polyethylene matrix [51]. Model calculations were performed in the same way, with a proportion of 25% non-crystalline material. An example is shown in Fig. 5.17, where the 1456 cm^{-1} peak appears to be unrelated to the CH_2 bending vibration. The relative intensities of different doublets do not seem to be reproduced by the calculated spectra as well as for PEH matrix samples, and there is clearly some difference in experimental and calculated singlet:doublet intensity ratios. However, the values for second moments are in good accord [51], and the agreement may generally be regarded as quite good. This result has important implications with regard to isotopic fractionation. Any significant fractionation would involve PEH crystallizing first, resulting in different chain conformations for the two species. The ability of a single molecular model to fit IR data for both PEH and PED matrix samples implies that such fractionation is minimal.

(b) Crystallization at low supercoolings

The range of single-crystal morphologies obtainable in polyethylene (Fig. 5.1) indicates a considerable potential for the mixed-crystal technique in probing the underlying fold arrangements, particularly when crystallization rates are fast enough to minimize isotopic fractionation. In fact, the variation in the CD_2 bending profile with crystallization temperature (through varying the solvent) has generally been found to be small [60]: an example is shown in Fig. 5.18. The interpretation of these results is still incomplete, but it appears that the range of group splittings obtained is more sensitive to the occupational probability along {110} directions than it is to the predominant fold direction.

A significant narrowing of those IR peaks in polyethylene which arise from crystal-field splitting was observed with the sample still in the form of a suspension, rather than a dried mat [61]: it is likely that crystal defects are fewer in the pyramidal crystals before their collapse. Although attempts have been made to utilize this effect in mixed-crystal suspensions, the further dilution of PED with solvent has so far prevented the observation of any significant line narrowing in the subtracted spectrum.

(c) Annealing solution-grown crystals

The influence of heat treatment on the morphology of semi-crystalline polymers, at temperatures below the melting point of the crystals, has been extensively studied. Features such as the increase in long period (l_x), both with temperature of annealing (T_a) and with the annealing time, are well established for polymers such as polyethylene (e.g. see reference 62). However, techniques such as X-ray diffraction, often applied to the

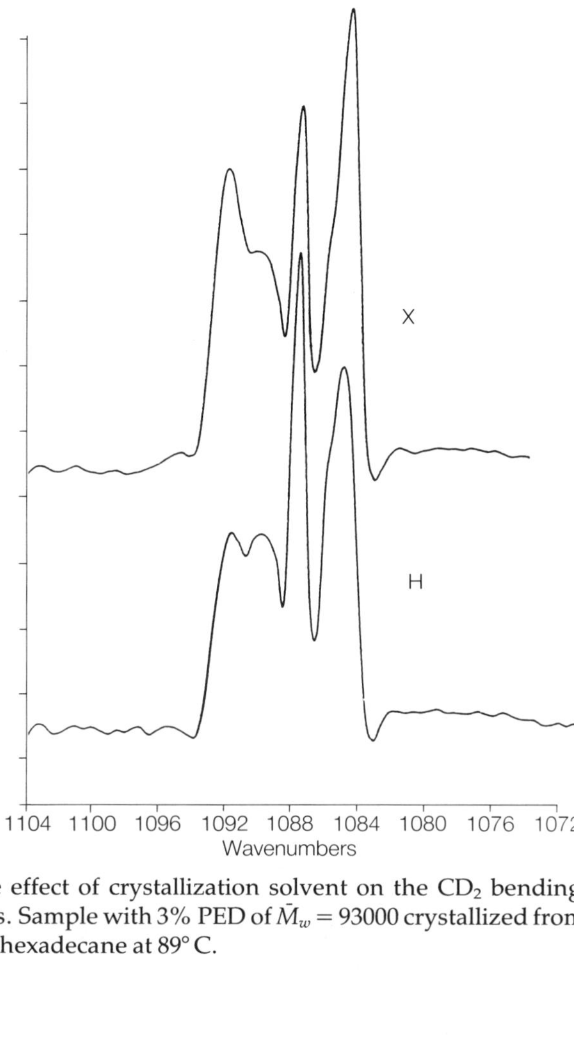

Fig. 5.18 The effect of crystallization solvent on the CD_2 bending spectrum of mixed crystals. Sample with 3% PED of $\bar{M}_w = 93000$ crystallized from (X) xylene at 70° C and (H) hexadecane at 89° C.

problem of annealing, measure averaged properties of lamellar stacks. The information obtainable on the molecular basis of annealing is necessarily limited. Mixed-crystal IR spectroscopy offers the possibility of studying changes in molecular conformation which occur during annealing, providing insight into the molecular processes involved. In particular, it is possible to make use of the molecular conformations already established for solution-grown polyethylene and described in the previous section as the starting point for models of annealing. The considerable mobility of polymer molecules below the crystal melting point is mirrored by a surprising degree of mobility in heated *n*-alkane mixtures, which have also been investigated spectroscopically using isotopic labelling. These 'model' systems will be mentioned first, before extending the discussion to polyethylene itself.

Using pairs of deuterated and normal *n*-alkanes, crystallized separately and then mixed, Ungar and Keller showed that changes occurred in both CD_2 bending and CH_2 rocking vibrations on annealing [56]. In each species, the major effect was a filling of the 'trough' in the centre of each mode, caused by molecules migrating into crystals of the other species. Those molecules which have migrated form a minority in crystals of the other isotopic species, and are not present in sufficient numbers to produce a doublet. A singlet component is therefore added to the spectrum, near the centre of each mode.

The effect of annealing on solution-crystallized polyethylene mats was first investigated by Ching and Krimm [63]. Using a dispersive spectrometer, with samples at room temperature, they showed that the CD_2 bending doublet changes shape during annealing. Measurements were predominantly concerned with cast films, which would have crystallized on cooling to 50°C. Generally, the CD_2 bending line-shape showed an enhanced singlet component with increasing T_a, with the spectrum tending towards that of a melt-crystallized sample for higher values of T_a. The progression towards a singlet line-shape was recognised as a significant indication of a low degree of isotopic fractionation in the original samples. Changes in the CH_2 rocking mode splitting with T_a were taken as evidence for a change from {110} to {200} folding with annealing. As noted earlier, this is erroneous, with the changes resulting from measurements made at room temperature rather than with liquid-nitrogen cooling. Nevertheless, the CD_2 bending results indicated that conformational changes occur during annealing.

This conclusion was supported in later work [64], which also made use of nitric acid etching to remove crystal folds and thus to influence molecular mobility. Again, the unetched sample showed a CD_2 bending mode tending towards a singlet with increasing T_a. Etching itself influences the spectrum, increasing the singlet intensity even without annealing. Both etched and unetched samples showed changes in band-shape, and hence in molecular conformations, for T_a values of 96°C and above.

The molecular processes involved in annealing were studied in detail by Spells and Sadler, using mixed-crystal IR spectroscopy [65] in conjunction with WANS measurements [66]. Measurements were focused on the range of T_a where the lamellar thickness increases irreversibly, but without the large-scale mobility which occurs at the highest T_a. This corresponds to the range $105 < T_a < 123$°C. A heating rate of 10 °C min^{-1} was used with an annealing time of 1 h. The starting point for the molecular conformation in unannealed single crystals was taken to be the model described in section 5.7.2.(a) (Fig. 5.8). Furthermore, the sheet-like

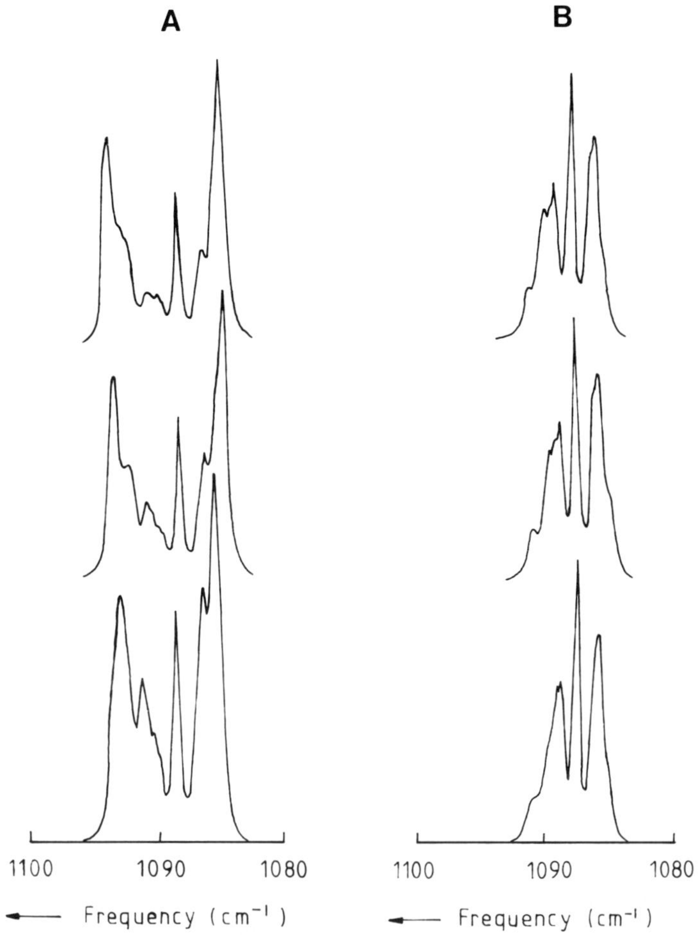

Fig. 5.19 Calculations showing the effect of annealing, using a model of random dilution of the original groups of stems, on the CD_2 bending spectrum. (A) Band shapes for as-crystallized mats using three-sheet (bottom), seven-sheet (middle) and 11-sheet (top) models, and the parameters listed in Table 5.3, to obtain the dependence on molecular weight. (B) Band shapes after using the random dilution model to represent annealing to the extent of doubling the crystal thickness for three (bottom), seven (middle) and 11 (top) sheet models. (Reprinted with permission from [65] copyright © 1989 American Chemical Society.)

conformation obtained for single crystals suggests that labelled stems are connected by folds in the plane of the sheet. The structure is assumed to have the maximum number of adjacent folds consistent with the stem arrangement. As annealing proceeds and the lamellae thicken, a molecular reorganization is anticipated, with changes in the nature of the groups

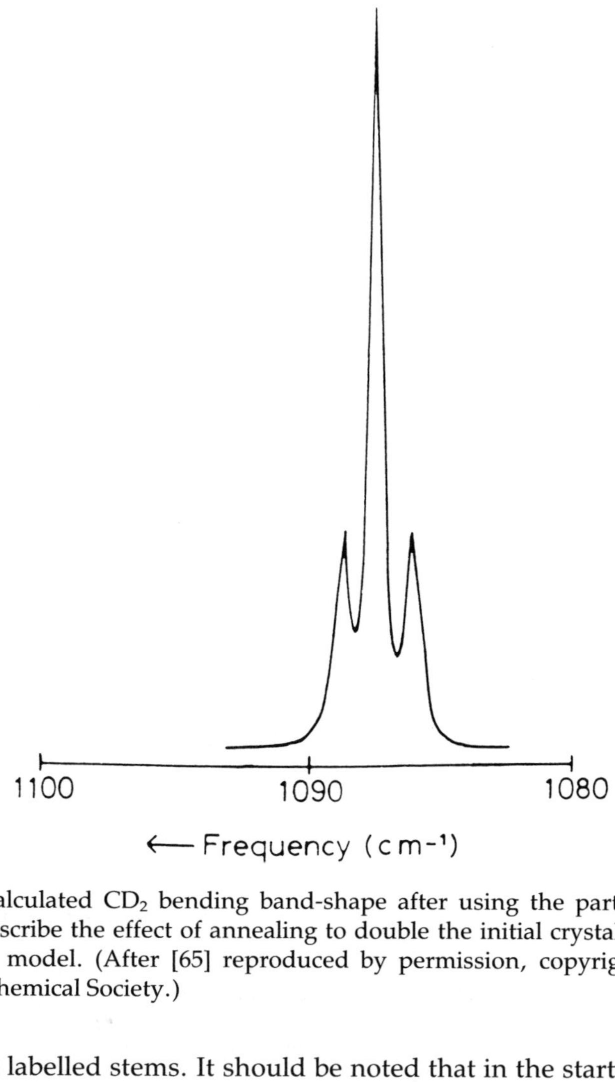

Fig. 5.20 Calculated CD_2 bending band-shape after using the partial melting model to describe the effect of annealing to double the initial crystal thickness. Three-sheet model. (After [65] reproduced by permission, copyright © 1989 American Chemical Society.)

of adjacent labelled stems. It should be noted that in the starting model the groups of adjacent stems within one sheet (i.e. the 'single-sheet groups') are generally small. An important distance scale, characterizing the annealing process, can then be introduced. L_a is taken to be a typical dimension of the region of the sample which is simultaneously molten for a particular value of T_a. Melting the polymer corresponds to L_a greater than the molecular dimensions. At low annealing temperatures, on the other hand, it is conceivable for L_a to be of the order of the stem diameter.

It is convenient that the highest value of T_a within the quoted range ($T_a = 123°C$) corresponds to a doubling of the original lamellar thickness. Model calculations are simplified if particular attention is paid to this value of T_a. The problem then reduces to that of selecting half of the labelled chain stems for removal from the lattice, since the (doubled) length of a chain stem has no effect on doublet splittings. Several models for this process were considered, but only the two extreme cases will be discussed here.

The simplest case ('random stem removal') involves the smallest possible rearrangement, with half of the original stems, chosen at random, being removed and the other half retaining their original positions. The distance L_a will, in this case, be smaller than or of similar magnitude to typical single-sheet groups. The sizes of groups of labelled stems will thus be reduced, accompanied by a reduction in the doublet splittings. The results of such calculations for various PED molecular weights (modelled using different numbers of sheets) are shown in Fig. 5.19. Larger doublet splittings for unannealed samples are clearly reduced by the stem removal, and the variation in spectra with number of sheets has become negligible for more than seven sheets. The other extreme can be represented by the 'partial melting' model, with L_a exceeding the molecular end-to-end distance. SANS data had previously shown that the radius of gyration within the lamellar plane is virtually unaffected by annealing within this temperature range [67]. This clearly imposes limits on the molecular models applicable, and the closest possible approach to a fully randomised structure is realised if the molecule occupies a 'puddle' of dimensions determined by the in-plane radius of gyration. With labelled stem positions chosen at random within the puddle, a structure is obtained with a low degree of stem adjacency and most stems within groups of three stems or fewer. This pattern of folding is similar to that proposed for melt-crystallized polyethylene [68]. Figure 5.20 shows the result of a calculation which neglects the 0.6% of stems within groups of five or more. Clearly, most stems are either isolated or in adjacent pairs (producing the predominant doublet with its very small splitting). At this stage, it was considered that any more detailed model than the two described here would not be justified, when taking into account the limitations of both the IR and neutron-scattering techniques.

Figure 5.21 shows IR measurements on annealed samples cooled with liquid nitrogen. Doublet splittings evidently decrease with increasing T_a, with the central singlet becoming more apparent. The same general behaviour was also observed for PED matrix samples, but with some evidence for isotopic fractionation at the highest annealing temperature ($T_a = 120°C$). The 'partial melting' model (Fig. 5.20) does not reproduce the large splittings satisfactorily, although the 'random stem removal' model (Fig. 5.19) does. In common with neutron-scattering measurements [66], these results indicated that, at least over the quoted range of T_a and for a heating rate of 10°C min^{-1}, a localized solid-state transformation was occurring, with the dimension L_a confined to no more than the

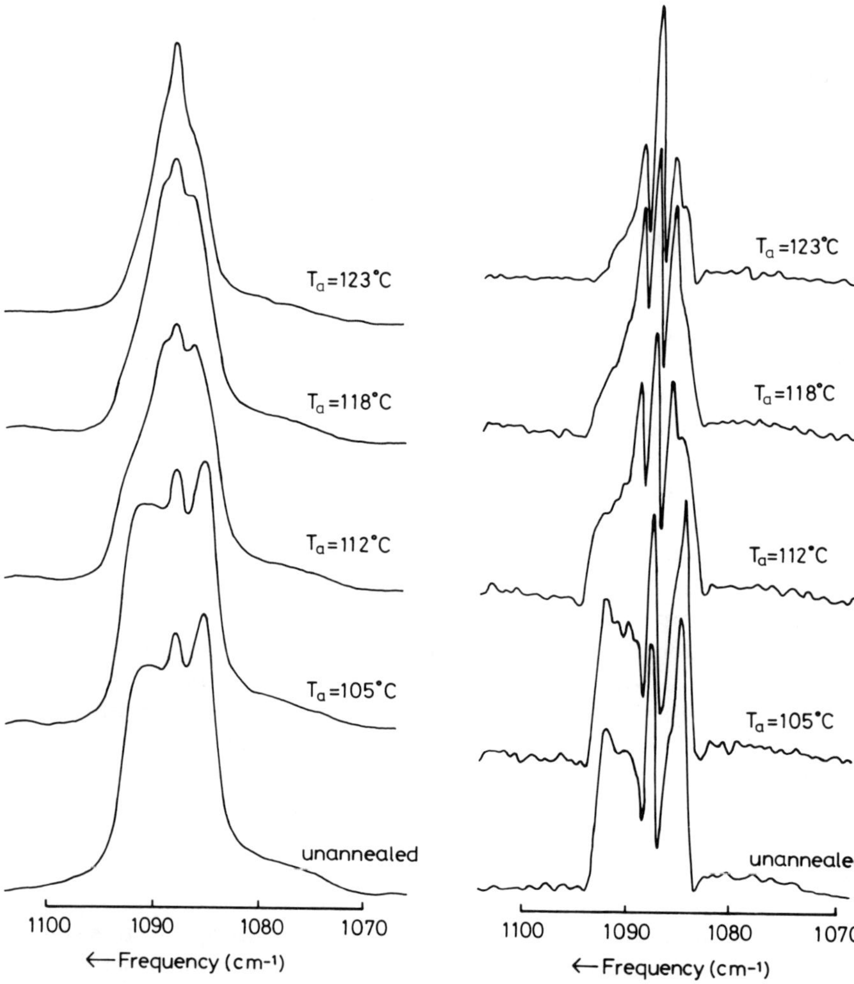

Fig. 5.21 IR spectra for 3% PED sample with $\bar{M}_w = 54000$, showing the effect of annealing. Left: raw data; right: deconvoluted data. (Reprinted with permission from [65] copyright © 1989 American Chemical Society.)

size of a single-sheet group (namely a few lattice spacings). The neutron-scattering results favoured the 'random stem removal' model, in terms of matching the scattered intensity distribution, over a wide range of molecular weights. For moderate molecular weights ($M_w < 100000$), the IR data also showed best agreement with this model, although for higher M_w a model involving preferential removal of stems from the ends of groups showed slightly better agreement. It should be stressed, however, that both techniques indicated a localized solid-state transformation, and that within the assumptions of the models used (e.g. monodisperse polymers, homogeneous annealing) it is probably unreasonable to expect more detail.

Clearly, annealing is a complex co-operative reorganization within single-crystal mats, and there remain many details to be investigated. The models used to date to analyse IR data have necessarily been simple, and in some cases topological constraints, such as the restrictions on mobility imposed by the chain trajectory, have been ignored. Nevertheless, the information obtained on the molecular rearrangements occurring during annealing is uniquely available from mixed-crystal measurements. Indeed, the full rapid scanning potential of FTIR spectrometers has yet to be utilized for collecting data during annealing itself.

(d) Deformation of solution-grown crystals

The mechanisms of molecular reorganization on deforming polymer films and fibres are generally poorly understood: although the morphology may be characterized for particular deformation conditions, direct evidence of molecular conformation is usually scarce. The mixed-crystal IR technique provides such direct evidence, specifically concerning the changes in size and orientation of blocks of adjacent labelled stems.

To date, the starting material has been restricted to polyethylene single-crystal mats, using the well-established chain conformation (see section 5.7.2.(a)). The brittleness of these mats limits the deformation techniques applicable, particularly if the annealing effects (section 5.7.2.(c)) which accompany measurements at elevated temperatures are to be avoided. Rolling the mats at room temperature resulted in a progressive reduction of the outer CD_2 bending doublet splittings, consistent with a reduction in group size [69] (see Fig. 5.22). However, this effect was found to be small at moderate roll ratios (roll ratio $\lambda \leq 5$), despite significant changes in crystallite orientation. This suggests that the lateral size of blocks breaking away from the original lamellae is initially large. Model calculations indicate that the block size is reduced to about 3.6 nm for $\lambda = 6.0$. The behaviour is thus characterized by progressively smaller blocks breaking from the original lamellae and rotating so that the molecular axis ultimately aligns in the roll direction. SANS measurements of the radius of gyration in directions parallel and perpendicular to the roll direction can be interpreted in terms of the separation of the detached blocks [70]. In addition, the CH_2 rocking vibration indicates that a proportion of the monoclinic phase is created on deformation [69], an effect which had previously been noted in mixed-crystal mats subjected to mechanical deformation [71].

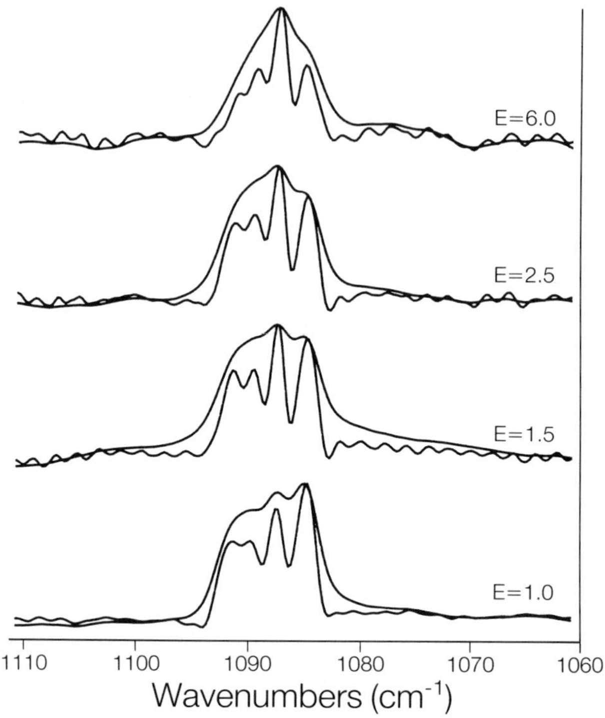

Fig. 5.22 The influence of rolling on mixed-crystal spectra. Sample with 3% PED of molecular weight 63000 as a function of roll ratio (*E*).

Clearly, the study of deformation mechanisms using mixed-crystal spectroscopy is at an early stage, and the full range of starting morphologies and deformation methods is still to be investigated.

5.7.3 Melt-crystallized polyethylene

Although the crystallization of polyethylene from the melt state is of greater immediate relevance to the majority of processing conditions than the solution crystallization already considered, melt crystallization poses considerable problems for the study of chain conformation by mixed-crystal techniques. These problems are largely associated with isotopic fractionation in melt-crystallized systems. This phenomenon, which has been discussed in section 5.4, has been extensively studied by neutron-scattering methods. The conclusion has been that in order to obtain samples with only minor levels of fractionation, it is necessary to restrict our attention to the rapid crystallization achieved by quenching molten samples, usually into cold water or liquid nitrogen. Subsequent neutron-scattering and IR work has been directed mainly towards quenched samples, although various methods for limiting the extent of fractionation have been devised.

The mixed-crystal IR technique was first used to study samples slowly cooled from the melt state by Ching and Krimm [26]. The CD_2 bandshape, obtained at room temperature using a dispersive spectrometer, was invariably a singlet, although possible variations in bandwidth were not investigated. Particular attention was paid to the matrix CH_2 rocking splitting, which was shown to vary with PED concentration, and also to differ from values for *n*-alkane mixtures. It was concluded that adjacent {200} folding was responsible for this behaviour. However, as noted earlier, this was erroneous, in that the CH_2 rocking results would be considerably modified by the use of liquid nitrogen cooling.

The observation of effects related to isotopic fractionation in neutron-scattering data has already been discussed. We simply summarize the conclusions here: fractionation is minimized by rapid crystallization, a small proportion of 'guest' polymer and a low molecular weight for the 'guest' polymer (preferably matched to that of the 'host' polymer in terms of crystal melting point).

By 1978, indications of a chain conformation quite different from that in polyethylene single crystals were emerging from neutron-scattering measurements on melt-quenched polymer. Kratky plots for the IANS region showed non-linear behaviour, unlike results for single crystals [30]. Straight lines could, however, be obtained from plots of Iq versus q^2, behaviour consistent with scattering from individual rods (the crystal stems), rather than from sheets of adjacent rods. Values for the rod dimensions, as calculated from neutron-scattering data, agreed well with accepted values. In contrast to the case of single crystals, this led to the suggestion that adjacent re-entry was not the major type of folding in melt-crystallized polyethylene. This result was clearly in contradiction to the conclusions of Ching and Krimm, based on IR measurements.

Further details of the chain conformation arising in samples quenched from the melt were provided by later neutron-scattering studies. The similarity between the radii of gyration obtained, using SANS data, from samples in the molten state and those quenched from the melt [28] suggested that little molecular movement accompanies crystallization. Using the concept that chain segments move the least possible distance on crystallization, a model was developed with crystal stem positions related to the spatial distribution of segments within a Gaussian coil in the melt state [68] (the 'freezing in' model). This model, although similar to the 'solidification' model proposed by Stamm and co-workers [72, 73] to explain neutron-scattering results on similar samples, was shown to be inadequate in reproducing IANS data. An alternative 'subunit' model was proposed, in which a large local rearrangement of the chain occurs in order for chain-folded crystallization to be accomplished, but the

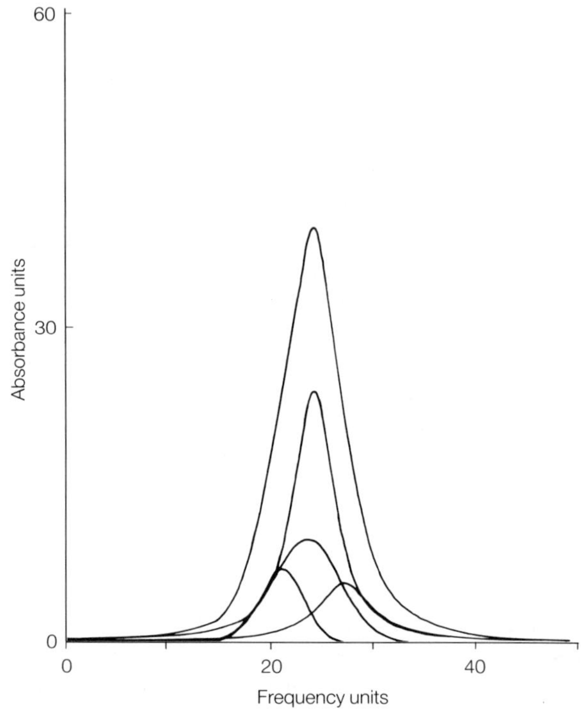

Fig. 5.23 Four-component fit to the CD_2 bending spectrum of a melt-crystallized 80 PEH/1 PED sample, with PED of $\bar{M}_w = 128000$. The top curve represents the sum of the four components, and agrees well with experimental data. Frequency scale: 0.0 = 1096.2 cm^{-1}; 50.0 = 1071.2 cm^{-1}. (After [47] reprinted by permission of John Wiley and Sons, Inc., copyright © 1983.)

distribution of **subunits** is now determined by the Gaussian chain in the melt state [68]. This 'subunit' model was found to give considerably improved agreement with data, for suitable choices of subunit. Both random and purely adjacent re-entry were ruled out, in terms of the fits with neutron data, but good fits were obtained for subunits consisting of short rows of labelled stems, on the scale of 1 nm or so. It was estimated that up to 40% of folds may be adjacent, but not in any specific crystallographic direction. These conclusions need interpretation in terms of the groups of adjacent labelled stems, considered in section 5.6.3, if IR measurements are to be used to test this model. To date, this has not been pursued, although it is apparent that the groups will generally be small and that numbers of stems per group of three or less will probably predominate.

The possibility of adjacent {200} folding, as concluded from early IR measurements [26], has implications for the WANS intensities. Wignall

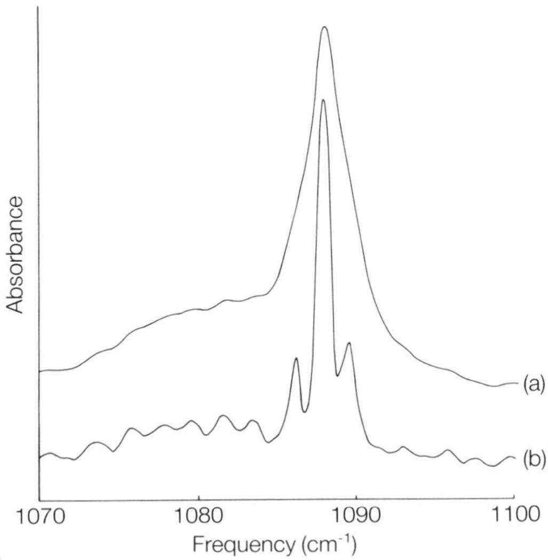

Fig. 5.24 (a) Experimental IR spectrum for a melt-quenched sample with 1% PED of $\bar{M}_w = 118000$, after subtraction of PEH reference spectrum. Resolution 1 cm^{-1}, (b) spectrum (a) after self-deconvolution. (Reprinted from [50] by permission of the publishers, Butterworth–Heinemann Ltd, © 1984.)

et al. investigated the neutron-scattering intensities [74], following the earlier suggestion that short sequences of adjacent or near-adjacent stems would cause a modulation of the WANS intensity [72]. No modification to the signal was observed, allowing an upper limit of about four stems to be assigned to the number of regularly folded stems in either the {100} or {010} planes, for both solution-crystallized and melt-quenched samples. Unfortunately, the method is insensitive to {110} adjacency, since the expected modulation overlaps the 110 reflection. The results are, however, consistent with the neutron-scattering measurements described earlier in indicating a low proportion of adjacency and very small group sizes in melt-crystallized polyethylene, in contradiction to the IR conclusions quoted.

The early IR measurements were later repeated, using liquid-nitrogen sample cooling [47]. Again, the CD_2 bending vibration was found to be a singlet, but of greater width than those observed in *n*-alkane mixtures. Curve fitting was applied to the data, with the conclusion that a crystalline singlet plus a non-crystalline singlet gave an inadequate fit: the area of non-crystalline peak necessary for a reasonable fit was considered to be too large in relation to the true sample crystallinity. Doublets were therefore introduced: an example of this curve fitting is shown in Fig. 5.23. The analysis is necessarily difficult, because the raw data show no direct indication of more than one component. The CH_2 rocking splittings were not used in this work. It was suggested that the crystalline singlet was associated with random re-entry, representing some 60–70% of the crystal stems. The remaining 30–40% of crystal stems were taken to be in adjacent arrangements, giving rise to the CD_2 bending splittings, with the splitting reaching a value of about 3.1 cm^{-1} for the highest molecular weights. This splitting approximates to that of a group with three adjacent {110} stems. Any fundamental disagreement between IR and neutron-scattering conclusions was thus removed.

The application of Fourier self-deconvolution to FTIR data from melt-quenched polyethylene is illustrated in Fig. 5.24. Some indication of the presence of more than one peak can be seen in the raw data (although not for low-molecular-weight samples). On deconvolution, two doublet components appear, with a separation of 3.4 cm^{-1}. This provides some confidence in the existence of doublets in melt-quenched spectra, but a quantitative analysis of such data, after spectral enhancement, is still awaited. It is noteworthy that the two methods of self-deconvolution and curve fitting lead to similar values for the doublet splitting, corresponding to a group of three stems.

5.8 SUMMARY AND OUTLOOK

Mixed-crystal infra-red spectroscopy, accompanied by neutron scattering and the theoretical basis to simulate IR spectra for a full range of conformational models, has provided details of the stem arrangement in polyethylene single crystals grown at high supercoolings, together with the rearrangements which occur on annealing or rolling single-crystal mats. For melt-grown crystals, the conformational details obtained are consistent with the more detailed model derived from neutron-scattering results.

Despite this significant body of information derived from mixed-crystal spectroscopy, there are several areas where the technique has yet to be fully exploited. First, deformation has only been studied for one sample type (single-crystal mats) and for one method of deformation (rolling): there is obvious scope for providing molecular details of deformation mechanisms for other conditions (e.g. melt crystallization). Second, further work on single crystals grown at higher crystallization temperatures should provide details of the different fold arrangements involved, information which must be consistent with a successful theory of polymer crystallization. Studies of *n*-alkane migration, using this technique [56], were noted earlier. Possible applications of this principle to the study of diffusion have not been fully exploited. Similarly, this chapter has primarily been concerned with experimental conditions where isotopic fractionation is minimized. The neutron-scattering community has recognised that fractionated systems can be put to use, particularly if the onset of fractionation can be clearly determined (e.g. reference 75), thereby indicating the mobility of molecular centres of mass. This change in behaviour may not be so obvious in IR results, but it has nevertheless been detected on annealing single-crystal mats [65] and in crystallization from one solvent [60].

Finally, the technique has, to date, only been applied to *n*-alkanes and polyethylene. Two of the requirements for a system to show sufficiently large correlation splittings are a crystal unit cell containing (at least) two chains and a large interaction (short separation) between the chains. It remains to be seen as to whether such conditions can be utilized in another polymer. However, it is clear that, for a wide range of problems, there is the potential for further significant contributions from mixed-crystal IR spectroscopy.

REFERENCES

1. Keller, A. (1957) *Phil. Mag*. **2**, 1171.
2. Organ, S.J. and Keller, A. (1985) *J. Mater. Sci*. **20**, 1571.
3. Organ, S.J. and Keller, A. (1985) *J. Mater. Sci*. **20**, 1602.
4. Hoffman, J.D., Davis, G.T. and Lauritzen, J.I. (1976) in *Treatise on Solid State Chemistry*, vol. 3, (ed. N.B. Hannay) Plenum Press, New York.
5. Sadler, D.M. (1983) *Polymer* **24**, 1401.
6. Sadler, D.M. (1981) in *Static and Dynamic Properties of the Polymeric Solid State*, (eds R.A. Pethrick and R.W. Richards) D. Reidel, Dordrecht.
7. Keller, A. (1962) *Polymer* **3**, 393.
8. Petraccone, V., Allegra, G. and Corradini, P. (1972) *J. Polym. Sci. C*, **38**, 419.
9. Bassett, D.C., Frank, F.C. and Keller, A. (1973) *Phil, Mag*. **8**, 1753.
10. Wittman, J.C. and Lotz, B. (1985) *J. Polym. Sci., Polym. Phys. Edn*, **23**, 205.
11. Frank, F.C. (1979) *Disc. Farad. Soc*. **68**, 7.
12. Barnes, J. and Fanconi, B. (1978) *J. Phys. Chem. Ref. Data*, **7**, 1309.
13. Tasumi, M. and Shimanouchi, T. (1965) *J. Chem. Phys*. **43**, 1245.
14. Tasumi, M. and Krimm, S. (1967) *J. Chem. Phys*. **46**, 755.
15. Tasumi, M. and Krimm, S. (1968) *J. Polym. Sci. A-2*, **6**, 995.
16. Krimm, S. (1974) in *Proc. Int. Symp. Macromol*. (ed. E.B. Mano), Elsevier, Amsterdam, p. 107.
17. Sadler, D.M. (1984) in *Structure and Properties of Crystalline Polymers*, (ed. I. Hall), Applied Science, London.
18. Guinier, A. and Fournet, G. (1955) *Small Angle Scattering of X-Rays* (trans. C.B. Walker), Wiley, New York.
19. Spells, S.J. and Sadler, D.M. (1984) *Polymer*, **25**, 739.
20. Hiebert, G.L. and Hornig, D.F. (1952) *J. Chem. Phys*. **20**, 918.
21. Hornig, D.F. and Hiebert, G.L. (1957) *J. Chem. Phys*. **27**, 752.
22. Bank, M.I. and Krimm, S. (1969) *J. Polym. Sci. A-2*, **7**, 1785.
23. Bank, M.I. and Krimm, S. (1970) *J. Polym. Sci., Polym. Letts*. **8**, 143.
24. Stehling, F.C., Ergos, E. and Mandelkern, L. (1971) *Macromolecules*, **4**, 672.
25. Krimm, S. and Ching, J.H.C. (1972) *Macromolecules*, **5**, 209.
26. Ching, J.H.C. and Krimm, S. (1975) *J. Appl. Phys*. **46**, 4181.
27. Schelten, J., Wignall, G.D., Ballard, D.G.H. and Schmatz, W. (1974) *Coll. and Polym. Sci*. **252**, 749.

28. Schelten, J., Wignall, G.D. and Ballard, D.G.H. (1974) *Polymer*, **15**, 682.
29. Schelten, J., Ballard, D.G.H., Wignall, G.D. *et al.* (1976) *Polymer*, **17**, 751.
30. Sadler, D.M. and Keller, A. (1977) *Macromolecules*, **10**, 1128.
31. Schelten, J., Wignall, G.D., Ballard, D.G.H. and Longman, G.W. (1977) *Polymer*, **18**, 1111.
32. Summerfield, G.C., King, J.S. and Ullman, R. (1978) *Macromolecules*, **11**, 218.
33. Buckingham, A.D. and Hentschel, H.G.E. (1980) *J. Polym. Sci., Polym. Phys. Edn.* **18**, 853.
34. Strazielle, C. and Benoit, H. (1975) *Macromolecules*, **8**, 203.
35. Atkin, E.L., Kleintjens, L.A., Koningsveld, R. and Fetters, L.J. (1982) *Polymer Bull.* **8**, 347.
36. Spells, S.J., Sadler, D.M. and Keller, A. (1980) *Polymer*, **21**, 1121.
37. Brockmeier, N.F. (1968) *J. Appl. Polym. Sci.* **12**, 2129.
38. Shen, M., Hansen, W.N. and Romo, P.C. (1969) *J. Chem. Phys.* **51**, 425.
39. Krimm, S. and Cheam, T.C. (1979) *Disc. Faraday Soc.* **68**, 244.
40. Cheam, T.C. and Krimm, S. (1981) *J. Polym. Sci., Polym. Phys. Edn*, **19**, 423.
41. Kauppinen, J.K., Moffatt, D.J., Mantsch, H.H. and Cameron, D.G. (1981) *Appl. Spectroscopy*, **35**, 271.
42. Kauppinen, J.K., Moffatt, D.J., Mantsch, H.H. and Cameron, D.G. (1981) *Anal. Chem.* **53**, 1454.
43. Kauppinen, J.K., Moffatt, D.J., Cameron, D.G. and Mantsch, H.H. (1981) *Appl. Optics*, **20**, 1866.
44. Gribov, L.A. (1970) *Opt. Spectrosc.* **29**, 465.
45. Zbinden, R. (1964) *Infra-red Spectroscopy of High Polymers*, Academic Press, New York.
46. Ungar, G. and Masic, N. (1985) *J. Phys. Chem.* **89**, 1036.
47. Jing, X. and Krimm, S. (1983) *J. Polym. Sci., Polym. Letters Edn*, **21**, 123.
48. Spells, S.J., Keller, A. and Sadler, D.M. (1984) *Polymer*, **25**, 749.
49. Sadler, D.M. and Keller, A. (1979) *Science*, **203**, 263.
50. Spells, S.J. (1984) *Polymer Commun.* **25**, 162.
51. Spells, S.J. (1985) *Polymer*, **26**, 1921.
52. Fanconi, B., McCrackin, F. and Sarazin, D. (1985) *Polymer*, **26**, 219.
53. Mathisen, H., Norman, N. and Pedersen, B.F. (1967) *Acta Chem. Scand.* **21**, 127.
54. Avitabile, G., Napolitane, R., Pirozzi, B. *et al.* (1975) *J. Polym. Sci., B*, **13**, 351.
55. Zerbi, G. (1982) in *Vibrational Intensity in Infra-red and Raman Spectroscopy*, (eds W.B. Person and G. Zerbi), Elsevier Scientific, New York.
56. Ungar, G. and Keller, A. (1979) *Colloid and Polym. Sci.* **257**, 90.
57. Jing, X. and Krimm, S. (1982) *J. Polym. Sci., Polym. Phys. Edn.* **20**, 1155.
58. Yoon, D.Y. and Flory, P.J. (1979) *Disc. Faraday Soc.* **68**, 288.
59. Yoon, D.Y. and Flory, P.J. (1983) *J. Amer. Chem. Soc. Prepr.* (August 1983).
60. Spells, S.J. and Sadler, D.M. (to be published).
61. Painter, P.C., Runt, J., Coleman, M.M. and Harrison, I.R. (1977) *J. Polym. Sci., Polym. Phys. Edn* **15**, 1647.
62. Fischer, E.W. and Schmidt, G.F. (1962) *Angew. Chem.* **74**, 551.
63. Ching, J.H.C. and Krimm, S. (1975) *Macromolecules*, **8**, 894.
64. Chang, C. and Krimm, S. (1983) *J. Polym. Sci., Polym. Letts. Edn*, **21**, 913.
65. Spells, S.J. and Sadler, D.M. (1989) *Macromolecules*, **22**, 3948.
66. Sadler, D.M. and Spells, S.J. (1989) *Macromolecules*, **22**, 3941.
67. Sadler, D.M. (1985) *Polymer Comm.* **26**, 204.
68. Sadler, D.M. and Harris, R. (1982) *J. Polym. Sci., Polym. Phys. Edn.* **20**, 561.
69. Okoroafor, E.U. and Spells, S.J. *Polymer* (in press).
70. Spells, S.J. and Okoroafor, E.U. *Polymer* (in press).
71. Kikuchi, Y. and Krimm. S. (1970) *J. Macromol. Sci., Phys.* **B4(3)**, 461.

72. Stamm. M., Fischer, E.W., Dettenmaier, M. and Convert, P. (1979) *Disc. Faraday Soc.* **68**, 263.
73. Dettenmaier, M., Fischer, E.W. and Stamm, M (1980) *Colloid and Polym. Sci,* **258**, 343.
74. Wignall, G.D., Mandelkern, L., Edwards, C. and Glotin, M. (1982) *J. Polym. Sci., Polym. Phys. Edn*, **20**, 245.
75. Sadler, D.M. and Barham, P.J. (1990) *Polymer*, **31**, 43.

6

Raman spectroscopy and mechanical properties

R.J. Young

6.1 INTRODUCTION

Over recent years there has been an upsurge of interest in using Raman spectroscopy to characterize the deformation of polymers, fibres and composites. In particular, it has been found that the technique of Raman spectroscopy in conjunction with an optical microscope (the 'Raman Microscope' or the 'Raman Microprobe') is particularly useful. The Raman effect is shown schematically in Fig. 6.1 where a beam of light of frequency ν_0 is scattered by a material. The majority of the radiation is scattered **elastically** at the same frequency (the Rayleigh scattering) but a small amount of the radiation is scattered **inelastically** at a different frequency $\pm\Delta\nu$ (the Raman scattering). It is this inelastically scattered light which carries information about molecular vibrations in the material and so is useful for the characterization of the materials.

This interest in the application of Raman spectroscopy is due to at least two factors. Firstly, there have been significant developments in the area of instrumentation with lasers being more widely available and easier to use (e.g. being air-cooled rather than water-cooled) and the introduction of highly sensitive photon detectors such as charge coupled device (CCD) cameras. Secondly, and probably most importantly, several new polymers have become available which are particularly amenable to characterization by Raman spectroscopy. High-performance fibres such as those obtained from liquid crystalline polymer solutions (e.g. Kevlar) have developed over recent years and have been shown to have particularly well-defined Raman spectra due to a combination of their highly oriented molecular structure and the aromatic nature of the molecular backbones. Also conjugated polymers with interesting and unusual optical and electronic properties such as substituted polydiacetylenes have

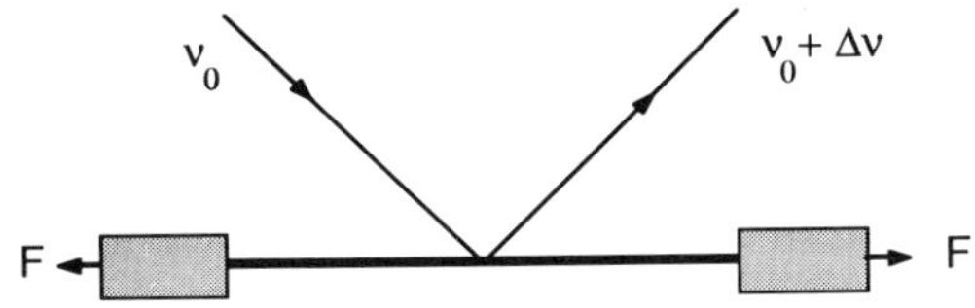

Fig. 6.1 Schematic representation of the Raman effect. The situation shown is for a fibre deformation experiment where the laser beam of frequency ν_0 is scattered from a fibre subjected to a force, F.

been developed and their molecular structure again leads to their having particularly well-defined Raman spectra.

This chapter will be concerned with the use of Raman spectroscopy for the characterization of the deformation of a wide range of both polymers and polymer-based composites. It is found that the frequency of the bands in the Raman spectra of some of these materials change under the action of stress, σ or strain, e. This change in the band position is characterized by either $d\Delta\nu/d\sigma$ or $d\Delta\nu/de$ which can be related directly to the deformation of the individual bonds in the materials. Following a short review of the instrumentation used in Raman spectroscopy the rest of the chapter will be concerned with its use in the analysis of deformation and mechanical properties.

6.2 INSTRUMENTATION FOR RAMAN SPECTROSCOPY

The original work concerning Raman spectroscopy and polymers was performed using dispersive spectrometers with photomultiplier detectors. Many of these systems incorporate microscopes for the alignment of the samples and collection of the spectra from regions as small as 2 μm in diameter. There have also been significant improvements in the hardware available for Raman spectroscopy recently. The introduction of array detectors has enabled a particular region of a spectrum to be acquired simultaneously rather than stepping through single points. This reduces drastically the time required to obtain a spectrum. A significant development in this area has been the introduction of charge coupled device (CCD) cameras [1] with quantum efficiencies of between 20 and 40% in the region 450–900 nm and a dark noise of less than 10^{-3} electrons per pixel. A typical Raman microprobe system incorporating a CCD detector is shown schematically in Fig. 6.2 and using such a system, the spectrum of Kevlar consisting of a single band in the region of 1580–1640 cm^{-1} can be obtained from areas of the order of 2 μm in diameter in a few seconds using a low-power 10 mW He–Ne laser [2].

More recently there have been reports [3,4] of the development of Fourier Transform (FT) Raman systems which have certain advantages

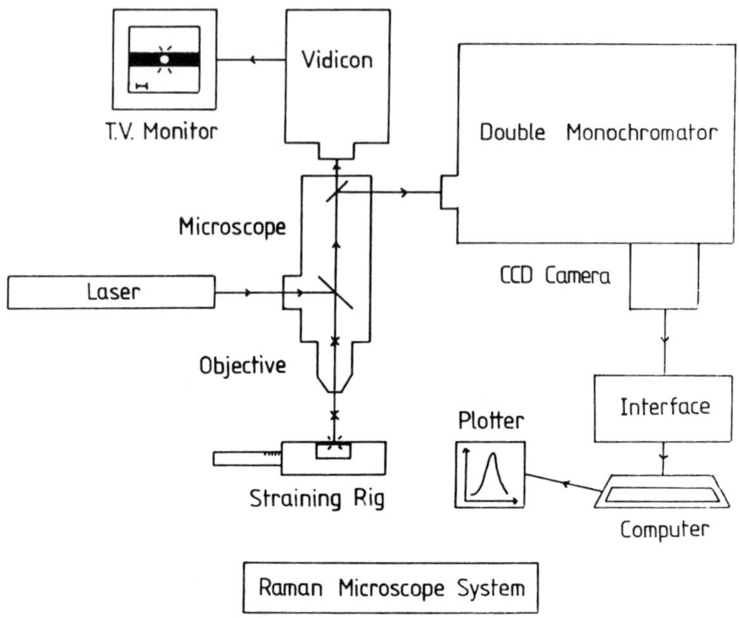

Fig. 6.2 Schematic representation of a Raman microprobe (microscope) system incorporating a CCD detector.

over conventional dispersive systems. These are potentially highly efficient with a high calibration accuracy. One particular advantage for polymers is that the use of IR excitation wavelengths greatly reduces the problems of fluorescence. Unfortunately, however, the intensity of the Raman scattering is reduced by an order of magnitude by the use of IR radiation rather than excitation in the visible region because the scattering intensity depends on the fourth power of the excitation frequency, ν_0^4. The choice between the use of either a dispersive or FT system depends upon the scattering characteristics of the material under investigation and the type of information being sought. Another significant advance is the development of the Raman microprobe into a true Raman microscope reported recently by Batchelder [5] whereby images of the Raman-scattered light may be obtained. There is clearly plenty of scope for the use of these new systems for the analysis of polymers and there is no doubt that there will be many such reports in the years to come.

6.3 POLYDIACETYLENE FIBRES

The first reported and one of the best examples of the use of Raman spectroscopy to follow deformation in materials is the case of substituted

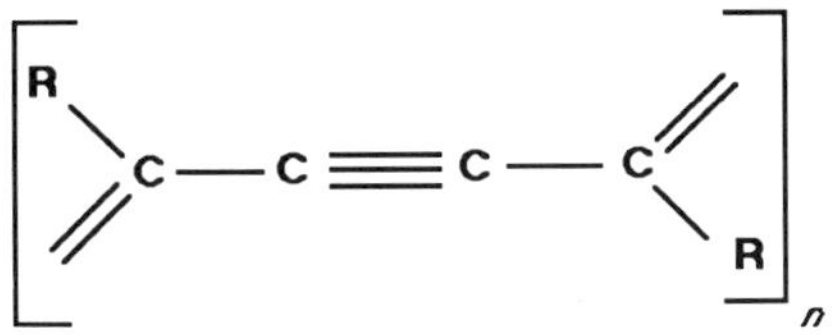

Fig. 6.3 Repeat unit of substituted diacetylene polymers. Typical R-groups are listed in Table 6.2.

polydiacetylene single crystals [6–12]. The macroscopic polymer crystals are produced by the solid-state polymerization of substituted diacetylene single-crystals monomers. The reaction is a topochemical solid-state polymerization [13] and the crystals produced have a high degree of perfection [14–16]. The polymer crystals can be generally in the form of lozenges or of fibres and the resulting polymer chains are extended with the structure given in Fig. 6.3. In the case of fibres, the molecules are aligned parallel to the fibre axis and since these fibres are essentially 100% crystalline they serve as excellent model materials for the study of the fundamental mechanical properties of ordered polymers. For example it has been found that such fibres exhibit their theoretical values of stiffness and high levels of strength [2, 17, 18]. This means that when the fibres are stretched the deformation is transferred directly to the deformation of the covalent bonds through bond stretching or bond-angle changes [7, 9].

Polydiacetylenes are found to undergo resonance Raman scattering [7] when the photon energy of the exciting laser beam is approximately equal to the energy of an electric-dipole allowed transition of the polymer. The intensities of Raman bands associated with totally symmetric vibrational modes which are strongly coupled to the electronic excited state can be enhanced by many orders of magnitude. If the laser wavelength is tuned to the maximum of the absorption band of the conjugated backbone, the vibrational modes of the backbone atoms are strongly enhanced relative to those of the side groups. Also, if the polymer is in a composite or another medium the bands are generally greatly enhanced relative to the vibrational modes of any matrix molecules. This gives rise to intense Raman scattering and relatively simple spectra from the polydiacetylene single crystals. Fortunately, there is a strong adsorption at about 2 eV (around 16000 cm^{-1}) which means that intense resonance Raman spectra can be obtained using, for example, relatively low-power (e.g. 10 mW) He–Ne lasers (excitation wavelength 632 nm or about 15820 cm^{-1}).

Table 6.1 lists the principal Raman bands observed for a variety of different fully polymerized polydiacetylene single crystals. The side groups (R) of these different crystals are listed in Table 6.2. The similarity of the band wavenumbers for the different polymers is striking even

Table 6.1 Summary of the principal Raman bands observed for fully polymerized polydiacetylene single crystals (After [9])

Polymer	T(*K*)	*Raman band wavenumbers* (*cm*$^{-1}$)				
TSHD	300	955	1205	1342	1464 1487	2088
TSHD	4	954 961	1203 1206 1269		1468 1486	2092
PUHD	300		1215	1345	1460	2072
DCHD	300	989	1214 1224	1341 1422	1452 1468 1493	2084

In addition, bands which are weak at room temperature were observed as follows:
TSHD(300 K) 145, 232, 260, 488, 665, 739, 842, 1632 cm^{-1};
TSHD(4 K) 31, 138, 234, 253, 490, 643, 666, 740, 845, 854, 1099, 1622 cm^{-1}.

though the side groups are chemically different and there can be of the order of 100 atoms per unit cell. This is because only vibrations which modulate the electronic states on the conjugated backbone giving rise to the 2 eV absorption band have significant Raman activity, due to the resonance effect. There are, however, some differences between the spectra obtained from different polydiacetylene derivatives due to the coupling of some vibrational modes with vibrations in the different side groups [9]. Attempts have been made to interpret the Raman spectra of polydiacetylenes and this has been done most successfully by calculating the in-plane vibrational modes for a polydiacetylene backbone using a simple set of valence force constants [9, 19]. The vibrational modes of the backbone were modelled with the main-chain carbon atoms represented

Table 6.2 Formulae and abbreviations for the diacetylene derivatives described in the text with general formula $R–CH_2–C{\equiv}C–C{\equiv}C–CH_2–R$ (After [12])

R	*Chemical name*	*Abbreviation*
$–OCONHC_2H_5$	2,4-Hexadiyne-1,6-diol-bis(ethyl urethane)	EUHD
$–OCONHC_6H_5$	2,4-Hexadiyne-1,6-diol-bis(phenyl urethane)	PUHD
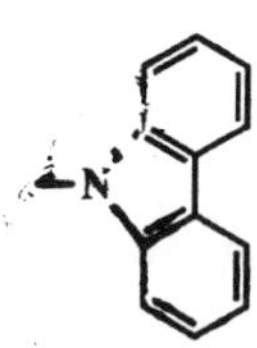	2,4-Hexadiyne-1,6-di(*N*-carbazol)	DCHD
$–OSO_2C_6H_4CH_3$	2,4-Hexadiyne-1,6-diol-bis(*p*-toluene sulphonate	TSHD

by point masses and the covalent bonds by springs. A set of force constants were produced by their model which gave a reasonable estimate of the Young's modulus of the materials. Four principal vibrational modes ν_1, ν_2, ν_3 and ν_4 were identified and they are listed in Table 6.4 along with the Raman frequencies. In some cases it was found that a particular mode could be coupled with vibrations of the side groups [9, 19].

Although the resonance Raman scattering from substituted polydiacetylene single crystals is of fundamental interest to the study of Raman spectroscopy of conjugated macromolecules, as far as this present chapter is concerned the effect of stress upon the Raman spectra is of most interest. In 1978 Cottle, Lewis and Batchelder [20] reported that the imposition of hydrostatic pressure caused an increase in frequency of the four Raman-active vibrational modes in polyTSHD (Table 6.2). They related these shifts to the decrease in the separation between the carbon atoms on the polymer backbone. A perhaps more significant discovery had been reported a year earlier by Mitra, Risen and Baughman [6] who found that there was an approximately linear decrease in two vibrational frequencies of single-crystal fibres of polyPUHD subjected to a tensile stress. They showed that this decrease could be explained in terms of the anharmonicity of the bonds between the carbon atoms on the chain. This study has been followed by a series of measurements of the dependence of the vibration frequencies upon stress and strain for several different substituted polydiacetylenes [7–12]. Batchelder and Bloor [7] performed an elegant series of experiment upon cleaved single-crystal fibres of polyTSHD fixed to the aluminium jaws of a small micrometer-driven straining rig by an epoxy resin adhesive. They showed that care had to be taken to obtain an accurate estimate of the strain in relatively short fibres during deformation and suggested that the exact values of strain reported by Mitra *et al.* [6] may be in error.

Table 6.3 Values of strain-induced Raman band shifts for ν_1 band of different polydiacetylene single crystals (Table 6.2)

Polymer	$d\Delta\nu/de(cm^{-1}/\%)$	*Ref.*
polyPUHD	20	6
polyTSHD	-20.3 ± 0.5	7
polyDCHD	-19.7 ± 0.4	85
polyEUHD	-18.8 ± 0.4*	8
	-19.0 ± 0.3†	8

* Phase 1 † Phase 2

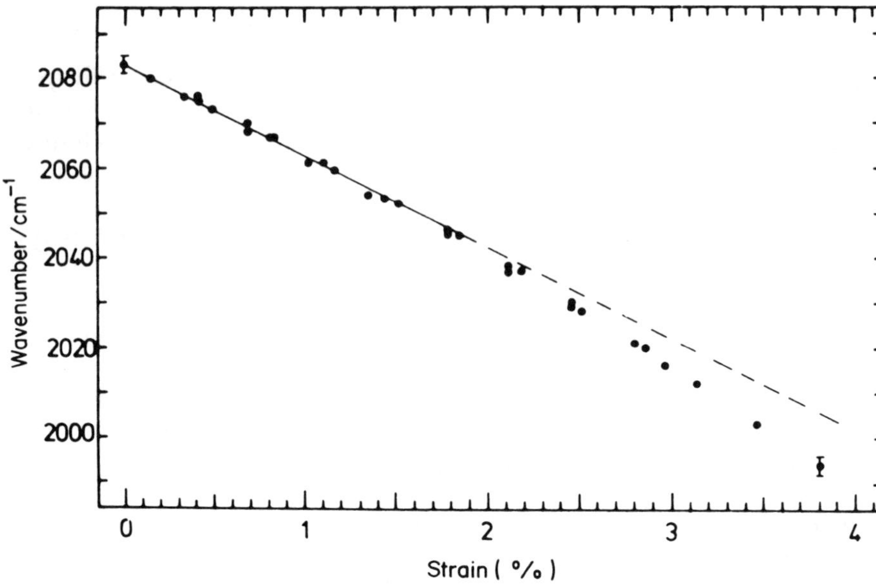

Fig. 6.4 Dependence of the wavenumber of the C≡C Raman band upon strain for a polyTSHD single crystal (After [9]).

The dependence of the wavenumbers for the Raman modes of several different substituted polydiacetylene single-crystal fibres has been measured by various groups of workers [8, 10–12]. Most attention has been paid to the behaviour of the ν_1 mode which is essentially the symmetrical stretching mode of the C≡C triple bond since this is the most sensitive to applied strain. The dependence of the wavenumber of this band upon applied strain for a polyTSHD single crystal is shown in Fig. 6.4 and the dependence of the positions of all the Raman bands for four polydiacetylenes with different side groups is summarized in Table 6.3. It is important to note that the rate of change of the wavenumber of the band with applied strain, $d\Delta\nu/de$, for the C≡C stretching ν_1 mode is about $-20\,cm^{-1}/\%$ and is virtually identical for all polydiacetylene single crystals even though they may have different modulus values [12]. This is because the structure of the backbone is very similar in each case and the shift of the Raman-band wavenumber with applied strain is a function only of the backbone structure whereas the Young's modulus depends upon both the properties of the backbone and the size of the side groups [12, 17, 18].

In the case of polyTSHD the behaviour of the ν_2 vibrational mode with strain was found to be more complex. It was found that the Raman band for this mode had a weak side band and that the two components shifted

Table 6.4 Calculated and experimental Raman band wavenumbers (cm^{-1}) for poly TSHD as a function of applied tensile strain (After [9])

	Strain (%)							
	0		*1*		*2*		*3*	
Mode	*Model*	*Expt*	*Model*	*Expt*	*Model*	*Expt*	*Model*	*Expt*
ν_1	2085	2083	2057	2061	2029	2042	2001	2015
ν_2	1486	1483	1482	1478	1480	1472	1478	1469
$+CH_2$	1463	1461	1459	1461	1454	1458	1449	1451
ν_3	1197	1202	1193	1197	1189	1192	1185	1185
ν_4	950	951	950.5	951	950.9	951	951.1	952

in a different way. This was accompanied by a reversal of the intensities of the two peaks with the side band becoming more intense at high strains. Batchelder and Bloor [7] interpreted this as a Fermi resonance between the ν_2 backbone vibrational mode and a scissors mode of the CH_2 unit in the side group (Table 6.2) with the two modes becoming degenerate at 2.1% strain and 1464 cm^{-1}.

It has been possible to use the vibrational model of Lewis and Batchelder [19] to model the effect of deformation upon the different vibrational modes and calculate the changes in the vibrational wavenumbers for polyTSHD [7]. The force constants were determined for each strain value using an empirical relationship between the force constant and bond length, and together with the model of Treloar [21] this enabled the changes in vibrational wavenumbers with strain to be determined. The results are given in Table 6.4 where it can be seen that there is generally good agreement between the measured and calculated wavenumbers. Additionally the model yields a value of Young's modulus in the chain direction of 46 GPa for polyTSHD which is close to the values measured mechanically for similar polydiacetylenes [12, 22].

A very comprehensive study has been made of the effect of deformation upon the Raman spectra of polyDCHD single-crystal fibres at different temperatures by Wu, Tashiro and Kobayashi [11]. They were also able to use FTIR spectroscopy to obtain infra-red spectra from the fibres. This procedure is normally very difficult but nevertheless, good spectra were obtained. The infra-red spectra were found to be relatively complex and many bands were found to come from the carbazole side groups. However, they found that when the fibres were subjected to stress in the FTIR spectrometer only the 1307 cm^{-1} band, which they assigned to asymetric C–C stretching, was found to shift significantly.

Combined with the relative difficulty of obtaining infra-red spectra from fibre-like samples, this is a clear demonstration of the advantage of using the Raman technique for measuring strain in high-modulus fibres. In a very careful series of experiments Wu, Tashiro and Kobayashi [11] also measured the dependence of the wavenumbers of twelve Raman-active bands for polyDCHD upon applied stress at room temperature and for four of the bands the measurements were made at three further temperatures up to 210° C. They found that at room temperature the bands associated with the ν_1, ν_2 and ν_3 modes underwent the largest stress-induced shifts as for polyTSHD [7]. They also found that the rate of shift, in terms of $d\Delta\nu/d\sigma$, tended to increase with increasing temperature. In the case of the ν_1 C≡C triple bond stretching mode it increased from $-40\,cm^{-1}$/GPa ($-18\,cm^{-1}$/%) at room temperature to a figure of $-56\,cm^{-1}$/GPa at 210° C. They found that over this temperature range the mechanically measured value of Young's modulus of the crystal also decreased by about 25%. They had difficulty in explaining the large increase in the rate of shift (with respect to stress) but when the reduction in the measured modulus is taken into account the rate of shift in terms of $d\Delta\nu/de$ is about $-18\,cm^{-1}$/% at 210° C, i.e. about the same as that at room temperature. This again confirms the point that the Raman technique follows directly the deformation of the main chain.

This work upon the strain dependence of the Raman modes of substituted diacetylene polymers has been of considerable interest to an improvement of our understanding of the fundamental physical properties of conjugated macromolecules. Additionally, it has enabled a whole series of model experiments to be devised to follow the deformation of polymers using optical methods. For instance model systems have been developed for the study of the micromechanics of deformation of fibre-reinforced composites and these are described in detail later.

6.4 POLYETHYLENE FIBRES

There has been considerable interest recently in preparing fibres of highly oriented polyethylene with very high values of stiffness and strength [23–27]. The techniques employed to achieve this end include ultra-drawing [23, 24], solid-state extrusion [25] and gel drawing or spinning [26, 27]. It has been possible by gel processing of ultra-high molar mass polyethylene to produce fibres with Young's modulus values of up to 200 GPa and strengths in the range 2–5 GPa and such materials are now available commercially under the trade names of Spectra® from Allied Signal in the USA and as Dyneema® from DSM in the Netherlands. These values of stiffness and strength are close to the theoretical limits for polyethylene which are about 300 GPa for the modulus [28] and 20–30 GPa for the strength [29]. Although real materials rarely have mechanical properties so close to their theoretical limits, it is of considerable interest to know how structural features control the mechanical properties of these high-modulus fibres and what factors lead to this shortfall in properties. For example, it is found that fibres produced under different conditions may have the same levels of orientation and degrees of crystallinity but very different levels of stiffness and strength [27]. There must clearly be structural differences in the fibres which leads to a difference in the molecular deformation processes in the fibres.

It has been relatively difficult to devise experiments to follow the deformation of these high-modulus fibres on the molecular level. Wide-angle X-ray scattering (WAXS) experiments upon fibres subjected to stress have been used extensively to monitor strains in the crystalline regions of high-modulus polyethylene fibres [30–32]. Such experiments can be relatively tedious requiring long data collection times (unless synchrotron radiation sources are employed [33,34]) and the interpretation of the data often requires assumptions to be made about the state of stress in the materials. It has been known for many years that the vibrational modes of polyolefin molecules are affected by the application of stress, particularly in the oriented state [35–37], and considerable effort has been put into studying the effect of stress upon the infra-red spectra of such materials. In general it has been found that some infra-red bands shift to lower frequency and increase in width, often also accompanied by the development of an asymmetric tail. However, difficulties are often encountered in obtaining infra-red spectra from fibres and it has been recognized recently that the measurement of Raman spectra during the stressing of highly oriented polyethylene fibres offers a unique opportunity to follow deformation of these materials on the molecular level [38–42].

It is found that well-defined Raman spectra can be obtained from highly oriented polyethylene fibres when the polarization direction of the laser beam is parallel to the fibre axis. Figure 6.5 shows a Raman spectrum in the region of 1080–1150 cm^{-1} for a high-modulus gel-spun polyethylene fibre before and after deformation [42]. It can be seen that there is a change in the position and shape of the 1128 cm^{-1} Raman band which has been assigned to the symmetric stretching of C–C single bonds [39]. It can be seen that following deformation the band develops a low-frequency tail and it is possible to fit the band to two Gaussian curves. In fact it has also been found that even before deformation the asymmetry of the band can be fitted to two peaks [42]. The variation in the position of the two peaks with strain is shown in Fig. 6.6 and it can be seen that they both move to lower frequency although the smaller peak moves more rapidly. This type of behaviour for polyethylene has been interpreted as being due to the presence of two populations of molecules in the

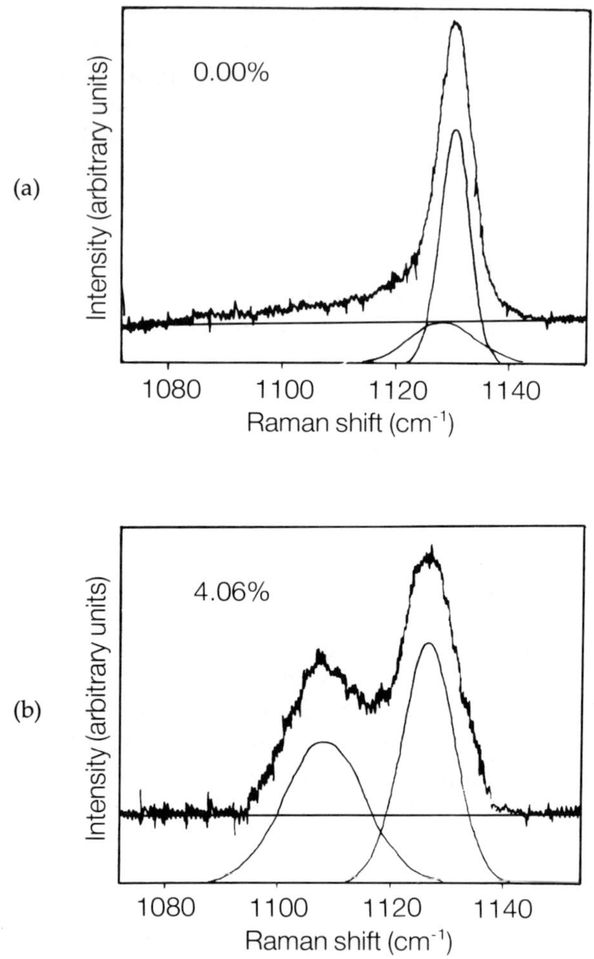

Fig. 6.5 Raman spectrum in the region 1080–1150 cm^{-1} for an experimental high-modulus gel-spun polyethylene fibre (After [42]). The Raman band has been fitted to two Gaussian curves. (a) Fibre undeformed and (b) at a strain of 4.06%.

microstructure of gel-spun polyethylene. During deformation they experience different levels of stress. It is found [41, 42] that the size of the rapidly moving peak scales with the Young's modulus of the fibre and hence it appears that it is related to the presence of microstructural features which give rise to the impressive mechanical properties of the fibres.

There was some controversy in the literature concerning the exact form of the stress-induced Raman shifts in polyethylene fibres. Some workers reported non-linear peak shifts with significant broadening of the bands

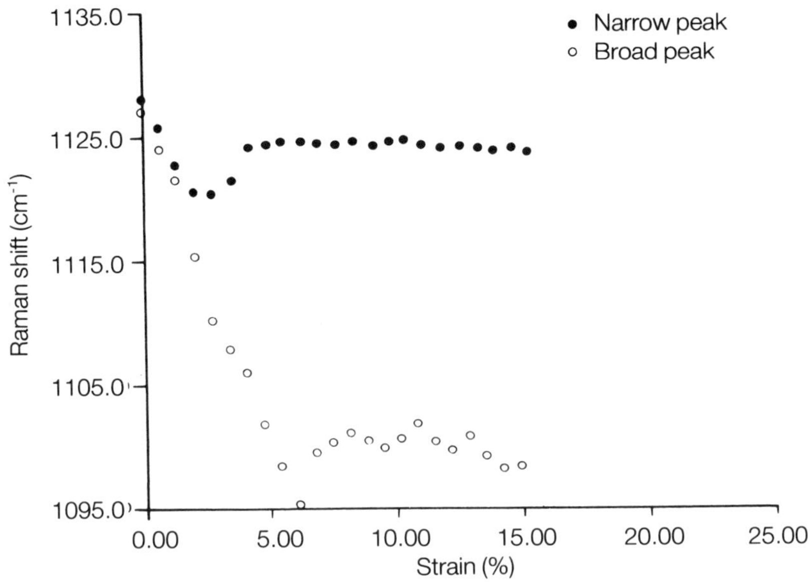

Fig. 6.6 Variation in the position of the two peaks in the Raman spectra of the high-modulus gel-spun polyethylene fibre in Fig. 6.5 with strain (After [42]).

and the development of a low-frequency tail [37, 39] whereas others apparently found linear peak shifts with little broadening [38, 40]. There is no doubt that in some instances the differences stem from differences in fibre structure due to variations in manufacturing route and testing conditions. Band splitting is certainly more apparent in gel-spun polymer which is deformed rapidly, and by obtaining spectra using a CCD camera (section 6.2). However, it appears that there is also a difference in the methods of data interpretation. If the Raman band is fitted to a single function (e.g. Gauss–Lorentz sum function [39]) then it is difficult to deal with any band asymmetry. For example, Prasad and Grubb [39] reported that although the shift of the band peak fitted to a symmetrical function is not linear with applied stress, the shift of the centre of gravity of the full band is linear. Such problems are overcome if the shifting Raman band is fitted to more than one function, i.e. if it is assumed that the band splits into several components as in Fig. 6.5. This clearly implies that the fibres must have either a two-phase structure or there must be different stresses on certain molecules in the structure. Prasad and Grubb [39] interpreted the low-frequency tail of the 1063 cm^{-1} C–C asymmetric stretching band of their spectra as non-crystalline overstressed tie molecules since they did not observe the development of any tail in their WAXS peaks. However, recent WAXS studies [41] have also indicated that the Bragg peaks in gel-spun polyethylene may split into two during deformation. Clearly more work is required to ascertain the relationship between Raman band movement and the behaviour of WAXS peaks under stress since this may enable a better insight to be obtained concerning the relationship between crystalline and molecular deformation processes.

Several models have been proposed to explain the mechanical properties of high-modulus polymer fibres [38, 43, 44] and they all assume that the fibres contain some highly stressed molecules or crystals contributing to the high level of modulus with other molecules or crystals taking lower levels of stress. It appears that for the first time Raman spectroscopy allows the deformation of the different components in the structure to be directly determined [38–42].

6.5 POLYOXYMETHYLENE FIBRES

There have been several attempts over recent years to produce high-modulus fibre of polyoxymethylene (POM) [45, 46]. Drawn samples of this material are often opaque and not suitable for optical scattering experiments but it has been possible to produce high-modulus samples of highly oriented POM using a variety of drawing techniques [38, 47]. Polyoxymethylene is a particularly interesting material for the study of the effect of deformation upon the Raman spectrum since unlike polymers such as polydiacetylenes or polyethylene the molecules have helical rather than planar zigzag configurations in the unit cell, and vibrational modes such as the internal rotations of the skeletal chain are possible in addition to bond stretching and bond-angle deformation modes.

It is found that there are several Raman bands in highly oriented POM for which there is a shift of the band to lower wavenumber upon the application of stress. The measured shifts $d\Delta\nu/d\sigma$ for the $920\,cm^{-1}$ C–O–C symmetrical stretching mode and the $539\,cm^{-1}$ C–O–C bond-angle bending mode in highly oriented POM with a modulus of 36 GPa are $-6\,cm^{-1}/GPa$ and $-19\,cm^{-1}/GPa$ respectively [47]. Tashiro and co-workers [47] have also measured shifts for the infra-red-active skeletal torsional mode at $231\,cm^{-1}$ and found a change of $-15\,cm^{-1}/GPa$. Additionally, they investigated the effect of stress upon the low-frequency vibrational bands at $39\,cm^{-1}$ and $65\,cm^{-1}$ and found that they shifted to higher frequency under tension. These bands are thought to be coupling modes between the skeletal torsional mode and the vibrational lattice mode and it was suggested that the shift to higher frequency may be due to the tension causing a lateral shrinkage due to the Poisson's contraction [47].

Although Tashiro and co-workers [47] commented that the shifts of the Raman bands in POM appeared to be relatively large, this may be due to

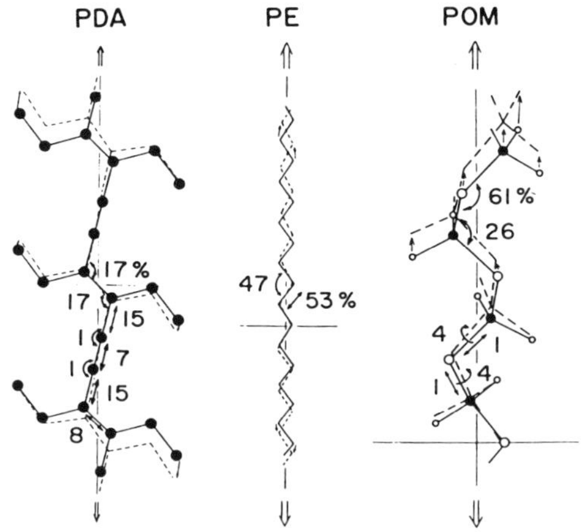

Fig. 6.7 Calculated potential energy distributions for the deformation modes of fibres of a polydiacetylene (PDA), polyethylene (PE) and polyoxymethylene (POM) (After [47]).

their shifts being expressed in terms of $d\Delta\nu/d\sigma$. The shift for the 920 cm^{-1} band in terms of $d\Delta\nu/de$ is only $-2.2\,cm^{-1}/\%$ whereas that for the 539 cm^{-1} band is $-6.8\,cm^{-1}/\%$. These values are of the same order as the band shifts in polyethylene [37–39] (expressed as $d\Delta\nu/de$) and considerably less than the shifts found in polydiacetylenes [6–12]. Nevertheless, the results do show that for helical molecules such as POM, mechanical deformation can produce considerable changes in the vibrational modes.

As well as observing changes in the Raman spectra, Tashiro and co-workers [38, 47] have examined in considerable detail the processes taking place in POM fibres during deformation both at the molecular level and in terms of crystallite deformation. They calculated the atomic displacements and the potential energy distribution for the POM chain during deformation to 10% strain and compared it with the behaviour of the planar zigzag polyethylene and polydiacetylene chains [47]. The results of their calculations are shown in Fig. 6.7 where it can be seen that the potential energy distribution is concentrated into the skeletal bending and torsional modes with some C–O–C stretching. In contrast, in the planar zigzag structures the potential energy distribution is concentrated into bond stretching and bond-angle bending in roughly equal proportions. These theoretical predictions are consistent with the experimental Raman and infra-red observations for the three polymers. In a second series of calculations [38] they modelled the behaviour of their fibres in terms of crystallite deformation using a combination of their Raman data and WAXS measurements under stress. The WAXS measurements can yield a 'crystal modulus' but this may require the assumption that the stress on the crystallites is the same as the bulk stress [30, 31] which may not be appropriate. They modelled their behaviour using Takayanagi-type models [31] and showed that the data from the two types of experiments produced complementary and consistent results. In particular, they showed that the broadening and asymmetry that developed in the 539 cm^{-1} Raman band during deformation could be interpreted as being due to the deformation of a second highly strained crystalline component as was suggested for polyethylene [38]. However, in the case of the POM the relative amounts of the two components was found to be comparable, particularly for samples with high draw ratios.

6.6 AROMATIC POLYAMIDE FIBRES

Fibres of aromatic polyamides (aramids) with high levels of stiffness and strength have been available commercially for several years and they are now used in a variety of applications, particularly in the areas of fibre-reinforced composites, protective clothing, tyre cord and ropes [48]. Aramid fibres are produced by spinning from liquid crystalline solutions using solvents such as sulphuric acid [49, 50]. The properties of the fibres can be improved by a brief heat treatment under tension at elevated temperatures [50]. Aramid fibres have very good performance characteristics which include high strength, high modulus, low creep, high thermal stability and good flame resistance [48]. They are available commercially under trade names such as Kevlar® (Du Pont) and Twaron (Akzo). Original materials such as Kevlar 49 were produced with values of filament modulus of the order of 120 GPa but recent developments have led to fibres with higher degrees of crystallinity such as Kevlar 149 which has a modulus of 170 GPa [50].

The first report of the Raman spectrum of an aramid fibre was made in 1979 by Penn and Milanovich [51] and a typical spectrum for a single filament of Kevlar 49 is given in Fig. 6.8. Aramid fibres generally give strong Raman scattering with little fluorescence and well-defined spectra can be obtained from single filaments using low-power laser beams [2, 52–56]. The structure of the most widely used aramid poly(*p*-phenylene terephthalamide) is shown below.

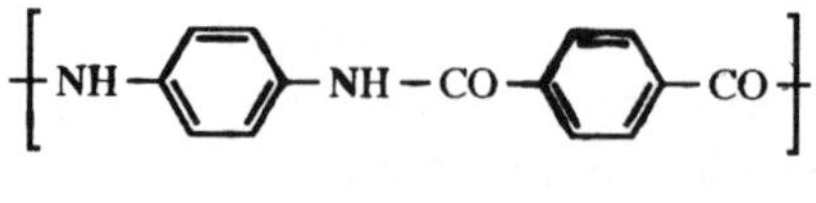

It has been pointed out [48] that the C–N bond has considerable double-bond character which, along with the *para*-substituted phenyl

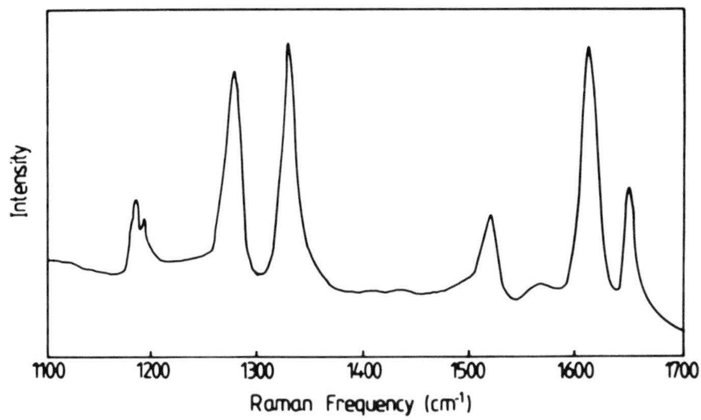

Fig. 6.8 Raman spectrum obtained from a single Kevlar 49 aramid fibre (After [56]).

group, leads to a resonance-conjugated system giving restricted rotation about that bond. These factors help to give rise to liquid crystalline solutions which allows the formation of highly oriented fibres. They also give rise to the characteristic yellow colouration of aramid fibres and probably increase the intensity of the Raman scattering due to resonance enhancement [9].

By comparison with model compounds, Penn and Milanovich [51] were able to assign some of the Raman bands to vibrations of structural groups in the poly(*p*-phenylene terephthalamide) molecule and the strong band at around 1610 cm^{-1} has been assigned to stretching of the *p*-phenylene ring. They also looked at the effect of deformation upon the Raman spectrum of Kevlar 49 and although they found that stress caused a change in the relative band intensities and depolarization ratios they did not find any measurable shift in the band wavenumbers with stress [51]. This last finding is completely at odds with the findings of Young and co-workers [52, 53] who have shown that significant and measurable frequency shifts of several Raman bands in Kevlar take place on the application of stress or strain. Indeed, Penn and Milanovich [51] appeared surprised that they did not find any frequency shifts, bearing in mind the earlier polymer deformation studies using infra-red [35] and Raman spectroscopy [6] and WAXS studies of deformed aramid fibres [57].

The situation was confused further by the recent publication by Edwards and Hadiki [54] who reported that no Raman band shifts were found when Kevlar fibres were deformed. They claimed that their work supported the findings of Penn and Milanovich [51] and disputed the findings of Young and co-workers [52, 53]. The situation was eventually

(a)

(b)

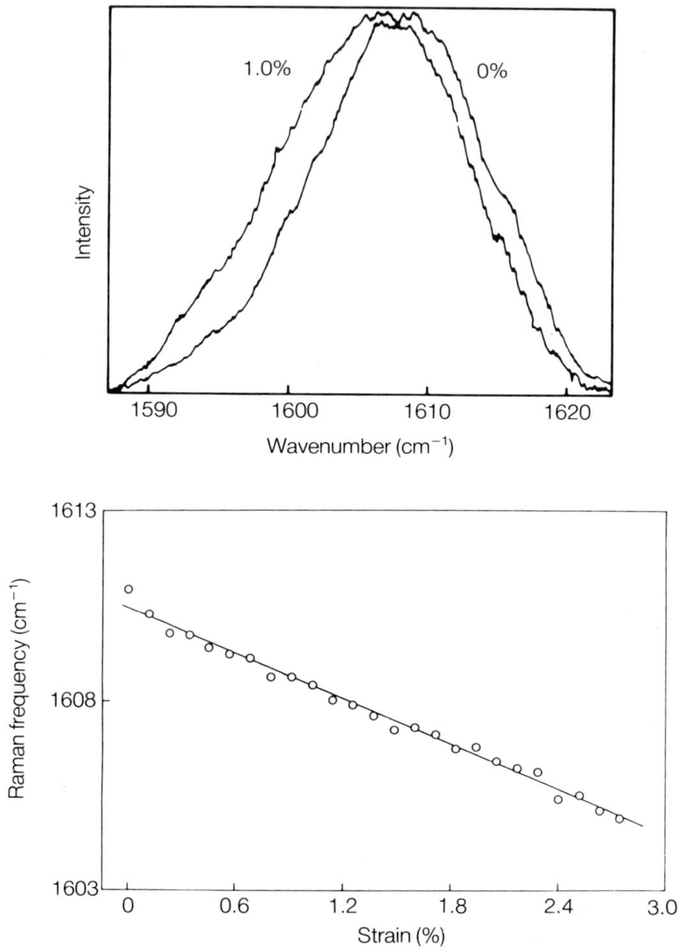

Fig. 6.9 (a) Shift of the 1610 cm^{-1} Raman band with strain for an aramid fibre. (b) Dependence of the peak position upon strain. (After [56]).

sorted out by a careful study by Young *et al.* [2] who repeated the experiments of both groups. Penn and Milanovich [51] and Edwards and Hadiki [54] had both used a relatively high-power argon ion laser whereas Young and co-workers [52, 53] had used a low-power He–Ne laser. It was demonstrated that the 488 nm line of argon ion lasers caused significant radiation damage to Kevlar fibres and caused the fibres to fracture at very low stresses and strains [2]. Hence the groups using argon ion lasers [51, 54] had damaged their fibres and careful inspection of their data showed that they had not been able to apply very high strains, at least not large enough to cause significant Raman band shifts. This behaviour is consistent with the general susceptibility of polyamides to degradation by short wavelength or UV light. In contrast it was shown that the 632 nm line of the He–Ne laser produced no radiation damage and so high strains could be applied and significant band shifts obtained [2]. Hence the dispute was resolved.

6.6.1 Molecular deformation processes

It is found that the position of the 1610 cm^{-1} Kevlar Raman band shifts to lower frequency under the application of a tensile stress as shown in Fig. 6.9(a). The peak position of the band is plotted as a function of strain in Fig. 6.9(b) and it can be seen that there is an approximately linear shift in peak position, $\Delta\nu$, with strain, e, up to failure at about 3% strain. The slope of the line is given by $d\Delta\nu/de$ and it is normally found [53, 56] that for different aramid fibres the slope is proportional to the Young's modulus, E, i.e.

$$d\Delta\nu/de \propto E \qquad (6.1)$$

Figure 6.10 gives a plot of $d\Delta\nu/de$ versus the fibre modulus, E, and it can be seen that there is an approximately linear relationship consistent with equation 6.1. This is an indication that the Raman technique is probing directly molecular stretching [56] and is consistent with the theoretical work of Northolt and co-workers [57–59]. They suggested that for the deformation of aramid fibres such as Kevlar the total strain is the sum of

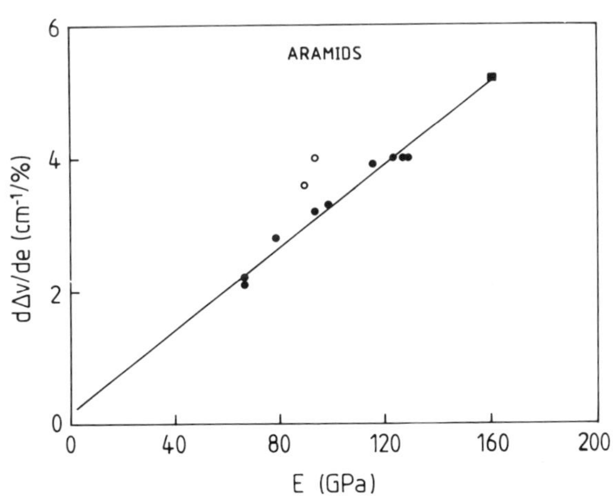

Fig. 6.10 Dependence of $d\Delta\nu/de$ with Young's modulus, E, for a series of commercial and experimental aramid fibres. The line is a least squares fit of the solid data points. The two fibres represented by open circles have a non-uniform skin/core structure (After [56]).

the strains due to two deformation processes, **stretching** and **rotation**, such that

$$e_{\text{total}} = e_{\text{stretch}} + e_{\text{rotation}} \tag{6.2}$$

It has been shown [56] that the Raman technique follows only the crystal and molecular stretching and, moreover, means that the change in the peak position, $d\Delta\nu$, is a measure of the molecular stress rather than the molecular strain [56]. The rate of shift per unit strain in the higher modulus fibres is a reflection of the higher levels of molecular stress in such fibres.

6.6.2 Skin/core structure

It appears that in the case of Kevlar fibres, the Raman spectrum is obtained essentially from the surface layers of the material. Measurements of the absorption characteristics of films of the aromatic polyamide have shown that 90% of the visible light is absorbed by a depth of the order of 1 μm of the material [60] and so it is clear that the spectrum shown in Fig. 6.8 comes from the skin regions of the 12 μm diameter Kevlar fibre.

An aspect of the deformation of the surface skin regions is reflected in the dependence of $d\Delta\nu/de$ upon the fibre modulus, E, as given by equation 6.1. It can be seen that most of the data points in Fig. 6.10 fall close to a straight line through the origin but that there are two points (open circles) that lie significantly above the line. Transmission electron microscopy [56] showed that these two fibres had significant differences in molecular orientation between the surface skin and core regions. Since the He–Ne laser beam used in the Raman spectroscopy penetrates only the order of 1 μm deep into the aramid fibres [60] the spectra were obtained from only the surface skin regions. Hence, for these two fibres the spectra come only from the highly oriented molecules in the surface regions which have higher Young's modulus values than for the core material. The values of $d\Delta\nu/de$ for these two fibres are higher than expected for a fibre of uniform structure and so the data points lie above the line in Fig. 6.10. This demonstrates clearly the usefulness of Raman spectroscopy in characterizing the deformation of surfaces in aramid fibres.

6.7 RIGID-ROD POLYMERS

Among recent developments of high-modulus polymer fibres, the fibres with the most impressive mechanical properties have been the heterocyclic rigid-rod polymer fibres produced through the USAF ordered-polymer programme [61–63]. The best fibres so far have been obtained using molecules such as poly(*p*-phenylene benzobisthiazole) (PBT),

Table 6.5 Chain repeat units for various rigid-rod and stiff-chain polymers used in high-performance fibres [64–66]

Polymer	*Chemical name*	*Repeat unit*
PBT	Poly(*p*-phenylene benzobisthiazole)	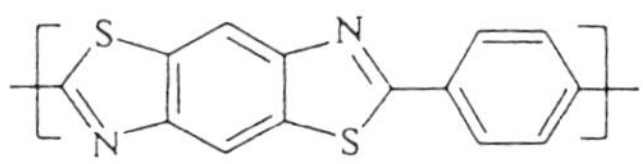
PBO	Poly(*p*-phenylene benzobisoxazole)	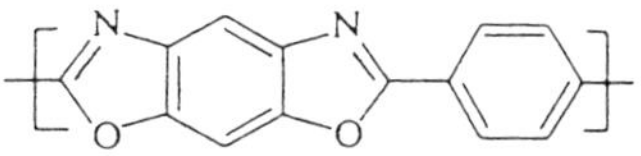
ABPBO	Poly(2,5(6)-benzoxazole)	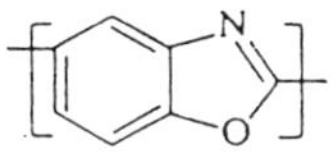

poly(*p*-phenylenebisoxazole) (PBO) or poly(2,5(6)-benzoxazole) (ABPBO) and the fibres are made in a similar way to aramid fibres by spinning from liquid crystalline solutions [61]. The chemical formulae of these molecules are given in Table 6.5. Fibres of PBT have a modulus of below 200 GPa in the 'as-spun' condition but this can be increased to 320 GPa after heat treatment [61, 62]. More recent work has shown that heat-treated fibres obtained from PBO can have even higher modulus values of up to 370 GPa [63]. Such fibres are also reported to have strengths of over 3 GPa and elongations to failure of up to 2%. Since these levels of stiffness and strength are combined with impressive oxidative and thermal stabilities [61] and low specific gravities, this means that they are extremely promising as reinforcing fibres for high-performance composites.

It has been found that well-defined Raman spectra can be obtained from single filaments of fibres such as PBT, PBO and ABPBO, although there is often some fluorescence which is generally worse for heat-treated fibres. As in the case of Kevlar it is likely that the conjugated electronic structure of the backbone of the rigid-rod molecules (Table 6.5) leads to resonance enhancement of the Raman scattering [9]. It is also found that the wavenumber of the Raman bands in such fibres is sensitive to the level of applied stress or strain [64–66]. Figure 6.11 shows examples of this effect in heat-treated fibres of PBT [64]. It can be seen that the position of the 1480 cm^{-1} band is particularly sensitive and shifts by over 20 cm^{-1} to lower frequency on the application of a strain of 2%. As in the case of aramids, this shift is accompanied by a broadening of the Raman band indicating a distribution of strain on the molecules in the structure. Similar shifts and broadening are also obtained with fibres of PBO [65] and ABPBO [66].

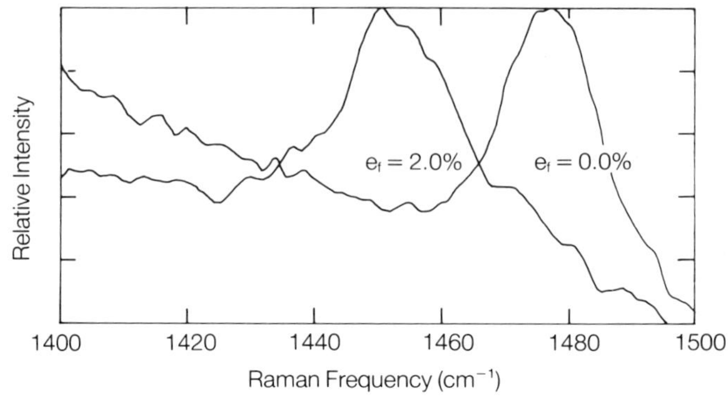

Fig. 6.11 Dependence of the peak position of the 1480 cm^{-1} Raman band for a heat-treated fibre of PBT upon strain. (After 64]).

The values of the strain-induced shifts in the wavenumbers of the different principal Raman bands for as-spun and heat-treated PBT, PBO and ABPBO fibres, $d\Delta\nu/de$ are summarized in Table 6.6 [64–66]. It can be seen in each case that the shift for a particular band tends to increase with the fibre modulus. It is likely that this is a reflection of improvements in molecular orientation leading to higher values of fibre modulus and higher strains on the covalent bonds along the polymer backbone. This is again similar to the behaviour of aramid fibres and means that the shifts

Table 6.6 Strain-induced Raman band shifts for different as-spun (AS) and heat-treated (HT) rigid-rod polymer fibres (Table 6.5)

Rigid-rod polymer	$\Delta\nu(cm^{-1})$	$d\Delta\nu/de(cm^{-1}/\%)$	*Ref.*
AS PBT	1175	0±1	64
	1480	−8.3±1	64
	1602	−6±1	64
HT PBT	1174	−1.3±0.5	64
	1477	−12.1±0.6	64
	1603	−8±1	64
AS PBO	1615	−2.1	65
HT PBO	1280	−7.9	65
	1537	−4.2	65
	1615	−5.4	65
AS ABPBO	1554	3.9±0.2	66
HT ABPBO	1558	5.9±0.3	66

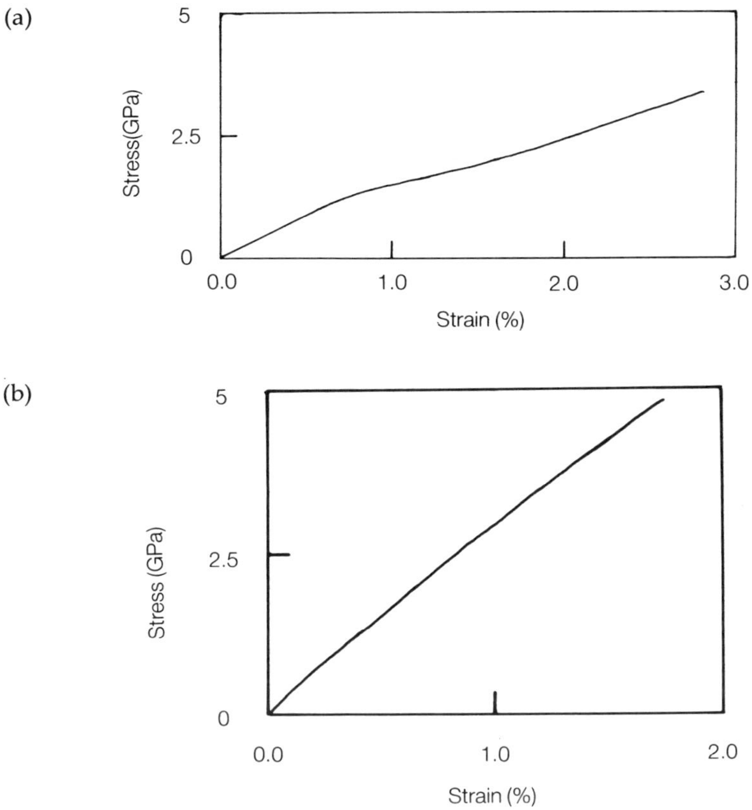

Fig. 6.12 Stress–strain curves for fibres of PBT (after [64]). (a) As-spun fibre and (b) heat-treated fibre.

for a particular band in terms of $d\Delta\nu/d\sigma$ will be approximately independent of fibre modulus. The values of $d\Delta\nu/d\sigma$ are then about $-3\,cm^{-1}/GPa$ for the $1600\,cm^{-1}$ band in PBT and similarly about $-3\,cm^{-1}/GPa$ for the $1615\,cm^{-1}$ band in PBO. The $1600\,cm^{-1}$ band in PBT has been tentatively assigned to the stretching of the phenyl rings although it may contain contributions from other vibrations [67]. It is likely that the $1615\,cm^{-1}$ band in PBO can be similarly assigned. The $1610\,cm^{-1}$ band in Kevlar which was shown in the last section to have a shift of $-3.3\,cm^{-1}/GPa$ is similarly thought to be associated with a phenyl-group stretching mode. Hence, it demonstrates that the Raman techniques gives a direct measure of the levels of stress and strain experienced by a particular group in the backbone and shows that similar groups in different molecules will experience approximately the same levels of stress. This has important implications for the design of molecules for high-modulus fibres.

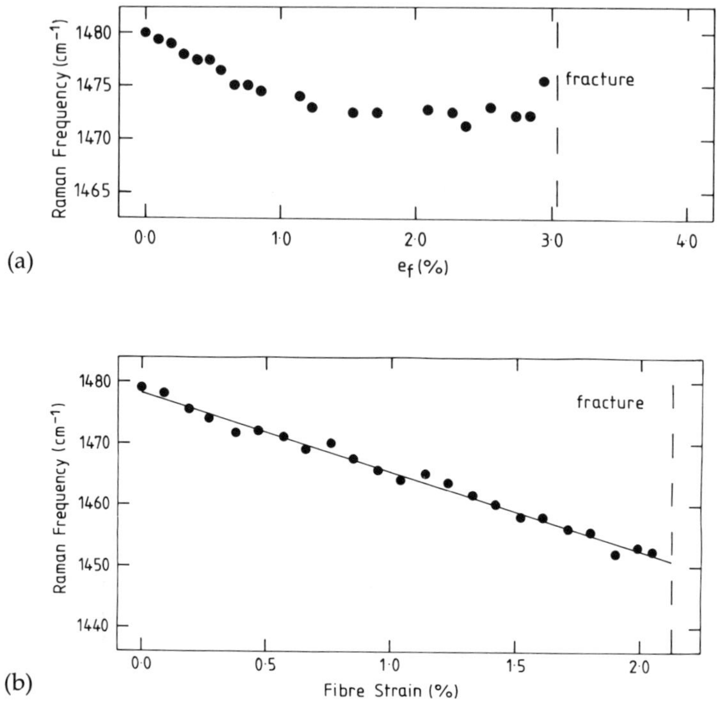

Fig. 6.13 Variation of the peak position of the 1480 cm^{-1} Raman band with strain for fibres of PBT (After [64]). (a) As-spun fibre and (b) heat-treated fibre.

Table 6.6 shows clearly that the mechanical properties of PBT, PBO and ABPBO fibres improve significantly following heat treatment. As well as an increase in the modulus and fracture strength of the fibres there is also a change in the stress/strain curve as can be seen for PBT in Fig. 6.12. The as-spun fibre tends to yield at about 0.8% strain whereas the heat-treated fibre does not yield and deforms in a linear elastic manner up to failure. This behaviour is mirrored in the variation of the position of the peak of the 1480 cm^{-1} Raman band with applied strain (Fig. 6.13). In the case of the heat-treated fibre there is an approximately linear shift with applied strain but for the as-spun fibre the position of the Raman band remains constant upon the application of further strain above about 0.8%. This is clearly a reflection of yield taking place in the fibre, probably by the translation of molecules past each other in the fibre direction. Odell and co-workers [68] concluded that PBT is only two-dimensionally crystalline with the chains packing parallel to each other but lacking register in the longitudinal direction. Heat treatment of the fibres is thought to increase the conjugation length in the polymer by improving the planarity of the molecules [69, 70]. This results in a better-defined molecular structure following heat treatment and improved molecular orientation. WAXS patterns from the fibres become better defined [64–69] and the improved structure is reflected in the higher modulus and the suppression of yield through molecular sliding detected from the Raman measurements. The studies of the deformation of PBT, PBO and ABPBO using Raman spectroscopy [64–66] have given a unique insight into the behaviour of rigid-rod molecules in fibres on a molecular level during deformation.

6.8 ORIENTED POLYMER FILMS

The work described earlier has shown that large shifts in the wavenumber of certain Raman bands assigned to main-chain vibrations can be found in highly oriented polymers. The shifts are often so large that Fina, Bower and Ward [70] reasoned that significant and measurable shifts might also be obtained for suitable Raman bands in samples with only moderate degrees of orientation such as hot-drawn poly(ethylene terephthalate) (PET) film. Previous studies [71, 72] using infra-red spectroscopy had shown that shifts of the order of 20 cm^{-1}/GPa could be obtained for the infra-red absorption band at 975 cm^{-1} on the deformation of oriented PET. Raman spectroscopy, however, gives the added possibility of providing information concerning molecular stresses along the backbone by using bands such as the Raman-active 1616 cm^{-1} main-chain phenyl group stretching vibration which is relatively strong in PET and well isolated from other bands. Additionally, the effect of deformation upon the wavenumber of this band has also been studied for fibres of aramids [2, 52, 53], PBT [64] and PBO [65] as discussed earlier, and so useful comparisons can be made concerning the response of the same band in different types of polymers.

Fina, Bower and Ward [70] prepared two samples of PET by hot drawing at 80° C. One had a draw ratio of 3.6 and a modulus of 6.6 GPa and the other a draw ratio of 4.6 with a modulus of 9.5 GPa. They found that there was a small but measurable shift of the 1616 cm^{-1} Raman band upon the application of stress and the largest shift (in terms of $d\Delta\nu/d\sigma$) was obtained for the sample of lower modulus. The actual values of the shifts are relatively small being typically less than 1 cm^{-1} but Fina and co-workers [70] showed that by careful experimentation such shifts were easily measurable. The shift in the peak position of the band was also accompanied by a band broadening. As in the case of most of the materials discussed so far this indicates that there is also a distribution of stresses on the molecules in the deformed PET samples.

Fina and co-workers [70] obtained most of their spectra with the polarizer and analyser parallel to the draw direction (and stressing direction) although they do report some results with them perpendicular to this direction. In this case they find that the shifts are only about one-third of those found when the polarizer and analyser are parallel to the direction of draw, although they did not report the direction of the shift, i.e. whether to higher or lower wavenumber.

It is interesting to note that the values of peak shift for oriented PET films are actually larger, by a factor of about two, than the shifts (in terms of $d\Delta\nu/d\sigma$) obtained for the similar band in aramids and rigid-rod fibres (sections 6.6 and 6.7) which were all of the order of $-3\,cm^{-1}/GPa$. The relatively high values of $d\Delta\nu/d\sigma$ for the PET films in comparison to the rather small actual shifts reflects the relatively low values of modulus of the PET samples. Fina *et al.* [70] pointed out that even though these different materials have different chain structures and different degrees of orientation the shift of the wavenumber of the main-chain phenyl-group stretching mode with applied stress is of the same order of magnitude for all the materials. This must clearly have implications for our understanding of molecular deformation processes in polymers. This work [70] has also shown that Raman spectroscopy could be a useful adjunct to the studies using conventional methods of the mechanical properties of polymers such as PET.

6.9 ISOTROPIC POLYMERS

It was shown in the last section that significant shifts in the positions of the Raman bands in PET with moderate degrees of orientation can be obtained and so the question that arises is whether measurable shifts in Raman bands can be obtained using isotropic polymers which contain little or no molecular orientation. In a little-quoted letter published in 1976, Evans and Hallam [73] reported measurable shifts to lower frequency in the wavenumbers of the bands in the Raman spectra of polymers such as polypropylene, polycarbonate, polystyrene and nylon 66 from samples which were presumably unoriented (although they did not give any details of specimen preparation). Shifts of the order of $1\,cm^{-1}$ were obtained for specimens deformed up to the point of specimen necking but the exact value of shift were found to depend upon the band in question [73]. They did not give any values of stress, but assuming a yield stress of their polypropylene of about 30 MPa, then the shifts they measured are of the order of up to $-30\,cm^{-1}/GPa$, which is quite significant. However, the relatively low yield stress of the material means that the magnitudes of the shifts will always be relatively small. Their data for polypropylene are reproduced in Table 6.7 [73] and it can be seen that between the point of yield and eventual specimen fracture, the behaviour of the Raman bands is relatively complex and that the wavenumber increases, decreases or stays the same for different bands. Unfortunately, they did not give sufficient details of their measurements to explain this effect. Evans and Hallam [73] pointed out the significant advantages of

Table 6.7 Frequencies (cm^{-1}) of selected bands in the Raman spectrum of mechanically loaded polypropylene (After [73])

	Stressed		
Unstressed	*Prior to neck*	*Prior to break*	*Band assignment*
2923.5	2923	2923	νC–H(CH_2*anti*)
2883	2883.5	2881.5	νC–H(CH_3*sym*)
2867.5	2867	2868.5	νC–H(CH_2*sym*)
2839.5	2839	2838	νC–H(CH_2*sym*)
1459	1458.5	1459.5	CH_2 bending
1435	1434	1435	CH_3 *antisym* bending
1151.5	1151	1152.5	Skeletal C–C
973.0	972.8	972.5*	Skeletal C–C
808.5	807.5	807.0	Skeletal C–C
320.0	320.0	321.5	Skeletal C–C bending
105.5	104.5	Absent	Lattice vibration?

* For comparison purposes this peak was studied on scale expansion and values are quoted to 0.1 cm^{-1} and were the mean of 10 runs. Other values are quoted to 0.5 cm^{-1} and were the mean of at least three runs.

the Raman technique over similar infra-red measurements and although a full publication of their work was promised, no record of its publication has been found.

Over recent years, a completely different approach has been adopted by Hu, Day, Stanford and Young [74–77] who have shown that, through the synthesis of specially designed co-polymers, it is possible to prepare isotropic polymers for which the deformation can be followed using Raman spectroscopy. They have demonstrated that such materials can be used for the study of polymer surface and interface deformation [76, 77] and their work in this area is reviewed later.

6.9.1 Urethane–diacetylene co-polymers

The approach that has been adopted [74–77] is to prepare a series of urethane–diacetylene co-polymers in which polydiacetylene units are incorporated into segmented co-polyurethanes. It was shown in section 6.3 that large strain-induced Raman band shifts can be obtained during the deformation of polydiacetylene single-crystal fibres [6–12]. In fact it is found that the largest shift measured so far ($-20\ cm^{-1}$/% strain) is for the C≡C triple-bond stretching band of polydiacetylenes (Table 6.3).

Polyurethanes constitute a versatile class of materials ranging from soft elastomers to glassy resins and are readily produced by a variety of processes into many different forms. Fibres, films, bulk sheets and

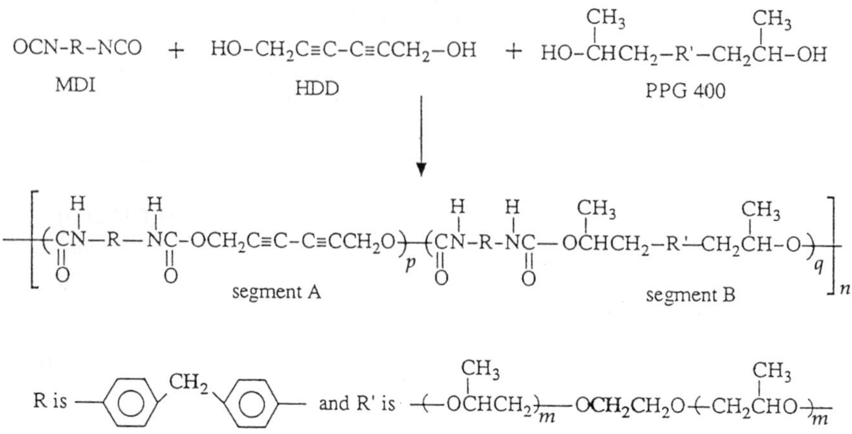

Fig. 6.14 Formation of linear-segmented co-polyurethanes comprising diacetylene–urethane hard segments, A, and polyether–urethane segments, B.

surface coatings can all be formed either from solution or in bulk. Segmented co-polyurethanes, because of their phase-separated structures, are particularly attractive since they enable the combination of disparate polymer properties to be obtained within a single material. Polydiacetylene single crystals are formed by the rapid solid-state polymerization [13–17] of substituted diacetylene monomer crystals on the application of heat or radiation. It is possible to induce similar solid-state reactions known as 'cross-polymerization' in the diacetylene groups of repeat units in certain co-polyurethanes and co-polyesters [78–80]. Cross-polymerization within the crystalline diacetylene regions produces a network structure in which the chains connecting the polydiacetylenes are analogous to the substituent side groups in the polydiacetylene single crystals [12]. In this way, linear co-polyurethanes containing phase-separated, diacetylene-containing domains, can be cross-linked *in situ* and transformed into insoluble and infusible materials [81–83]. These previous studies were concerned only with elastomeric materials formed via a two-stage solution process [81–83] at only one composition. The more recent studies [74–77] have been concerned with the development of more rigid co-polymers formed by a relatively simple one-shot bulk polymerization route.

A reaction scheme that can be used to prepare the co-polyurethanes is shown in Fig. 6.14. The reactants used are 4,4′-methylenediphenylene di-isocyanate (MDI), 2,4-hexadiyne-1,6-diol (HDD) and a polypropylene glycol (PPG400). It is possible to vary the structure and consequent properties of the materials by varying the relative proportions of HDD and PPG400. The exact details of the reaction conditions are given elsewhere [75] and linear polymers can be produced with molar masses of the order of 10000 g mol^{-1}. These linear segmented urethane co-polymers are soluble in a variety of solvents and can therefore be processed into a variety of forms such as surface coatings [76].

It can be seen from the reaction (Fig. 6.14) that the structure of the material consists of diacetylene–urethane hard segments, A, and poly-ether–urethane segments, B. The development of this alternating segmented structure results in phase separation due to incompatibility of the chemically distinct segments (A and B). Thermodynamic incompatibility depends primarily on the interaction parameter between the diacetylene- and polyether-based segments (determined by their intrinsic solubility parameters), and their sequence lengths (degrees of polymerization). The development of hydrogen bonding and potential crystallinity of the hard segments further enhance the driving force for phase separation. The linear co-polyurethanes thus form as essentially a two-phase morphology consisting of rigid, highly hydrogen-bonded hard segment domains (with a distribution of sizes) dispersed in a ductile, polyether–urethane phase. The formation of linear diacetylene-containing co-polyurethanes provides the distinct advantage, in subsequent applications, of enabling the co-polymers to be processed from solution. During or after removal of solvent, the phase-separated co-polymers can be rapidly cross-linked *in situ*, using heat or radiation, either of which causes cross-polymerization of diacetylene units within the hard segment domains. The basic chemical reaction of the diacetylene unit is shown simplistically in Fig. 6.15 in the terms of a single full-extended chain.

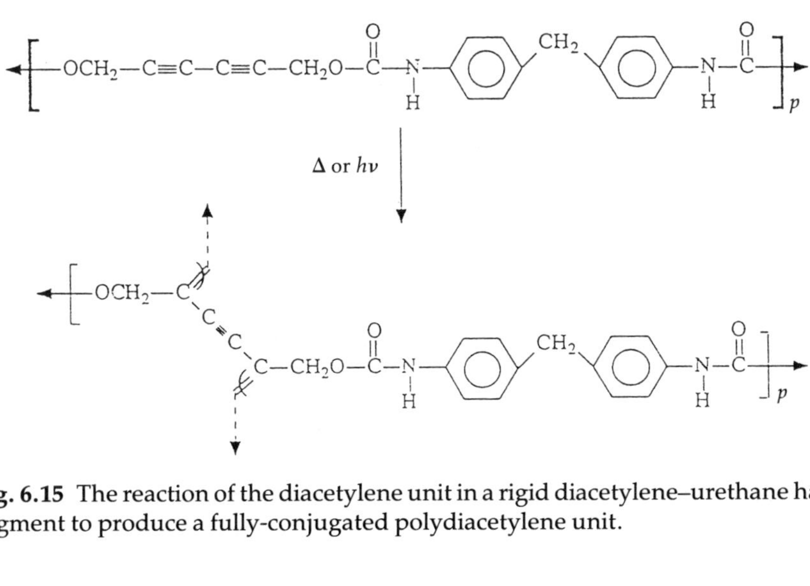

Fig. 6.15 The reaction of the diacetylene unit in a rigid diacetylene–urethane hard segment to produce a fully-conjugated polydiacetylene unit.

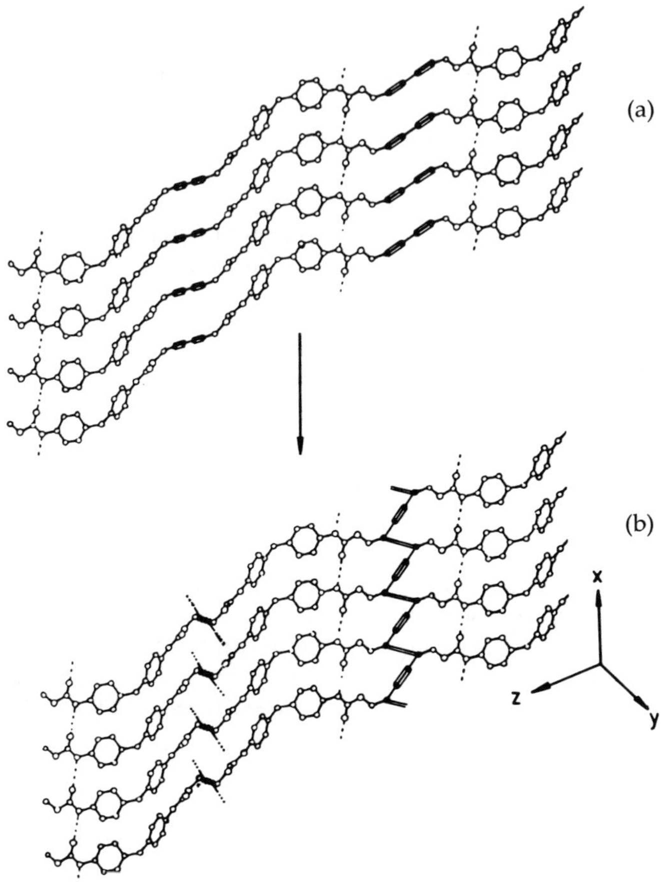

Fig. 6.16 Schematic representation of the solid-state topochemical polymerization of the diacetylene–urethane hard segments. (a) Linear-segmented block co-polymer; (b) cross-polymerized material. (After [75]).

In practice, however, the solid-state topochemical reaction involves many such chains packed within the hard segment domains, and the resulting cross-polymerization occurs three-dimensionally as depicted in Fig. 6.16. The diacetylene–urethane hard segments are assumed to be crystalline and have fully extended conformations in which the HDD unit is all-*trans*, and the chains are staggered so that adjacent chains are linked by straight C=O .... H–N hydrogen bonds in both directions perpendicular to the urethane chain axes. It is found [76] that these hard segment domains are organized in the form of spherulitic entities which are seen to be of the order of 1 μm in diameter using transmission electron microscopy.

The idealized structure in Fig. 6.16, however, is unlikely to be totally representative of the overall structure actually obtained for the hard segment domains, although regions of such three-dimensional order must exist within domains dispersed throughout the co-polyurethanes in order to achieve overall cross-polymerization. The formation of fully conjugated polydiacetylene (PDA) chains within the phase-separated co-polyurethanes produces dramatic colour changes (white → red → deep-purple) and transforms the co-polymers into completely insoluble and infusible materials. The extent of cross-polymerization that is achieved depends upon a number of factors such as the time and temperature of heating, the concentration of hard segment domains and the degree of order within the domains [75].

The relationship between chemical composition, structure and properties for the co-polymers has been described in detail elsewhere [76, 77]. In general it is found that the glass transition temperature, T_g and Young's modulus, E increase with hard segment content and heat treatment temperature. It was found that the material with the optimum composition and properties had a value of T_g of about 80° C and a Young's modulus (isotropic) of about 1.7 GPa, both of which are typical of a conventional glassy polymer.

6.9.2 Deformation studies

A Raman spectrum for a sample of the cross-polymerized co-polyurethane is shown in Fig. 6.17 and it can be seen that it has four main

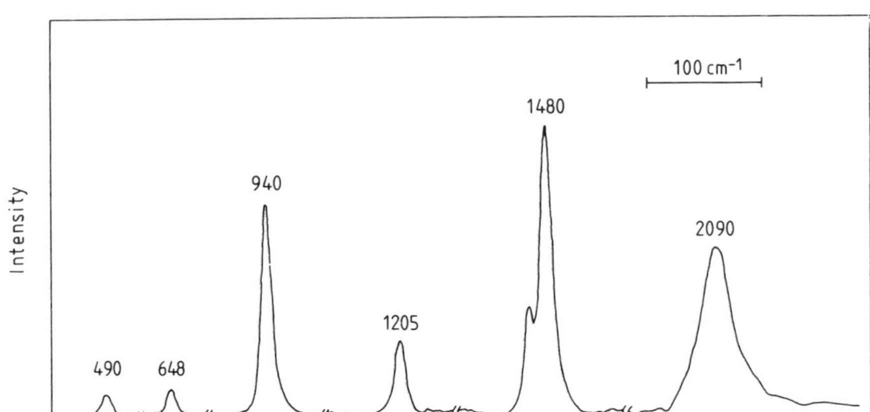

Fig. 6.17 Full Raman spectrum for the cross-polymerized diacetylene–urethane co-polymer. (After [75, 76]).

(a)

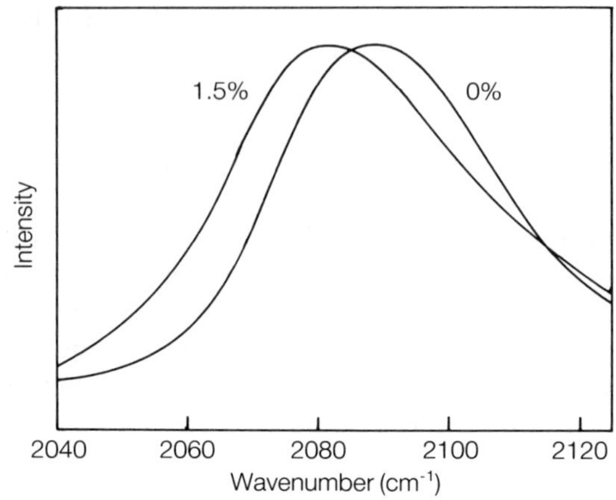

(b)

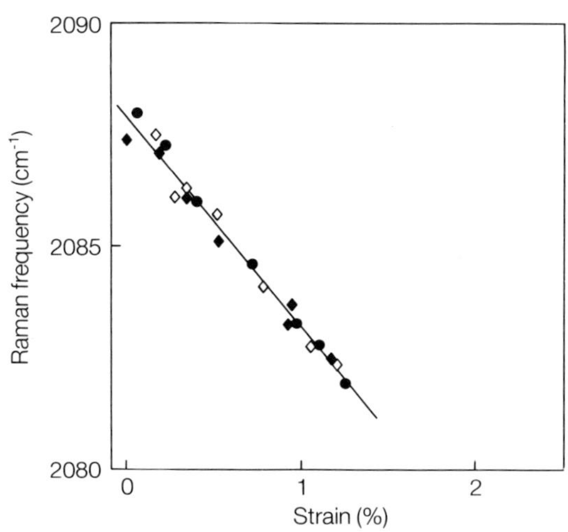

Fig. 6.18 (a) Shift of the 2090 cm^{-1} Raman band with strain for the cross-polymerized urethane–diacetylene co-polymer. (After [76]). (b) Dependence of the peak position of the 2090 cm^{-1} Raman band upon strain (♦–1st loading, ●–unloading, ◊–reloading). (After [76]).

scattering bands [75, 76]. The spectrum is remarkably similar to that obtained from a polydiacetylene single-crystal fibre [9]. For the fibres, a strong spectrum is only obtained when the fibre axis is parallel to the direction of polarization of the laser beam whereas it is found that the spectrum from the co-polymer is identical for all orientations of the sample [75, 76]. This shows clearly the isotropic nature of the co-polymer. It was found [75, 76] that the intensities of the bands in the Raman spectrum varied with both the hard segment content and heat treatment temperature and moreover it was demonstrated that this variation could be used to follow the cross-polymerization reaction. In particular the C≡C triple-bond stretching band at about 2090 cm^{-1} is not present before cross-polymerization and is indicative of the formation of the poly-diacetylene chains in the structure.

Figure 6.18(a) shows the position of the C≡C stretching Raman band for one of the co-polymers in the undeformed and deformed states [76]. Upon deformation there is a pronounced shift to lower frequency and slight broadening of the Raman band. This shift is a clear indication of stress transfer from the polyether–urethane matrix to the diacetylene–urethane hard segments and shows that this stress is translated into direct deformation of the polydiacetylene chains. Broadening of the bands suggests that a distribution of stresses is developed within the non-uniform hard segments domains during deformation.

The dependence of the band position upon tensile strain is shown in Fig. 6.18(b). It can be seen that it is approximately linear up to about 1% strain with a slope, $d\Delta\nu/de$, of the order of about $-5\,cm^{-1}/\%$ strain and is also completely reversible on both unloading and reloading. The slope of the line in Fig. 6.18(b) is significantly lower than that for polydiacetylene single-crystal fibres (Table 6.3) but is comparable to that of high-performance fibres such as those based upon aromatic polyamides or rigid-rod polymers.

It has been found that the relationship shown in Fig. 6.18(b) can be used to measure strain in a wide variety of situations. It is known that polydiacetylenes absorb visible light very strongly and so, for the bulk co-polyurethane, the spectrum is only obtained from material in the surface regions. Hence any strain measurements will be only for surface material. Moreover, since it is possible to focus the laser beam in the spectrometer to a spot of the order of 2 μm in diameter it is possible to obtain considerable spatial resolution.

Various examples of using these materials for surface-strain mapping have been presented in a recent publication [77]. The determination of stress concentrations around defects such as holes or notches in a deformed plate of the co-polyurethane is shown in Fig. 6.19. A circular hole and a notch of pre-determined dimensions were accurately machined into a 3 mm thick specimen of the co-polymer. The specimen was deformed in tension in the Raman spectrometer and the change in position of the C≡C stretching band at 2090 cm^{-1} with co-polymer strain (measured remotely with a resistance strain gauge) was determined at different positions around the defects, as shown schematically (inset) in Fig. 6.19. The slope ($d\Delta\nu/de$) of each line, relative to that for the remote applied deformation data (open O) is proportional to the stress concentration at each position. For the hole, the highest stress concentration, as expected is at the equator (solid ●) and is essentially zero at the pole (solid ◆). For the notch, the stress concentration (open ◊) increases sharply

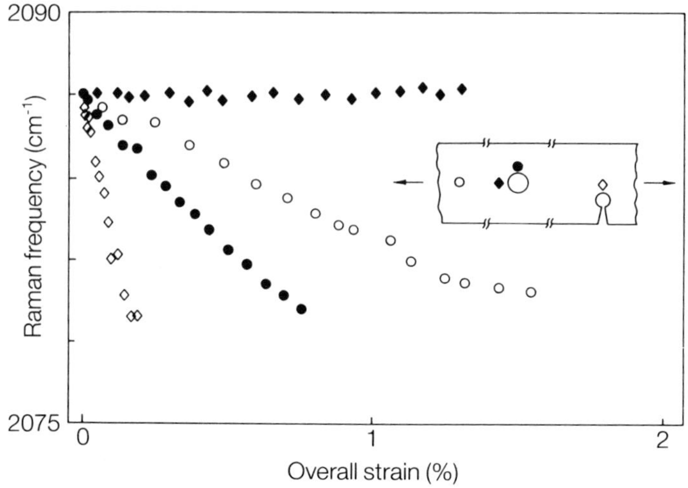

Fig. 6.19 The effects of stress concentrations on the peak position of the C≡C Raman band for a 3 mm thick cross-polymerized diacetylene–urethane plate deformed in tension. The data were obtained from spectra obtained at the different positions indicated. (After [77]).

depending on the notch tip radius and distance from the tip. The results obtained for the various defect geometries [77] show the stress concentration values measured by Raman spectroscopy to be very similar to those determined from conventional stress analyses [84].

The good solubility and adhesive characteristics of the diacetylene-containing co-polyurethanes make them attractive materials for subsequent use as surface coatings that can be applied with controlled thickness to a variety of substrates. Subsequent cross-polymerization, *in situ*, would then convert the coatings into cross-linked materials with strain-sensitive properties that can be determined quantitatively, in conjunction with the substrate, using Raman spectroscopy. To illustrate this use [76], a solution of the co-polyurethane was applied at 0.05 mm coatings to the following substrates: (a) a sheet of highly cross-linked (non-diacetylene containing) polyurethane resin, (b) an inorganic glass filament (≈25 μm diameter) and (c) a sheet of aluminium. The solvent was removed by evaporation and the coatings were thermally treated at 100° C for 40 hours. The coated substrate specimens were deformed in tension in a Raman spectrometer and the shift in the position ($\Delta\nu$) of the C≡C triple-bond stretching band was monitored as a function of the overall specimen strain, e. Excellent linearity between $\Delta\nu$ and e was obtained in each case. The strain sensitivities of the three coated substrates determined from the slopes of the plots, are given in Table 6.8 and the values of $-5\,\mathrm{cm}^{-1}/\%$ strain for $\mathrm{d}\Delta\nu/\mathrm{d}e$ are almost identical to that

Table 6.8 Strain sensitivities of the Raman frequencies of the C≡C triple bond stretching band for the diacetylene–urethane co-polymer and for the co-polymer coated on different substrates (after [76])

Substrate	$d\Delta\nu/de$ (cm^{-1}/%)
Pure co-polymer	5.3 ± 0.4
Polyurethane	5.5 ± 0.4
Glass fibre	5.7 ± 0.4
Aluminium	5.5 ± 0.4

of the bulk sheet material. Clearly, these results demonstrate that the co-polymers can be used as coating to monitor accurately the deformation of a substrate, which is particularly useful if the substrate is of complex geometrical shape or is not readily accessible for direct measurements. As such, the polydiacetylene-containing co-polyurethanes are shown to behave as optical strain gauges.

6.10 COMPOSITES

It has been demonstrated that for a very wide range of high-modulus polymeric fibres there is an approximately linear shift of the wavenumbers of bands in their Raman spectra upon the application of stress and strain. Although this has been used to give a detailed understanding of the molecular processes taking place during the deformation of the fibres, the phenomenon has an equally important application in the understanding of deformation micromechanics in composites reinforced with such fibres. This is an area which is currently attracting considerable attention in several laboratories and it is likely to be a subject of active research for several years to come. Nevertheless, there are already a number of publications in the literature concerning the application of Raman spectroscopy to composites and this work is discussed in detail below.

6.10.1 Single-fibre composites

The first investigations of the application of Raman microscopy to the analysis of composite micromechanics was undertaken by Young, Galiotis, Batchelder and co-workers [85, 86] and followed an earlier study [87] where they had shown that the technique could be used to determine strains within an adhesive bond between two polydiacetylene single-crystal fibres. They looked in detail at the deformation of single-fibre composite specimens consisting of polydiacetylene (polyDCHD–Table 6.2) single-crystal fibres in the centre of epoxy resin tensile testing bars with their axes parallel to the direction of the application of the tensile stress. It is found that well-defined Raman spectra can be obtained at different points on the fibre using a focused laser beam or preferably a Raman microscope system. The polyDCHD fibres give strong resonance Raman scattering whereas epoxy resins give only weak Raman scattering and are generally free from significant fluorescence unless they contain additives or modifiers. Hence, good spectra can be obtained using relatively low-powered lasers (e.g. 5 mW He–Ne laser [85, 86]) from polydiacetylene fibres covered by several millimetres of epoxy resin.

It was shown in Fig. 6.4 that the wavenumbers of the Raman-active bands in polydiacetylene single-crystal fibres are particularly sensitive to strain, with the band for C≡C triple bond stretching mode shifting by $-20\,\text{cm}^{-1}/\%$. Since the positions of the band peaks can be determined to better than $\pm 1\,\text{cm}^{-1}$, plots such as those shown in Fig. 6.4 can be used as strain gauge calibration curves for the fibres in a composite and fibre strain can be measured to a precision of better than ±0.05%. Since it is possible using a microscope system to focus the laser beam to a spot size of as small as 1 μm in diameter, then strain can be measured from point-to-point along a fibre in very small steps. This is demonstrated in Fig. 6.20 where that variation of fibre strain with position along the fibre is given at three levels of the applied matrix strain, e_m (measured using a conventional resistance strain gauge) for a single-fibre epoxy composite specimen cured at room temperature [86]. It can be seen that when the matrix is deformed the strain increases from the end of the fibre and then saturates along the central portion of the fibre such that the fibre and matrix strains are equal ($e_f = e_m$). The length over which the fibre strain rises from the value at the end to that of the central region is equivalent to half of the 'critical length', l_c. This length is closely related to the critical length determined from fibre pull-out tests [88] although the Raman technique enables it to be measured elastically without fracturing the fibres.

Figure 6.20 is a vivid illustration of the power of the Raman technique to measure strain (or stress if the material is calibrated in terms of $d\Delta\nu/d\sigma$) distributions in fibres in composites. The fundamental mechanism of reinforcement in composites is the transfer of stress from the matrix to the fibre through shear stresses at the interface [88]. Previous investigations of stress transfer in model composites had been concerned virtually exclusively with the analysis of the stresses in the matrix using techniques such as photoelastic stress analysis [89] whereas the Raman technique gives a direct measure of the stress or strain in the fibre to as high degree of accuracy and spatial resolution. The exact position at which the strain is

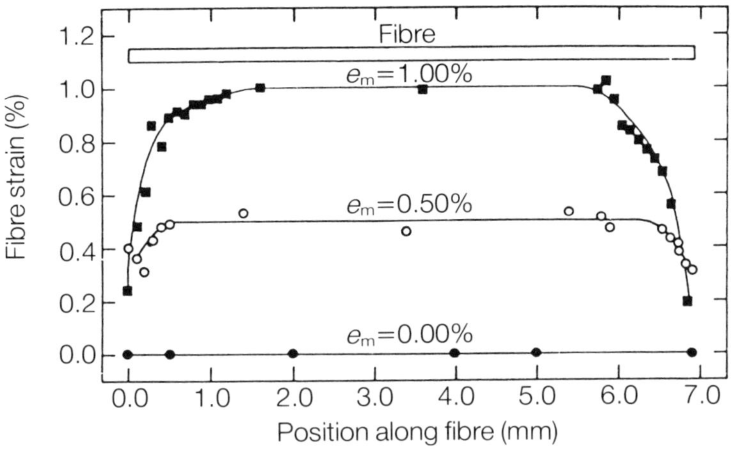

Fig. 6.20 Variation of strain, measured using Raman spectroscopy, with position along the fibre for a single fibre of polyDCHD in an epoxy resin bar subjected to the indicated levels of matrix strain e_m. (After [86]).

measured depends upon the optical skin depth of the fibre materials. Because the effective skin depth for Raman scattering in the polyDCHD fibres used to obtain Fig. 6.20 is around 10 nm for the He–Ne laser line employed, the Raman technique in this case measures only strains at the surface of the fibres. In less absorbing materials, strains will be measured in material at greater depths.

Because of the previous difficulties in measuring fibre stresses and strains directly there have been several attempts to model the transfer of stress from the fibre to matrix in discontinuous fibre-reinforced composites. One of the most widely used analytical approaches is the so-called 'shear-lag' model of Cox [88, 90]. The development of advanced computers has enabled numerical methods such as finite element analysis [91] or the finite difference approach [92] to be employed to model stress distributions in both the fibres and matrix. It is found that the results of the direct measurements of the strain distributions in fibres [85, 86] are in approximate agreement with the calculated distributions [90–92] but differ considerably in detail. For example in the Cox model [90], it is assumed, to simplify the mathematics, that there is no adhesion across the fibre ends and so the model predicts that the stress at the ends of the fibres falls to zero. It can be seen from Fig. 6.20 that the strain in the deformed fibres has a finite value at the fibre end due to traction across the end faces of the fibres. In contrast the numerical analyses, which do not require the simplifying assumption to be made, do predict the correct form of behaviour. Another discrepancy between the experimental data and the models is the increase in l_c with e_m shown in

Fig. 6.20. Such behaviour is not expected from any of the analyses. However, they all assume linear elastic deformation of both the fibre and matrix and do not take into account matrix yielding. It is found that there are considerable stress concentrations in the matrix in the vicinity of the fibre ends and so it is likely that shear yielding will take place in the matrix at even modest levels of e_m. This would be expected to lead to the increase in l_c with e_m which is observed in Fig. 6.20.

The variation of critical length l_c with fibre diameter is shown in Fig. 6.21. It has been shown [85] that both the model of Cox [90] and dimensional analysis predict the linear relationship which is obtained. The finite difference calculations of Termonia [92] have enabled the variation of l_c with fibre diameter to be determined as a function of the ratio of fibre modulus to matrix modulus and this is given as the theoretical line in Fig. 6.21 for the modulus values of the polyDCHD/ epoxy resin system employed. It can be seen that the theoretical analysis predicts a higher slope than is found experimentally. The discrepancy is not surprising since the finite difference method assumes linear elastic behaviour and no yielding. Clearly the theoretical analyses will have to be modified to take into account non-linear and plastic deformation before real composite systems can be modelled properly.

Robinson *et al.* [86] were careful to point out that the critical length l_c measured using the Raman technique is not the same as the critical length measured from pull-out tests [93] which is often given the same symbol. In the latter case fibres are either pulled out from a matrix or fractured and the pull-out length depends upon the ratio of the fibre strength to the interfacial shear strength. The value of l_c measured using the Raman technique as seen in Fig. 6.20 is really an elastic transfer length and is principally a function of the values of the fibre and matrix modulus. The situation may be somewhat different when the interface between the fibre and matrix starts to break down, for example when specimens are cycled to high strains. In this case [85] it is found that a debonded region develops at the end of the fibre, and l_c and the distribution of the strain in the fibre changes.

Hsu and co-workers [94, 95] have reported the results of a set of experiments upon epoxy resin samples containing polyPUHD (Table 6.2) polydiacetylene single-crystal fibres which are very similar to those reported originally by Young, Galiotis, Batchelder and co-workers [85, 86]. Their results are essentially the same but they also looked in detail at the effect of test temperature and the geometry of the fibre ends upon the distribution of strain within the fibres. They showed that the negative thermal expansion coefficient of their polyPUHD fibres could lead to the generation of thermal stresses in the fibre within the epoxy resin bar on changing temperature [94]. In their more recent publication [95] they measured the strain distributions within fibres having different end

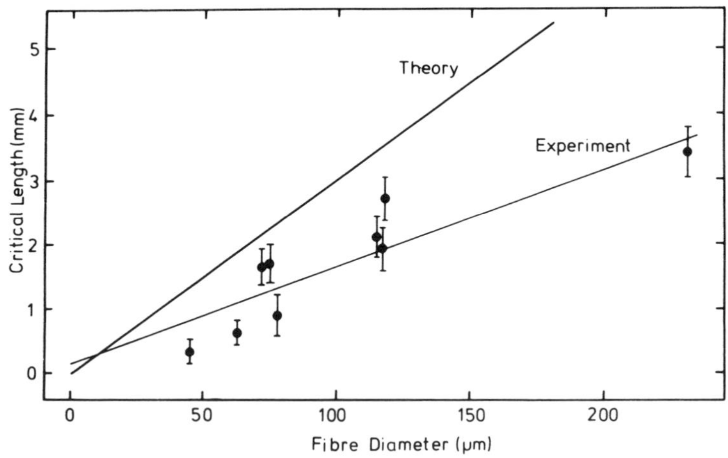

Fig. 6.21 Variation of critical length with fibre diameter for single fibres of polyDCHD in an epoxy resin bar (cf. Fig. 6.20). The experimental line is a least squares fit of the data (after [86]) and the theoretical line is due to Termonia [92].

geometries. They showed that when the fibre had a blunt end, the behaviour was as shown in Fig. 6.20 following a Cox-type distribution of strain at the end of a fibre with a tapered end. In this case they found that the strain was uniform along the whole of the fibre and did not decrease at the end. This is not surprising since it can be seen from Fig. 6.21 that the critical length decreases as the fibre diameter decreases and will become vanishingly small as the fibre tapers to a point.

It has been found that matrix shrinkage during curing has an important effect upon the micromechanics of the deformation of single-fibre composites. Most of the observations described earlier have been obtained in systems where the matrix has been cured at room temperature. For room-temperature-cured specimens the fibre strain, $e_f = e_m$ in the centre of the fibre but as the post-cure temperature is increased, e_f becomes significantly less than e_m. This mismatch in strain is found to correlate with the appearance of stress-induced twins in the fibres [96] which are caused by the shrinkage of the resin from the cure temperature down to room temperature. The degree of resin shrinkage can be determined to a high degree of accuracy [96] and this promises to be an excellent method of screening composite matrix resins to evaluate the levels of shrinkage at different cure temperatures.

Although the examples given earlier are primarily concerned with polydiacetylene single-crystal fibres in an epoxy resin, the study of model single-fibre composites is not necessarily restricted to such systems. Galiotis and co-workers [97] have recently reported the analysis of strain distributions in model composites consisting of single Kevlar 49 aramid fibres or HMS4 high-modulus carbon fibres in an epoxy resin. They found that in the case of the aramid, deformation to low strains produced the classical Cox-type strain distribution along the fibre (*cf*. Fig. 6.20). When the specimen was cycled to 1.6% strain the strain distribution along the fibre changed dramatically with l_c increasing significantly and the strain e_f increasing approximately linearly to a maximum at the fibre centre. They attributed this change in behaviour to the breakdown of bonding between the fibre and matrix.

In a second series of experiments, Galiotis and co-workers [97] followed the strain distribution in an HMS4 carbon fibre during a fragmentation test [98]. This is when single brittle fibres are loaded in a resin until they fracture, normally at fairly regular intervals along their length. Using the Raman technique it was shown [97] that debonding can take place at the ends of the fibres in the vicinity of cracks and that care must be taken in the interpretation of conventional fragmentation test data.

6.10.2 Fibre compression

The compressive properties of high-modulus polymer fibres are generally quite poor and fibres such as Kevlar can undergo failure by a kinking process at relatively low strain [99]. This is a severe limitation to their use in high-performance composites [100]. The kinking process is very similar to the compressive twinning that polydiacetylene fibres are found to undergo [96] and single-fibre composite specimens deformed in uniaxial compression have been used to follow the development of twins during the compressive deformation of polydiacetylene single-crystal fibres [101] and of kink bands in rigid-rod polymer fibres [102].

It is possible to employ the Raman technique to follow the full tensile/compressive deformation behaviour of single polymer fibres and Fig. 6.22 shows the variation of the peak position of the 1610 cm^{-1} Raman band for a Kevlar 49 fibre deformed in a resin in both tension (+ve strain) and compression (−ve strain) [103]. Since it was pointed out in section 6.6.1 that $d\Delta\nu$ is essentially a measure of molecular stress, Fig. 6.22 represents a molecular stress/strain curve for the Kevlar fibre. It can be seen that the behaviour is approximately linear in tension and at low strain in compression, although it has been found [103] for other aramid fibres that there can be significant curvature of the plots in the elastic region. What is most interesting is that in the compressive region there is a critical strain above which there is no further change in peak position. This represents the point at which the fibres undergo compressive failure by kink band formation [103]. It is extremely difficult to obtain such accurate information on the compressive behaviour of fibres using any other technique.

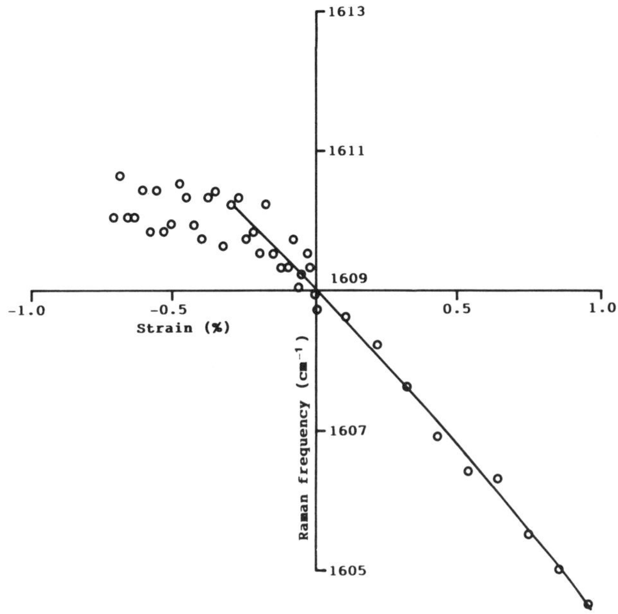

Fig. 6.22 Variation of the peak position of the 1610 cm^{-1} Raman band for a single fibre of Kevlar 49 subjected to tension and compression. (After [103]).

6.10.3 Interfacial micromechanics in polymer fibre reinforced composites

The transfer of stress across the fibre/matrix interface is a fundamental problem in the understanding of composite micromechanics. It is the subject of a large number of theoretical investigations [88, 90–93] and the Raman technique now allows the transfer of stress to be measured directly for real composite systems. An example of how this may be done will be reviewed for rigid-rod polymer fibres in an epoxy resin.

It was demonstrated in section 6.7 that for rigid-rod polymer fibres such as PBT, PBO and ABPBO (Table 6.5) the bands in their Raman spectra shift to lower frequency under the action of tensile stress or strain [64–66]. This gives invaluable information upon the way in which rigid-rod polymer molecules deform in the fibres, leading to the fibres having such impressive mechanical properties [64–66]. This section will describe the use of Raman spectroscopy to measure the point-to-point variation of stress within these three types of fibres in an epoxy resin composite [104, 105]. Examples will be given for six fibres, i.e. both 'as-spun' (AS) and 'heat-treated' (HT) forms of PBT, PBO and ABPBO in a cold-setting epoxy resin matrix. Full details of the structure and mechanical properties of the fibres are given elsewhere [64–66].

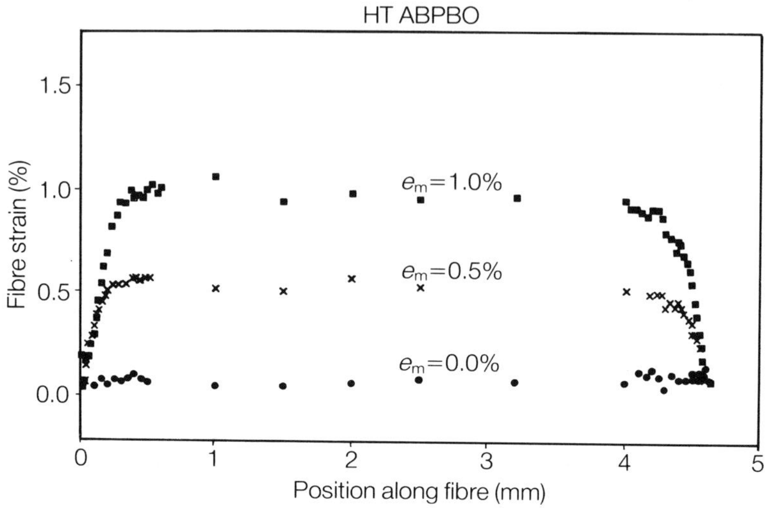

Fig. 6.23 Variation of the fibre strain with position along the fibre for a single-fibre HT ABPBO epoxy resin composite. The results are given for three different levels of matrix strain, e_m. (After [104]).

Single-fibre composite specimens were prepared consisting of individual fibres aligned in the axial direction in the centre of an epoxy resin tensile bar, cured at room temperature to avoid thermal stresses [86], as described in detail elsewhere [104, 105]. The dependence of the positions of the Raman bands for the six different fibres upon strain are summarized in Table 6.6 [64–66]. In general, it is found that for the HT fibres there is an approximately linear decrease in the frequency ($\Delta\nu$) of the band peak positions with strain, e, up to fracture whereas for the AS fibres there is a linear decrease up to fibre yield with the frequency remaining approximately constant when the fibre is strained beyond yield.

The dependence of the fibre strain upon position along an HT ABPBO fibre in the single-fibre composite is shown in Fig. 6.23. It can be seen that the behaviour is of the 'classical' Cox [90] shear-lag type with the strain rising from zero at the fibre ends to equal the matrix strain in the central region of the fibre. In addition, it can be seen that in the central region of the fibre, the fibre strain equals the matrix strain. This type of behaviour is typical of all the rigid-rod polymer fibres in the composites at low strain (<1%) [104, 105]. However, at higher strain levels the behaviour is quite different and can be summarized as follows.

a. In the AS fibres, **fibre yielding** occurs for $e > 1\%$ and it is found that the fibre strain tends to lag behind the matrix strain at higher strains.

b. In the HT fibres, **fibre fracture** tends to take place for $e > 1.3\%$ giving characteristic triangular distributions of strain in fibre fragments [98] – very different from the shear-lag profiles.
c. At high strain, **matrix yielding** and **interfacial failure** take place giving rise to trapezoidal distributions of strain along the fibres.

6.10.4 Interfacial shear stresses

Once the variation of the tensile stress and strain with position along the fibres is determined it is relatively easy to calculate the variation of the interfacial shear stress, τ_i, along the fibres [104, 105]. In the work of

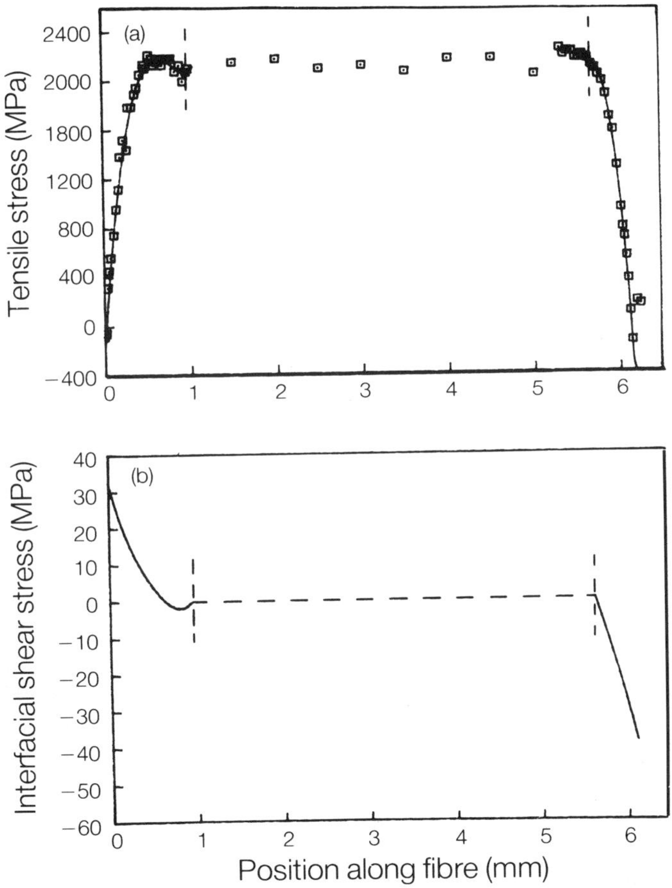

Fig. 6.24 (a) Variation of tensile stress with position along an HT PBT fibre in a single-fibre composite subjected to an overall strain of 0.8%. (b) Variation of interfacial shear stress with position along the fibre in (a) derived using equation 6.3. (After [105]).

Table 6.9 Maximum interfacial shear stresses ($\tau_i \pm 10\%$) for the different fibres (after [105])

Fibre	*AS PBT*	*HT PBT*	*AS PBO*	*HT PBO*	*AS ABPBO*	*HT ABPBO*
τ_i(MPa)	40	36	50	15	54	23

Ang [104] described here, this was done by fitting the variation of stress, σ, with position, x, to a 5th order polynomial and obtaining the derivative $d\sigma/de$. It can be readily shown [104, 105] that

$$2\tau_i = r\, d\sigma/dx \tag{6.3}$$

where r is the fibre radius. The results of this analysis for an HT PBT fibre in a single-fibre composite deformed to 0.8% strain are shown in Fig. 6.24. It can be seen that the variation of fibre stress with position is again classical [88–93]. In addition the variation of τ_i with position is also presented and it can be seen that τ_i is a maximum at the end of the fibres and then falls to zero over the central regions of the fibres. This is exactly the form of behaviour expected from the Cox-type shear-lag analysis [88–93]. It appears that the fibre-matrix adhesion for the HT PBO fibres was not as good as for the HT PBT ones as can be seen from Fig. 6.25. It can be seen that there is a trapezoidal distribution of tensile stress along the fibres and that τ_i peaks at some distance away from the end of the fibres, due probably to breakdown of the fibre/matrix interface. If it were due to matrix yielding, similar behaviour would have been expected for the PBT fibre composite in Fig. 6.24.

Young and Ang [105] performed a large number of experiments for composites containing the six types of fibres and deformed to different levels of overall strain. In each case the distribution of fibre stress along the fibres was measured and τ_i determined. The maximum values of τ_i for the six different fibres are listed in Table 6.9. As far as is known, none of the fibres were given any special surface treatments. It can be seen from the Table that τ_i is generally higher for the AS fibres than for the HT fibres. The low value of τ_i for HT PBO is consistent with the behaviour shown in Fig. 6.25.

It has been demonstrated that Raman spectroscopy is a powerful method of following the interfacial micromechanics of the deformation of polymer fibres within composites. It is possible to do this quantitatively and determine fundamental parameters such as the interfacial shear stress, τ_i. Although the technique has been demonstrated only for model single-fibre composites, it is clearly of considerably more general use and could be applied to a very wide range of high-performance fibres in a large number of matrix systems.

(a)

(b)

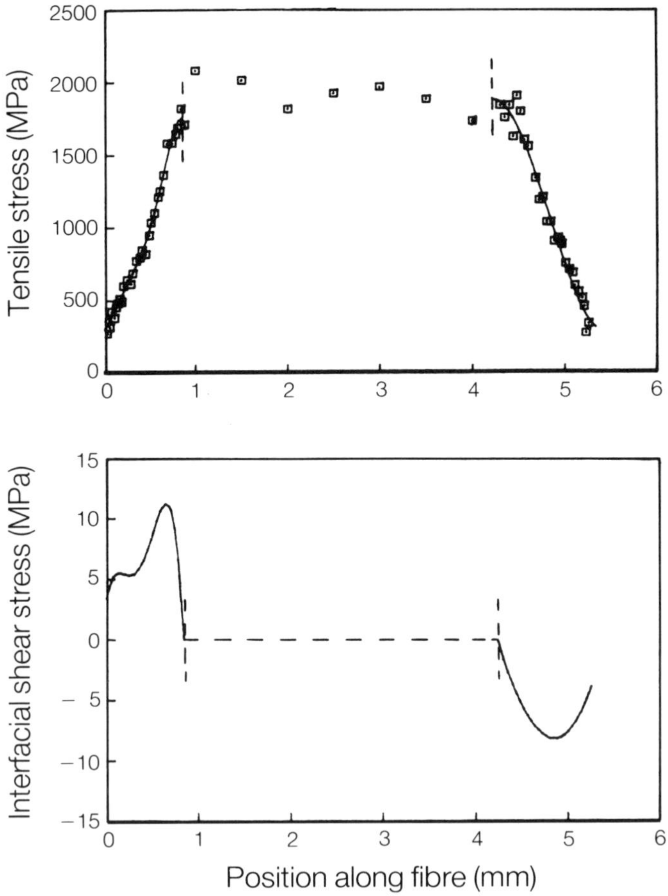

Fig. 6.25 (a) Variation of tensile stress with position along an HT PBO fibre in a single-fibre composite subjected to an overall strain of 0.8%. (b) Variation of interfacial shear stress with position along the fibre in (a) derived using equation 6.3. (After [105]).

6.10.5 High-volume fraction composites

The previous sections have been concerned exclusively with the behaviour of model single-fibre composites, but for engineering applications the composite systems of interest contain high-volume fractions (typically of the order of 60%) of fibres. The Raman technique, however, is also a powerful method of measuring strain in such systems although only a small amount of work has been published in this area so far.

One of the first examples of the potential was demonstrated by Robinson and co-workers [106], who showed that the strain in a glass-fibre reinforced epoxy resin could be monitored using a polyDCHD fibre placed among the glass fibres. The polydiacetylene fibre acted as an internal 'strain gauge'. They also showed that in this case l_c was different from the value for the same type of fibre in a single-fibre composite. A more sophisticated approach was adopted by Underwood, Sharpe and Batchelder [107] who placed polydiacetylene fibres within the plies of a carbon-fibre composite laminate during the lay-up procedure. Because the carbon fibres are not transparent they guided the light to their polydiacetylene fibre using a fibre optic and so were able to measure the internal ply strains at a specific point in the laminate.

The direct measurement of strain in fibres in a high-volume fraction composite using the Raman technique was first demonstrated by Young and co-workers [108] for Kevlar fibres in a cold-setting epoxy resin. The laser beam was somewhat defocused so that the average strain over several fibres was measured and very good agreement was found between the strains measured using the Raman technique and those measured with a resistance strain gauge. It is not difficult to think of many further applications of the technique for mapping strains in different types of composites.

A recent application of the technique has been to follow fibre deformation and determine the microscopic thermal stresses in carbon-fibre reinforced PEEK composites [109–111]. There is now considerable interest in carbon-fibre reinforced PEEK being used as a high-performance composite material [112, 113]. It is somewhat different from conventional polymer composite systems since it employs a high-temperature semi-crystalline thermoplastic matrix which gives the material significant advantages in certain applications. There are reports [114] of significant levels of residual macroscopic internal stresses in carbon-fibre reinforced PEEK, because the material is exposed to relatively high temperatures (of the order of 400° C) during fabrication. On cooling to room temperature, the PEEK matrix contracts its volume significantly due to both thermal contraction and crystallization. Since carbon fibres have a negative thermal expansion coefficient, this thermal mismatch can lead to significant macroscopic internal stresses in cross-ply laminates [114] and microscopic thermal stresses within the fibres and matrix in the plies. Since the Raman technique is able to measure fibre strains for carbon fibres [115], it can be potentially employed to measure strains in individual fibres within a ply in the composite. Care, however, must be exercised in analysing the data [116] since Raman peaks from the peak matrix superimposed on the carbon bands can give rise to apparent shifts not caused by internal stresses.

6.10.6 Molecular composites

Macroscopic composites normally consist of relatively large high-modulus fibres of the order of 5–12 μm in diameter in a low-modulus matrix and the structures produced are normally limited to plies of aligned fibres in uniaxial or cross-ply laminates [88]. The fabrication of such materials is generally tedious and the geometries of structures that can be produced are limited. There is currently considerable interest in producing 'molecular composites' which consist of a reinforcing phase of rod-like molecules dispersed on the nanometre scale in a matrix of flexible molecules [117–121]. It is found that good molecular composites are only formed when there is a fine dispersion of the reinforcing rigid-rod molecules on the molecular level [119–121]. Such materials can potentially be processed into structures which give one-, two- or even three-dimensional reinforcement in a wide variety of shapes and extend the basic principles of fibre reinforcement to the molecular level. A good fibre/matrix interface can be ensured through co-polymerization if necessary.

In spite of the attraction of such materials, only two successful molecular composite systems have been reported widely. One was developed in Japan and is based upon aramid molecules dispersed in a matrix of flexible molecules such as a nylon polymer [118]. Problems were encountered, however, with the aggressive solvents required for the aramid decomposing the matrix polymer, and the mechanical properties of such materials were found to be only relatively modest. A considerably more successful system has been developed by Helminiak and co-workers [119–121] under the USAF ordered-polymer programme. The system is based upon rigid-rod polymer molecules such as PBT (section 6.7) dispersed in a semi-flexible coil, high-temperature matrix resin such as poly(2,5(6)benzimidazole) (ABPBI). Fine dispersions of the rigid-rod molecules in the matrix polymer can be obtained by careful control of the properties of ternary solutions of the molecules in a solvent [119]. Molecular composites produced by such techniques can have very impressive mechanical properties and films of the composite have been reported to have modulus values of the order of 80 GPa [120] which compares well with aluminium, which has a modulus of 70 GPa.

The excellent mechanical properties of the molecular composite stem from the efficient transfer of stress from the flexible matrix to the reinforcing rigid-rod molecules. Since the deformation of molecules such as PBT can be monitored in high-modulus fibres using Raman spectroscopy [64], then it is possible to apply the same technique to follow their deformation in molecular composites. This has recently been demonstrated by Young, Day and Ang [122] for a 70/30 PBT/ABPBI molecular composite film. It was found that well-defined Raman spectra can be obtained from the PBT molecules in the molecular composite. The spectrum is very similar to that obtained from the same molecules in a high-modulus fibre. Mechanical deformation causes a shift in the wavenumber of the 1480 cm^{-1} Raman band to lower frequency and a broadening of the band [122]. This behaviour is similar to that of PBT high-modulus fibres [64], but in the case of the molecular composite $d\Delta\nu/de$ is only of the order of $-1\,cm^{-1}/\%$ for the material and so considerably lower than $d\Delta\nu/de$ for PBT fibres. However, the molecular composite film has an in-plane modulus of about 80 GPa and so $d\Delta\nu/d\sigma$ is of the order of $-1.25\,cm^{-1}/GPa$, which is about a quarter of the rate of shift of the $1480\,cm^{-1}$ band in PBT fibres (about $-4.8\,cm^{-1}/GPa$ [64]).

It is possible to make an initial attempt at modelling the mechanical behaviour of molecular composites using theories developed for macroscopic composite systems [88] since the film can essentially be thought of as a flexible matrix reinforced with 'fibres' in the form of the rigid-rod molecules or small clusters of molecules [119–121]. It is known that if the modulus of a composite consisting of uniaxially-aligned fibres is E_c then the modulus of the same volume fraction of fibres in an equivalent random in-plane composite would be $3E_c/8$ and that of the equivalent random three-dimensionally oriented composite would be $E_c/8$ [88]. If the molecular composite film is envisaged to be similar in structure to an in-plane random composite (fibre volume fraction, $V_f = 0.7$) and the PBT fibre similar to a uniaxially aligned composite ($V_f = 1.0$), then a modulus of the order of 80 GPa would be predicted for the molecular composite, which is very similar to the values found in practice [120, 121].

It is somewhat more difficult to model the shifts in the wavenumber of the Raman bands [122]. Since the direction of polarization of the laser beam was oriented parallel to the stressing direction, the spectra will come essentially from molecules oriented with their chain directions parallel to the stressing direction. It might therefore be expected that if the strain in the 'fibres' is the same as that in the specimen as a whole, then $d\Delta\nu/de$ should be the same for the PBT/ABPBI molecular composite as for the PBT fibres. This is certainly not the case [122] since Raman measurements show much larger frequency shifts for the PBT molecules in the fibres than for the molecules in the molecular composite. The reinforcing units, however, will have anisotropic elastic properties with a high modulus parallel to the molecular axis and a much lower transverse modulus. This makes stress analysis of the situation rather difficult to model in the molecular composite and it is clearly an area where more experimental and theoretical work is required.

6.11 CONCLUSIONS

It has been demonstrated that Raman spectroscopy is a very powerful technique for following the deformation of high-performance fibres and composites and may also be of use in the study of the deformation of isotropic polymers. It has been shown that the shift in the Raman bands is related directly to deformation of the bonds in the polymers and so the technique offers a unique method of following molecular deformation in polymers. The relationship between the shift in a particular Raman band and stress or strain can be used to map deformation in a wide variety of systems. For example the point-to-point variation of strain in the high-performance fibres in a composite can be determined with a spatial resolution of the order of 2 μm which enables the micromechanics of composite deformation to be studied with a resolution and precision which was hitherto unobtainable. Smart polymer coatings have been developed which allow strain mapping in the substrates with a similar level of resolution. It is clear that the Raman technique is in its infancy and over the years to come there will be significant developments in instrumentation and its application to different materials which will allow further significant advances to be made.

ACKNOWLEDGEMENTS

The author is grateful to a large number of colleagues and research students in London and Manchester who have helped with his work upon the development of the Raman spectroscopy for the study of mechanical properties in materials. Part of this chapter was prepared during study leave at EPFL in Lausanne, Switzerland and he would like to thank Professor H.H. Kausch for his hospitality and helpful discussion.

REFERENCES

1. Batchelder, D.N. (1988) *Euro. Spectrosc. News*, **80**, 28.
2. Young, R.J., Lu, D. and Day, R.J. (1991) *Polym. Int.* **24**, 71.
3. Zimba, C.G., Hallmark, V.M., Swalen, J.D. and Radbolt, J.F. (1987) *Appl. Spec.* **41**, 721.
4. Purcell, F.J. (1990) *Spectrosc. Int.* **1**, 33; (1989) *Spectroscopy*, **4**, 24.
5. Batchelder, D.N., Cheng, C. and Pitt, G.D. (1991) *Advanced Materials*, **3**, 566.
6. Mitra, V.K., Risen, Jr., W.M. and Baughman, R.H. (1977) *J. Chem. Phys.* **66**, 2731.
7. Batchelder, D.N. and Bloor, D. (1979) *J. Polym. Sci., Polym. Phys. Ed.*, **17**, 569.
8. Galiotis, C., Young, R.J. and Batchelder, D.N. (1983) *J. Polym. Sci., Polym. Phys. Edn*, **21**, 2483.
9. Batchelder, D.N. and Bloor, D. (1984) 'Resonance Raman spectroscopy of conjugated macromolecules', in *Advances in Infrared and Raman Spectroscopy*, Vol. 11, (eds R.J.H. Clark and R.E. Hester), Wiley Heyden Ltd., Chichester.
10. Galiotis, C. (1982) 'Polydiacetylene single crystal fibres', Ph.D. Thesis, University of London.
11. Wu, G., Tashiro, K. and Kobayashi, M. (1989) *Macromolecules*, **22**, 188.
12. Young, R.J. (1987) 'Polymer single crystal fibres', in *Developments in Oriented Polymers – 2*, (ed. I.M. Ward), Applied Science, London.
13. Wegner, G. (1977) *Pure Appl. Chem.* **49**, 443.
14. Wegner, G. (1969) *Z. Naturforsch.* **24b**, 824.
15. Bloor, D., Ando, D.J., Preston, F.H. and Stevens, G.C. (1974) *Chem. Phys. Lett.* **24**, 407.

16. Bloor, D., Koski, L., Stevens, G.C., *et al.* (1975) *Mater. Sci.* **10**, 1678.
17. Galiotis, C. and Young, R.J. (1983) *Polymer*, **24**, 1023.
18. Galiotis, C., Read, R.T., Yeung, P.H.J. *et al.* (1984) *J. Polym. Sci., Polym. Phys. Edn.*, **22**, 1589.
19. Lewis, W.F. and Batchelder, D.N. (1979) *Chem. Phys. Lett.* **60**, 232.
20. Cottle, A.C., Lewis, W.F. and Batchelder, D.N. (1978) *J. Phys. C*, **11**, 605.
21. Treloar, L.R.G. (1960) *Polymer*, **1**, 95.
22. Baughman, R.H., Gleiter, H. and Sendfeld, N. (1974) *J. Polym. Sci., Polym. Phys. Edn*, **12**, 1511.
23. Capaccio, G. and Ward, I.M. (1974) *Polymer*, **15**, 233.
24. Capaccio, G., Gibson, A.G. and Ward, I.M. (1979) in *Ultra-High Modulus Polymers*, (eds A. Ciferri and I.M. Ward), Applied Science, London.
25. Zachariades, A.E., Mead, W.T. and Porter, R.S. (1979) in *Ultra-High Modulus Polymers* (ed. A. Ciferri and I.M. Ward), Applied Science, London.
26. Smith, P. and Lemstra, P.J. (1979) *Makromol. Chem.* **180**, 2983.
27. Smith, P. and Lemstra, P.J. (1981) *J. Mater. Sci.* **15**, 505.
28. Holliday, L. and White, J.W. (1971) *Pure Appl. Chem.* **26**, 545.
29. Kinloch, A.J. and Young, R.J. (1983) *Fracture Behaviour of Polymers*, Applied Science, London.
30. Sakurada, I., Ito, K. and Nakamae, K. (1966) *J. Polym. Sci.* **C15**, 75.
31. Clements, J., Jakeways, R. and Ward, I.M. (1978) *Polymer*, **19**, 639.
32. Nakamae, K., Nishino, T. and Ohkubo, H. (1991) *J. Macromol. Sci – Phys.* **B30**, 1.
33. Grubb, D.T. and Liu, J.J.-H. (1985) *J. Appl. Phys.* **58**, 2822.
34. Prasad, K. and Grubb, D.T. (1990) *J. Polym. Sci.: Part B: Polym. Phys.* **28**, 2199.
35. Roylance, D.K. and Devries, K.L. (1971) *J. Polym. Sci., Polym. Lett.* **9**, 443.
36. Wool, R.P. and Statton, W.O. (1974) *J. Polym. Sci., Polym. Phys. Edn*, **12**, 1575.
37. Wool, R.P., Bretzlaff, R.S., Li, B.Y. *et al.* (1986) *J. Polym. Sci. Polym. Phys. Edn*, **24**, 1039.
38. Tashiro, K., Wu, G. and Kobayashi, M. (1988) *Polymer*, **29**, 1768.
39. Prasad, K. and Grubb, D.T. (1989) *J. Polym. Sci., Polym. Phys. Edn*, **27**, 381.
40. Kip, B.J., van Eijk, M.C.P. and Meier, R.J. (1991) *J. Polym. Sci.: Part B: Polym. Phys*, **B29**, 99.
41. Kip, B.J., Tabaksblat, R., Roovers, W.A.C. *et al.* (1991) Rolduc Polymer Meeting, Rolduc, The Netherlands.
42. Wong, W.F. and Young, R.J. (unpublished work).
43. Barham, P.J. and Arridge, R.G.C. (1977) *J. Polym. Sci., Polym. Phys. Edn*, **15**, 1177.
44. Gibson, A.G., Davies, G.R. and Ward, I.M. (1978) *Polymer*, **19**, 683.
45. Brew, B. and Ward, I.M. (1978) *Polymer*, **19**, 1338.
46. Nakagawa, K., Konaka, T. and Yamakawa, S. (1985) *Polymer*, **26**, 84.
47. Wu, G., Tashiro, K. and Kobayashi, M. (1989) *Macromolecules*, **22**, 758.
48. Tanner, D., Fitzgerald, J.A., Riewald, P.G. and Knoff, W.F. (1989) in *High Technology Fibers – Part B*, (eds M. Lewin and J. Preston), Marcel Dekker, New York.
49. Schaefgen, J.R. in *The Strength and Stiffness of Polymers*, (eds A.E. Zachariades and R.S. Porter), Marcel Dekker, New York.
50. Kwolek, S.L., Memeger, W. and van Trump, J.E. (1988) in *Polymers for Advanced Technologies*, (ed. M. Lewin), VCH Publishers Inc., New York, p. 421.
51. Penn, L. and Milanovich, F. (1979) *Polymer*, **20**, 31.
52. Galiotis, C., Robinson, I.M., Young, R.J. *et al.* (1985) *Polymer Comm.* **26**, 354.

53. van der Zwagg, S., Northolt, M.G., Young, R.J. *et al.* (1987) *Polymer Comm.* **28**, 276.
54. Edwards, H.G.M. and Hakiki, S. (1989) *Brit. Polym. J.* **21**, 505.
55. Day, R.J., Robinson, I.M., Zakikhani, M. and Young, R.J. (1988) in *Integration of Fundamental Polymer Science and Technology – 2*, (eds P.J. Lemstra and L.A. Klientjens), Elsevier Applied Science, p. 571.
56. Young, R.J., Day, R.J., Dong, L. and Knoff, W. (1992) *J. Mater. Sci.* **27**, 5431.
57. Northolt, M.G. and van Aartsen, J.J. (1978) *J. Polym. Sci., Polym. Symp.* **58**, 283.
58. Northolt, M.G. (1980) *Polymer*, **21**, 1199.
59. Northolt, M.G. and van der Hout, R. (1985) *Polymer*, **26**, 310.
60. Lu, D. (1991) 'Structure/property relationships in aromatic polyamide fibres', Ph.D. Thesis Victoria University of Manchester.
61. Allen, S.R., Filippov, A.G., Farris, R.J. and Thomas, E.L. (1983) in *Strength and Stiffness of Polymers*, (eds A.E. Zachariades and R.S. Porter), Marcel Dekker, New York.
62. Allen, S.R., Farris, R.J. and Thomas, E.L. (1985) *J. Mater. Sci.* **20**, 2727.
63. Krause, S.J., Haddock, T.B., Vezie, D.L. *et al.* (1988) *Polymer*, **29**, 1354.
64. Day, R.J., Robinson, I.M., Zakikhani, M. and Young, R.J. (1987) *Polymer*, **28**, 1833.
65. Young, R.J., Day, R.J. and Zakikhani, M. (1990) *J. Mater. Sci.* **25**, 127.
66. Young, R.J. and Ang, P.P. (1992) *Polymer*, **33**, 975.
67. Venkatesh, G.M., Shen, D.Y. and Hsu, S.L. (1981) *J. Polym. Sci., Polym. Phys. Edn*, **19**, 1475.
68. Odell, J.A., Keller, A., Atkins, E.D.T. and Miles, M.J. (1981) *J. Mater. Sci.* **16**, 3309.
69. Minter, J.R., Shimanwa, K. and Thomas, E.L. (1981) *J. Mater. Sci.* **16**, 3303.
70. Fina, L.J., Bower, D.I. and Ward, I.M. (1988) *Polymer*, **29**, 2146.
71. Vettegren, V.I. and Novak, I.I. (1973) *J. Polym. Sci., Polym. Phys. Edn.*, **11**, 2135.
72. Mocherla, K.K.R. and Statton, W.O. (1977) *J. Appl. Polym. Sci., Appl. Polym. Symp.* **31**, 183.
73. Evans, R.A. and Hallam, H.E. (1976) *Polymer*, **17**, 838.
74. Stanford, J.L., Young, R.J. and Day, R.J. (1991) *Polymer*, **32**, 1713.
75. Hu, X., Stanford, J.L., Day, R.J. and Young, R.J. (1992) *Macromolecules*, **25**, 672.
76. Hu, X., Stanford, J.L., Day, R.J. and Young, R.J. (1992) *Macromolecules*, **25**, 684.
77. Hu, X., Day, R.J., Stanford, J.L. and Young, R.J. (1992) *J. Mater. Sci.* **27**, 5958.
78. Wegner, G. (1970) *Makromol. Chem.* **134**, 219.
79. Day, D. and Lando, J.B. (1981) *J. Polym. Sci., Polym. Lett. Edn.*, **19**, 227.
80. Patil, A.O., Desphande, D.D., Talwar, S.S. and Biswas, A.B. (1981) *J. Polym. Sci., Polym. Chem. Edn*, **19**, 1155.
81. Rubner, M.F. (1986) *Macromolecules*, **19**, 2114.
82. Liang, R.S. and Reiser, A.J. (1987) *J. Polym. Sci., Polym. Chem. Edn.*, **25**, 451.
83. Nallicheri, R.A. and Rubner, M.F. (1991) *Macromolecules*, **24**, 517.
84. Williams, J.G. (1973) *Stress Analysis of Polymers*, Longman, London.
85. Galiotis, C., Young, R.J., Yeung, P.H.J. and Batchelder, D.N. (1984) *J. Mater. Sci.* **19**, 1640.
86. Robinson, I.M., Young, R.J., Galiotis, C. and Batchelder, D.N. (1987) *J. Mater. Sci.* **22**, 3642.
87. Galiotis, C., Young, R.J. and Batchelder, D.N. (1983) *J. Mater. Sci. Lett.* **2**, 263.

88. Hull, D. (1981) *An Introduction to Composites*, Cambridge University Press, Cambridge.
89. Tyson, W.R. and Davies, G.J. (1965) *Brit. J. Appl. Phys.* **16**, 199.
90. Cox, H.L. (1952) *Brit. J. Appl. Phys.* **3**, 72.
91. Cararra, A.S. and McGarry, F.J. (1968) *J. Comp. Mater.* **2**, 222.
92. Termonia, Y. (1987) *J. Mater. Sci.* **22**, 504.
93. Kelly, A. (1970) *Proc. Roy. Soc.* **A319,** 95.
94. Kim, P.K., Xu, Y.Y., Chang, C. and Hsu, S.L. (1986) *Polymer*, **27**, 1547.
95. Fan, C.F. and Hsu, S.L. (1989) *Macromolecules*, **22**, 1474.
96. Robinson, I.M., Yeung, P.H.J., Galiotis, C. and Young, R.J. (1986) *J. Mater. Sci.* **21**, 3340.
97. Galiotis, C., Jahankhani, H., Melanitis, I. and Batchelder, D.N. (1989) *ECCM III*, (eds A.R. Bunsell, P. Lamicq and A. Massiah), Elsevier, London, p. 765.
98. Favre, J.-P. and Jacques, D. (1990) *J. Mater. Sci.* **25**, 1373.
99. DeTeresa, S.J., Allen, S.R., Farris, R.J. and Porter, R.S. (1984) *J. Mater. Sci.* **19**, 57.
100. Greenwood, J.M. and Rose, P.G. (1974) *J. Mater. Sci.* **9**, 1809.
101. Robinson, I.M. (1987) 'Optomechanics of polymer deformation', Ph.D. Thesis, University of London.
102. Day, R.J., Ang, P.P. and Young, R.J. (1989) in *Interfacial Phenomena in Composite Materials '89*, (ed. F.R. Jones), Butterworths, London, p. 121.
103. Lu, D. and Young, R.J. (Unpublished results).
104. Ang, P.P. (1991) 'Deformation micromechanics of high modulus polymer fibres and composites', Ph.D. Thesis, Victoria University of Manchester.
105. Young, R.J. and Ang, P.P. (1991) in *Interfacial Phenomena in Composite Materials '91*, (eds I. Verpoest and F.R. Jones), Butterworth–Heinemann, Oxford, p. 49.
106. Robinson, I.M., Young, R.J., Galiotis, C. and Batchelder, D.N. (1987) *ICCM VI & ECCM II*, (eds F.L. Matthews, N.C.R. Buskell, J.M. Hodgkinson and J. Morton), Elsevier, Amsterdam, p. 1–333.
107. Underwood, F., Sharpe, D. and Batchelder, D.N. (1989) *ECCM III*, (eds A.R. Bunsell, P. Lamicq and A. Massiah), Elsevier, London, p. 759.
108. Galiotis, C., Robinson, I.M., Batchelder, D.N. and Young, R.J. (1988) in *Engineering Applications of New Composites*, (eds S.A. Paipetis and G.C. Papanicolaou), Omega, Oxford, p. 409.
109. Young, R.J. and Day, R.J. (1989) *Brit. Polym. J.* **21**, 17.
110. Young, R.J., Day, R.J., Zakikhani, M. and Robinson, I.M. (1989) *Comp. Sci. Tech.* **34**, 243.
111. Galiotis, C., Melanitis, N., Batchelder, D.N. *et al.* (1988) *Composites*, **19**, 321.
112. Crick, R.A., Leach, D.C., Meakin, P.J. and Moore, D.R. (1987) *J. Mater. Sci.* **22**, 2094.
113. Davies, P., Cantwell, W.J. and Kausch, H.H. (1989) *Comp. Sci. Tech.* **35**, 301.
114. Jeronimidis, G. and Parkin, A.T. (1988) *J. Comp. Mater.* **22**, 401.
115. Robinson, I.M., Zakikhani, M., Day, R.J. *et al.* (1987) *J. Mater. Sci. Lett.* **6**, 1212.
116. Galiotis, C. (Private Communication).
117. Takayanagi, M. (1985), in *High Modulus Polymers and Composites*, (ed. C.L. Choy), The Chinese University Press, Hong Kong, p. 141.
118. Takayanagi, M. (1983) *Pure Appl. Chem.* **55**, 81.
119. Wiff, D.R., Helminiak, T.E. and Hwang, W.-F. (1988), in *High Modulus Polymers*, (eds A.E. Zachariades and R.S. Porter), Marcel Dekker, New York.
120. Krause, S.J., Haddock, T.B., Price, G.E. *et al.* (1986) *J. Appl. Polym. Sci., Polym. Phys. Edn*, **24**, 1991.

121. Krause, S.J., Haddock, T.B., Price, G.E. and Adams, W.W. (1988) *Polymer*, **29**, 195.
122. Young, R.J., Day, R.J. and Ang, P.P. (1990) *Polymer Comm.* **31**, 47.

7

NMR spectroscopy of solid polymer systems

C. Dybowski and A.J. Brandolini

7.1 INTRODUCTION

Over time, nuclear magnetic resonance (NMR) spectroscopy has proven to be a remarkably versatile tool for the description of molecular structure and dynamics. This versatility arises from the wide range of nuclear interactions affecting the NMR signal. Because these interactions depend on different nuclear and molecular parameters, one can manipulate them to extract very specific information. Since the development, in the mid-1970s, of solid-state pulse techniques, it has been possible to tailor an NMR experiment to highlight the specific interaction of interest [1–4], leading to a finer description of chemical structure, morphology, orientation, and chain dynamics [5–7].

In NMR, each individual interaction has particular advantages for the study of solid polymer systems. The origin of nuclear magnetic resonance lies in the effect of a strong static magnetic field on the weak intrinsic magnetic moment of a nucleus with non-zero nuclear spin. This Zeeman interaction determines the average resonance frequency of a particular nuclear species, permitting observation of specific elements without interferences from others present in the sample. Relatively smaller interactions than the Zeeman interaction, due to the coupling of nuclear moments with electronic angular momentum or neighboring nuclear spins, cause slight shifts in this frequency or splittings of the resonance, making NMR such a useful technique for chemical and structural identification. Direct dipolar couplings among like (homonuclear) or dissimilar (heteronuclear) spins reflect both the proximity and the relative motions of these spins. These couplings also enable spins in various regions of a sample to transfer information from one to another. Dipolar couplings can provide specific insights into the details of molecular motion, and changes in them can often be correlated with thermodynamic or mechanical transitions observed in a bulk material. The chemical shift causes distinct resonance positions for spins in unique electronic environments. It arises from coupling of nuclear moments to the orbital angular momentum of nearby electrons. It is inherently anisotropic since the local electronic structure is not spherically symmetric, endowing the chemical shift with the capacity to describe directionality of motion or structure. This anisotropy can be used to describe supramolecular features, such as crystallite orientation, that have a directional character. If the anisotropy is suppressed experimentally, distinct resonances appear for spins in chemically or crystallographically unique electronic environments. The isotropic chemical shift makes NMR spectroscopy a useful tool for the identification of chemical structure and morphology, and – in some cases – for focusing on particular nuclei of interest without a troublesome background signal. Certain nuclei having a non-spherical charge distribution (nuclear spin $I>1/2$) exhibit quadrupolar coupling to local electric-field gradients. When present, the effects of the electric quadrupole interaction on the NMR signal sensitively reflect both the rate and geometry of molecular motion.

When excited in a pulse experiment, the detectable magnetization has characteristic persistence constants, or relaxation times, that describe the kinetics of approach to the equilibrium state or the flow of energy into or out of a magnetic-energy reservoir. For example, at equilibrium in a magnetic field, all magnetization is parallel to the magnetic field direction; thus, the non-equilibrium state in which a magnetization exists that is perpendicular (or transverse) to the magnetic field will disappear, and this process takes a finite time. If the disappearance is exponential, one may describe the process by a relaxation time, in this case labelled T_2. Other magnetic parameters approach equilibrium on different time scales, and each has its characteristic relaxation time. The relaxation times, like other kinetic parameters, depend on the nature of the interaction that induces relaxation and on the thermal motion of the molecule. Thus, investigation of these NMR parameters provides insight into the rate and/or amplitude of molecular dynamics, and may also reflect the molecular basis of bulk properties.

The way in which some nuclear interactions are altered by the imposition of a gradient of the magnetic field can be exploited to provide spatially resolved images of the distribution of nuclei in a sample. By use of techniques to enhance the contrast of the images based on the values of NMR parameters like relaxation times, one may discern macroscopic distributions of mobile and relatively immobile components of a sample. NMR imaging, which is so highly refined as a medical technique, is still in its infancy as a tool in materials research. Spatial resolution is not yet good enough to allow routine observation of microscopic features, but several interesting and useful macroscopic studies have been made.

In this chapter, we discuss each interaction individually and provide several appropriate examples of its applications. These examples have been drawn from recent publications only as illustrations; we have not attempted to make a comprehensive survey of the solid-state NMR literature; several such reviews are available [8–15]. We have not presented detailed discussions of the theoretical bases and experimental considerations for solid-state NMR experiments; this information is available elsewhere [1–7]. Our aim is to highlight the information that can be extracted from the manipulation and observation of the effects of each interaction on the magnetization. The development of solid-state NMR has been a process of learning how to overcome the strong nuclear interactions inherent in a solid. The achievement of relatively well-resolved, 'liquid-like' resonances is made by discarding a great deal of information contained in these interactions. While these high-resolution solid-state spectra are very useful for identifying polymers or for gaining information about chemical structure, they are not necessarily an end in themselves. Through careful and judicious selection of experimental criteria, selected interactions can be observed independently. Through even more imaginative combinations of experiments, correlations between interactions can be represented in a two-, three- or even four-dimensional format [16–17].

7.2 ZEEMAN INTERACTION

One of the strengths of NMR spectroscopy, both in solution and in the solid state, is the ability to focus on one given nucleus at a time. Because of the selectivity of the Zeeman interaction, each distinct nucleus resonates in a narrow range around a frequency called the Larmor frequency, ω_0, as given in equation 7.1.

$$|\omega_0| = \gamma B_0 \tag{7.1}$$

In this equation, γ is a constant called the gyromagnetic ratio, and B_0 is the magnetic field strength. Not all isotopes are NMR-active. For example, the most abundant carbon nucleus, ^{12}C, does not show a nuclear magnetic resonance. The only NMR-observable isotope of carbon, ^{13}C, has a natural abundance of approximately 1.1%, resulting in a low signal from a sample of natural abundance. Many other chemically important, NMR-active nuclei such as ^{15}N are even rarer. Natural-abundance spectra can be recorded for such nuclear species, but synthesis of enriched polymers offers another possible, and often more practical, route to detection of NMR in these cases. In addition to these factors, nuclei with spins greater than 1/2 have a nuclear quadrupole moment, which generally leads to

Table 7.1 Properties of some commonly studied nuclei

Nucleus	*Spin*	*Natural abundance (%)*	*Frequency (MHz)*
^{1}H	1/2	99.99	100.0
^{19}F	1/2	100	94.1
^{31}P	1/2	100	40.5
^{7}Li	3/2	92.6	38.9
^{119}Sn	1/2	8.58	37.3
^{129}Xe	1/2	26.4	27.7
^{23}Na	3/2	100	26.5
^{27}Al	5/2	100	26.1
^{13}C	1/2	1.11	25.1
^{29}Si	1/2	4.70	19.9
^{2}H	1	0.02	15.4
^{17}O	5/2	0.04	13.6
^{15}N	1/2	0.37	10.1

broader spectra. Broad spectra correspond to a very rapid decay of the transverse magnetization, and such signals can be difficult to detect. Table 7.1 lists relevant properties for some of the more commonly studied nuclei [18]. Of the common nuclei, ^{1}H has the highest resonance frequency at a given field strength. Many chemically important nuclei such as ^{13}C, ^{29}Si and ^{27}Al have resonances in a band around a frequency near one quarter that of ^{1}H. From a practical standpoint, this means that the radio-frequency electronics for excitation and detection of these nuclei are similar, and resonances may often be detected with the same instrumentation.

Most published NMR experiments record either ^{1}H or ^{13}C signals, but in some applications, information can be obtained more readily from the spectrum of another nucleus, e.g. ^{31}P or ^{19}F. The nucleus may be incorporated into the polymer, contained in additives, introduced during modification, or infused into the solid polymer as a probe molecule. Many heteroatom-containing polymers have been studied by NMR: ^{31}P in polyphosphazenes [19] ^{15}N in polyamide [20–23] and silk [24] and ^{29}Si in polysiloxanes [25–27]. Since, in these examples, the heteroatom is located along the polymer chain, the spectra are particularly sensitive to backbone structure and dynamics. Thus, NMR spectra of these nuclei are more informative than ^{13}C spectra of pendant carbon-containing groups. Certain behaviors, such as cation clustering in ionomers (specifically, styrene-co-styrene sulfonate and ethylene-co-methacrylic acid) may be studied directly by observing the resonance of the counterion, such as ^{23}Na [28]. In favorable cases, features such as cross-link points can be investigated, as in thiophosphate-cross-linked propylene oxide networks [29]. The polymer contributes no background signal because the only phosphorus is located at the cross-link points; this considerably simplifies interpretation. Polymer additives, such as trioctyl phosphate plasticizer in poly(2,5-dimethyl-1,4-diphenylene oxide) and polystyrene [30] and organophosphite antioxidants in polyester blends [31] have also been investigated. Certain nuclei, most notably ^{129}Xe [32] are useful as probe molecules. In such studies, xenon gas is infused into the polymer under high pressure. Variable-temperature studies clearly illustrate features such as the increased polymer chain mobility in amorphous materials [33–35] and the phase structure of block copolymers [36].

7.3 DIPOLAR COUPLING

Nuclear magnetic moments may be thought of as bar magnets, in one approximation. At points in the vicinity of a nucleus, there is a component of the magnetic field whose origin is the dipolar field of this 'magnet'. Another nuclear spin in its neighborhood therefore experiences a coupling to the first spin through Zeeman interaction with this field. The magnitude of this interaction depends on the distance between the two spin sites, r, and the orientation of the internuclear vector relative to the direction of the magnetic field. Since the magnitude of the coupling is proportional to r^{-3}, measurement of the dipolar coupling is a precise way to determine internuclear distances. In addition, the dipolar coupling allows a mechanism for the exchange of energy with other modes of the molecule, particularly rotation, in both solid and solution states. Thus, measurement of relaxation times dominated by thermally induced modulation of the dipolar coupling provides a means for determining activation energies for rotational motion. In solution, it is fortunate that the rapid isotropic motion of small molecules produces an averaging of this interaction to zero, so that there is neither splitting of the resonance nor a shift in the signal's position due to dipolar coupling.

There are two types of dipolar coupling: homonuclear (like spins, primarily ^{1}H–^{1}H or ^{19}F–^{19}F) and heteronuclear (dissimilar spins, such as ^{13}C–^{1}H or ^{29}Si–^{1}H). Each type of interaction can be exploited to gain information about the spin system, but different experimental techniques are required to deal with each type.

Homonuclear dipolar couplings of abundant nuclei, such as ^{1}H or ^{19}F, are the most commonly studied. For rare spins, homonuclear interactions (for example, ^{13}C–^{13}C or ^{15}N–^{15}N) are less common unless the materials being studied are highly labeled [37]. These nuclei usually are predominantly affected by heteronuclear dipolar couplings to abundant nuclei in the neighborhood.

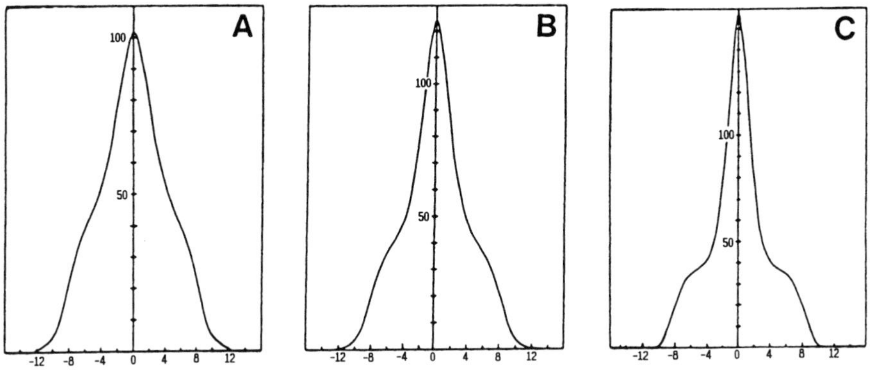

Fig. 7.1 Dipolar-coupled ^{19}F NMR spectra of polyvinylidene fluoride at (A) room temperature; (B) 50° C; (C) 80° C. (Reprinted from [38] by permission of Butterworth–Heinemann Publishers Ltd, © 1985.)

For abundant spins, a nucleus is subject not to a single dipolar coupling, but to dipolar couplings with many different nearby spins. Therefore, abundant-spin NMR frequently can only give average information on the strength of the dipolar coupling, meaning that only average distances can be specified from the NMR spectrum. One important aspect of homonuclear dipolar couplings among abundant nuclei is spin diffusion. If magnetization can be created in one part of a system that is spatially distinct from other parts, for example by creation of magnetization in only the amorphous parts of a polymer, then the natural trend is for the system to redress this gradient by a flow of magnetization to areas of depletion. This flow happens through the dipolar couplings among the spins, and the rate of this diffusion is determined by the magnitude of these couplings, which depend on the distances involved, the effects of relative nuclear motion and the strength of the coupling (as determined by the size of the gyromagnetic ratio). Thus, measurement of the rate of spin diffusion can give indirect information on these important parameters.

For semi-crystalline polymers, such as poly(vinylidene fluoride) (PVDF), the broad-line spectrum of abundant nuclei (i.e. a spectrum produced with no attempt to suppress couplings) consists of a broad component characteristic of the more rigid crystalline domains and a narrow peak due to the mobile amorphous regions. A typical ^{19}F spectrum is shown in Fig. 7.1 [38]. With increasing temperature, there is a distinct narrowing of the amorphous resonance and a reduction in the area of the crystalline peak. This is associated with conversion of rigid chains into mobile ones, and these line shapes can be deconvoluted to give the proportion of rigid and mobile chains. The line shape depends on

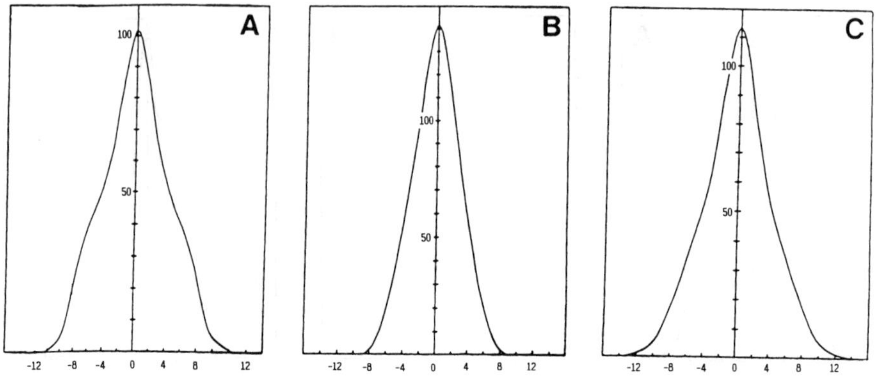

Fig. 7.2 Dipolar-coupled ^{19}F NMR spectra of drawn polyvinylidene fluoride ($\lambda = 6$) with various orientations of the draw axis relative to the magnetic field: (A) 0°; (B) 45°; (C) 90°. (Reprinted from [38] by permission of Butterworth–Heinemann Publishers Ltd, © 1985.)

orientation, and this effect can be observed for drawn samples, as shown in Fig. 7.2. By analyzing the line shapes, one can derive orientation distribution functions for the sample. This approach can also be applied to the chemical-shift anisotropy interaction, as will be discussed in section 7.4.

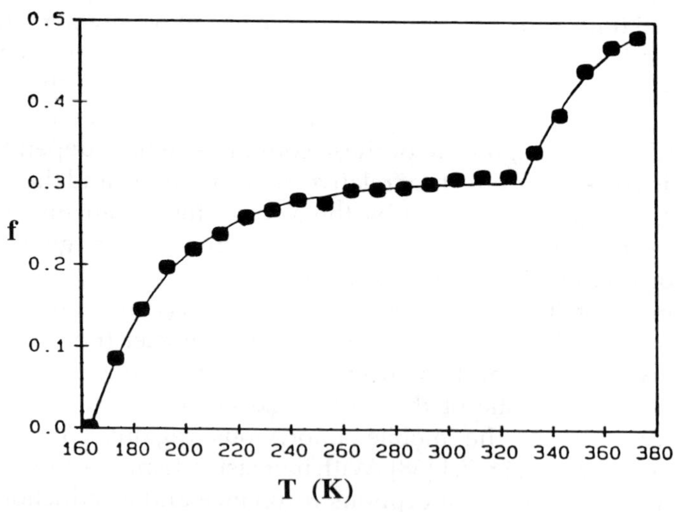

Fig. 7.3 Plot of narrow-component fraction, *f*, in ^{1}H spectra of solid PE as a function of temperature. (Reprinted with permission from [39], copyright John Wiley & Sons, Inc., © 1989.)

The ^1H spectrum of polyethylene (PE) can similarly be separated into crystalline and amorphous components, and these also vary with temperature, resulting in spectra very similar to those shown earlier for PVDF [39]. The fraction of spins in the narrow component exhibits a complex dependence on temperature, as shown in Fig. 7.3. At any temperature, the fraction of the NMR signal in the narrow resonance is not necessarily the amorphous fraction as measured by classical techniques such as differential scanning calorimetry (DSC); however, the asymptotic limiting fraction at high temperature does correspond well with the amorphous fraction determined by DSC. There is an indication of two motional processes with onset transition temperatures of 163 K and 328 K, which correlate with the α and γ transitions for linear PE. The diffusion of magnetization from one region to the other can be studied (using a pulse technique developed by Goldman and Shen) [40–41], from which one may make an estimate of the average domain size. The time constant for spin diffusion leads to an estimated crystallite size of >100Å, consistent with domain sizes deduced from X-ray diffraction measurements.

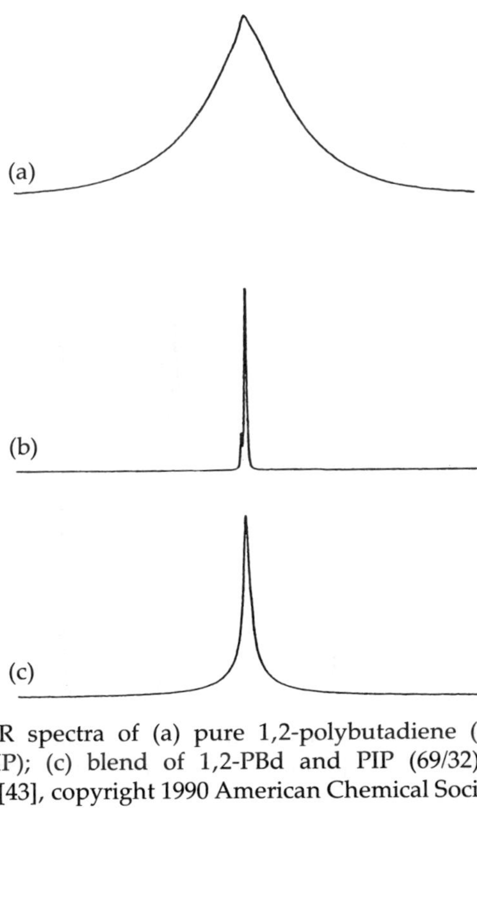

Fig. 7.4 ^1H NMR spectra of (a) pure 1,2-polybutadiene (1,2-PBd); (b) pure polyisoprene (PIP); (c) blend of 1,2-PBd and PIP (69/32). (Reprinted with permission from [43], copyright 1990 American Chemical Society.)

Blends have been fairly extensively studied by ^{1}H solid-state NMR [42]. If the components of a blend are completely miscible, the ^{1}H spectrum of the blend does not give the compound line shape typical of a two-phase system, as shown in Fig. 7.4 [43]. The spectrum of pure poly(vinylethylene) is quite broad, as expected for a relatively rigid material. Polyisoprene, as befits an elastomer, exhibits a very narrow resonance. The miscible blend's line width is relatively narrow, indicating compatibility at the molecular level. The two-dimensional approach can be used to differentiate between miscible and immiscible blends, as in solvent-cast blends of polystyrene (PS) and poly(vinyl methyl ether) (PVME) [44, 45]. There is no evidence of spin diffusion between PS and PVME when the blend is cast from a chloroform solution, pointing to an immiscible system. On the other hand, solvent-casting from a toluene solution produces strong correlations, which reveal that this system results in intimate phase mixing.

Heteronuclear dipolar couplings (for example, coupling of ^{13}C and ^{1}H in organic materials) also result in broadening of the resonance. Inherent in this spectral feature is the ability to make measurements of internuclear distances, although again the presence of multiple couplings means that only an average distance is measured. Heteronuclear couplings also vary in strength depending on the extent of averaging by molecular motion. As for decoupling of nuclei in solution, one must irradiate the sample at the resonance frequency of the interfering nucleus to produce high

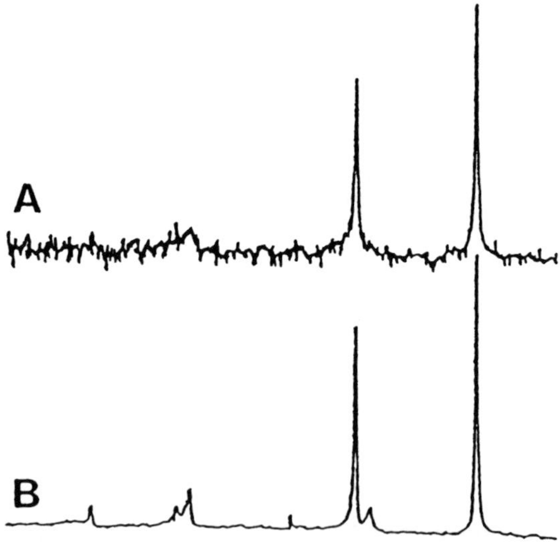

Fig. 7.5 ^{13}C spectra of Hytrel® (A) with scalar decoupling; (B) with dipolar decoupling. (Reprinted with permission from [46], copyright 1981 American Chemical Society.)

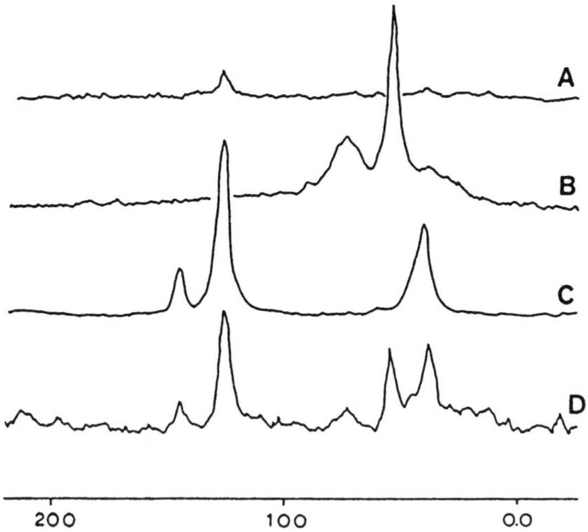

Fig. 7.6 ^{13}C NMR spectra of (A) perdeuterated polystyrene PS-d_8 (with CP); (B) polyvinyl methyl ether (PVME) (no CP); (C) polystyrene (with CP); (D) blend of PS-d_8 and PVME (65/35) (with CP). (Reprinted with permission from [48], copyright 1987 American Chemical Society.)

resolution. The extent of decoupling depends on the strength of the dipolar interactions, allowing one to selectively decouple regions based on their strength. Figure 7.5(a) illustrates that low-power (scalar) ^1H irradiation selectively decouples the softer segments of Hytrel, a segmented thermoplastic elastomer [46]. Since motion in these regions is more 'liquid-like', the dipolar interactions are easier to remove by lower-level irradiation. The power level required to observe the butylene terephthalate hard segment is double that used in the scalar decoupling experiments. In essence, one must make the irradiation the dominant effect for decoupling to be efficient; in the case of low-power irradiation, only weakly coupled spins will satisfy this condition. In the case of higher-power irradiation, this condition is satisfied for all of the carbons.

The cross-polarization (CP) phenomenon used to enhance rare signals by exchange of magnetization between abundant and rare spins [47] takes advantage of static dipolar couplings between ^1H and another nucleus (most commonly, ^{13}C or ^{29}Si). While this technique is often exploited primarily for signal enhancement, the dynamics of polarization transfer can reveal information about the proximity of species in blends and composites. Since CP depends on a ^1H spin reservoir, deuterated materials do not exhibit this phenomenon; the ^{13}C resonance is, therefore, greatly attenuated for carbons near deuterons. Dipolar interactions operate through space, and carbons that are proximate to protons will be cross-polarized, even if they are not covalently bonded. Figure 7.6

illustrates this point clearly [48]. Even though virtually no signal is observed for perdeuterated PS under CP conditions, miscible PS/PVME blend systems show a significant PS signal. Immiscible blends, such as polystyrene/poly(methyl methacrylate) do not exhibit any signal arising from the deuterated moiety [49], except at the interface between the two components.

The same approach can be used to study the effectiveness of compatibilizing agents, as in blends of sulfonated PS and methyl methacrylate-co-4-vinylpyridine [50]. The compatibilized system shows clear evidence of greater cross polarization to the deuterated component, indicating enhanced miscibility. Cross polarization can be used to probe the interface between polymer and filler in a composite material. Figure 7.7 shows the ^{29}Si CP spectrum of silica, compared with the spectrum recorded by the direct polarization (DP), or Bloch decay, experiment [51]. The CP spectrum of this material [Fig. 7.7(a)] shows one major peak, which corresponds to silanol groups receiving their magnetization from nearby hydroxyl protons. The DP spectrum of silica [Fig. 7.7(b)] also exhibits a peak due to bulk silicon atoms. When poly(vinyl alcohol) (PVOH) is added to silica as a 1:1 composite, cross polarization becomes more efficient, which has been ascribed to a decrease in mobility of the silanol groups caused by the presence of the polymer [Fig. 7.7(c)]. The restricted motion increases the static ^{1}H–^{29}Si heteronuclear couplings, making the ^{29}Si nuclei more readily polarized, or said another way, making the cross-polarization time constant shorter. This hypothesis is supported by an interrupted decoupling, or dipolar dephasing, experiment, in which a short time delay is inserted between the cross-polarization step and the onset of high-power irradiation to implement decoupling of the ^{29}Si from ^{1}H. During this period strongly coupled spins lose magnetization by the effect of strong dipolar interactions, resulting in the loss of signal from these spins. Weakly coupled spins, which retain their polarization, are observed. In the PVOH/silica system, the composite gives no signal if the decoupling is interrupted, further confirming the results of the CP and DP experiments. Such a dipolar-dephasing experiment can be used to great advantage to edit spectra to enhance the intensity of protonated carbons relative to non-protonated carbons [52–54].

Homo- and heteronuclear dipolar couplings can be exploited simultaneously in an experiment to study blends [55]. To study mixing at the molecular level, spin diffusion is established between chemically distinct protons in a blend of an aromatic poly(etherimide) (PEI) and a poly(aryl-ether ketone) (PAEK). The resulting ^{1}H spectrum could be observed directly, but the broader chemical-shift dispersion of ^{13}C is exploited by using the post-diffusion ^{1}H magnetization to cross-polarize ^{13}C's. In this way, the particular carbons affected by the diffusion process are made

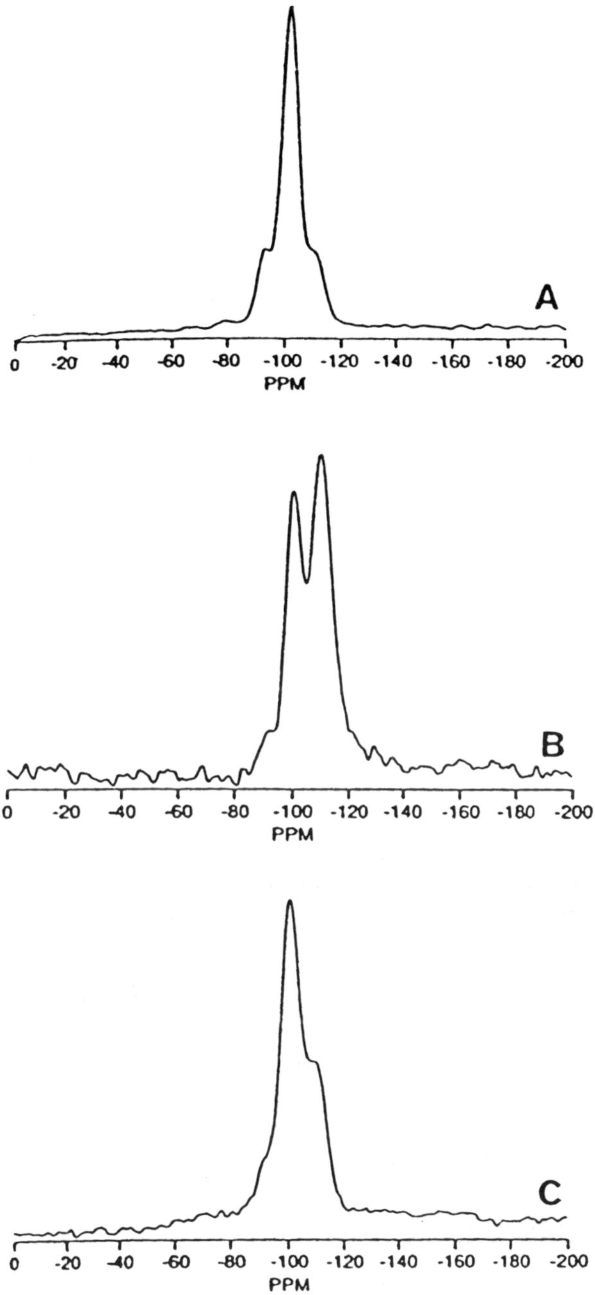

Fig. 7.7 ^{29}Si NMR spectra of (A) silica (with CP); (B) silica (no CP); (C) poly(vinyl alcohol) (PVOH) and silica (1/1) (with CP). (Reprinted with permission from [51], copyright 1991 American Chemical Society.)

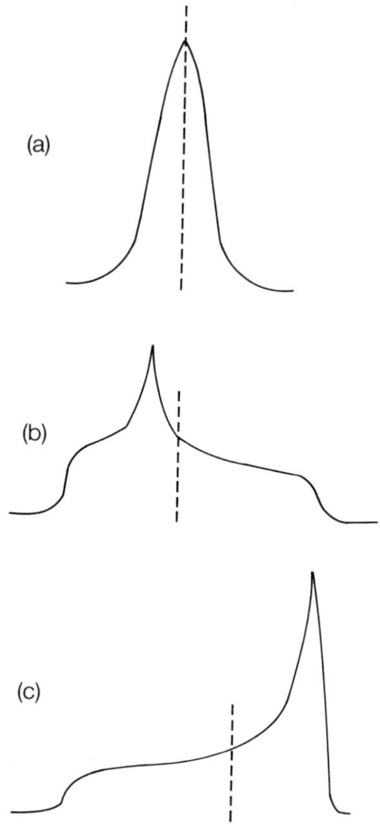

Fig. 7.8 Typical chemical-shift powder patterns with various symmetries: (a) spherical; (b) non-axial; (c) axial. The dashed line marks the position of the isotropic chemical shift.

obvious. When the PEI protons are excited, ^{13}C peaks due to PAEK appear only when the spin-diffusion process is allowed to proceed for times longer than 400 μs. Knowing the strength of the ^{1}H–^{1}H dipolar coupling, a measurement of the time constant for spin diffusion permits one to obtain an estimate for the scale on which the two components are mixed, in this case 1 nm or less.

7.4 CHEMICAL-SHIFT ANISOTROPY

The inherently anisotropic character of the chemical shift is masked in spectra of solutions by the rapid, relatively isotropic tumbling of molecules. In solids, however, motions are sufficiently slow and/or restricted that the shift exhibits a pronounced dependence on the orientation, relative to the static magnetic field, of the bonding structure. Thus, for example, the resonance position of carbon monoxide in the solid state depends on the orientation of the C–O bond relative to the magnetic field. An NMR chemical-shift spectrum of a solid may be looked upon as a plot of the number of molecules in the sample having the particular orientation (or set of orientations) corresponding to that resonance frequency. Thus, NMR spectra of spin-1/2 nuclei recorded so as to accentuate this interaction result in a spectrum containing information about the directionality of molecular motion or structural features.

Samples of powders, in which all orientations are equally likely, give rise to characteristic 'powder patterns', the result of superposition of resonances from nuclei in various possible orientations of the local axes (known as the principal axes) relative to the magnetic field direction. These patterns appear to be broader and less informative than the spikes of solution-state spectra, but a careful interpretation of the line shape can reveal significant details about the electronic structure in the vicinity of the nucleus, as well as about the molecule's geometry of motion. Generally there are three categories of powder-pattern symmetry: (a) spherical; (b) axial; and (c) non-axial. They are illustrated in Fig. 7.8. In the first case, the electronic structure around the nucleus is spherically symmetric and no information on orientation can be gained. In the second, it is cylindrically symmetric about an axis (such as for carbon monoxide) and the NMR spectrum is a map of the orientations of this axis relative to the magnetic field. In the third case, there is no unique direction to the local electronic structure. In this case, the NMR spectrum is a complex mapping of this electronic structure and any general point on the spectrum may correspond to many orientations of the electronic structure relative to the magnetic field. In some cases, rapid anisotropic motion about an axis can result in an effective axially symmetric pattern (for example, in the case of a rotating methyl group); in this case, one may obtain information about the distribution of orientations of the axis. This is clearly shown in Fig. 7.9, which decomposes the static ^{13}C powder pattern of polyethylene into its crystalline and amorphous components [56]. Even though there is obviously no significant chemical difference between the two phases, the crystalline component exhibits non-axial symmetry, with a width of 266 Hz; the amorphous region has axial symmetry with a width of only 95 Hz. The relative narrowness of the amorphous resonance is explained, to a first approximation, by the greater molecular motion known to occur in this region. The difference in symmetry requires somewhat more explanation. The symmetry of the amorphous phase is consistent with relatively rapid segmental reorientations occurring about an axis parallel to the chain backbone. If these are fast on the NMR frequency scale (>2 kHz in this case), then the motion will appear to be axially symmetric. In the crystalline phase, these

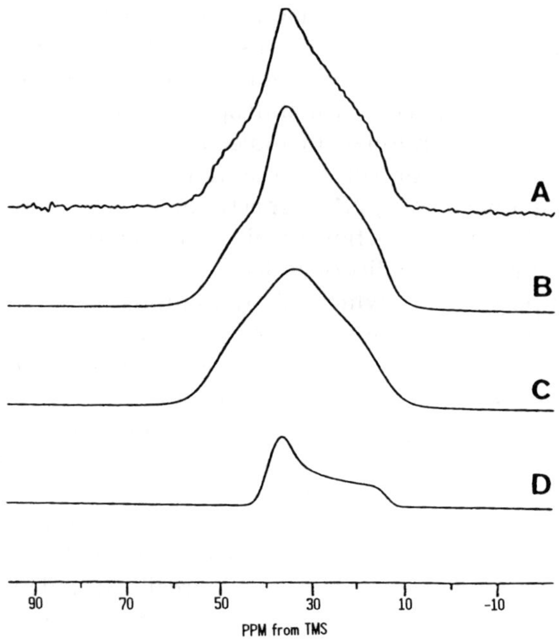

Fig. 7.9 ^{13}C chemical-shift powder patterns for linear polyethylene: (A) observed spectrum: (B) calculated best-fit spectrum; (C) crystalline component; (D) amorphous component. (Reprinted with permission from [56], copyright 1989 American Chemical Society.)

reorientations are slower, so that the two axes perpendicular to the backbone are not averaged by the motion.

When a non-random distribution of orientations of the local axes is induced in a polymer sample, for example by elongation, the spectrum is a perturbed powder spectrum, reflecting the orientation distribution. If spectra are recorded as a function of the angle between the draw axis of a uniaxially deformed material and the static magnetic field, the resulting spectra can be analyzed by calculating the moments of the spectrum, which are related to the Legendre polynomials describing the orientation function $N(\theta)$ [57–59]:

$$N(\theta) = \sum \frac{2l-1}{8\pi} < P_l(\cos(\theta) > P_l(\cos(\theta)) \qquad (7.2)$$

Figure 7.10 shows ^{19}F spectra for drawn polytetrafluoroethylene (PTFE). As the angle between the draw axis and the magnetic field is varied, the shape of the spectrum changes. Analysis of similar spectra for various

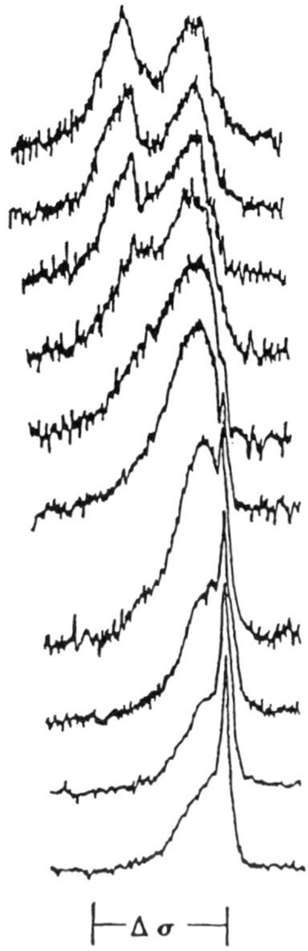

Fig. 7.10 ^{19}F NMR spectra of oriented polytetrafluoroethylene (PTFE) ($\lambda = 2.4$) for various angles between the draw axis and the magnetic field. (Reprinted from [58, 59], copyright John Wiley & Sons, Inc., © 1983.)

draw ratios leads to the distribution functions shown in Fig. 7.11. The marked change that occurs in the shape of the curve between $\lambda = 1.4$ and $\lambda = 2.4$ has been attributed to the yield point, which occurs between these two strains for PTFE. The process of disorientation which occurs during annealing has been studied by this approach [60].

Even more complex orientation patterns, such as biaxial order, for polymers with more intricate chemical structures, are also amenable to chemical-shift tensor analysis [61]. Biaxially oriented poly(ethylene terephthalate) film clearly shows a dependence of line shape on orientation with respect to the magnetic field, as shown in Fig. 7.12. The line shape is complicated, however, by the four overlapping ^{13}C tensors (carbonyl, protonated aromatic, non-protonated aromatic, and methylene), but the application of the two-dimensional approach allows these resonances to be separated. If these spectra are taken at various orientations, a fairly complete description of three-dimensional orientation can be developed.

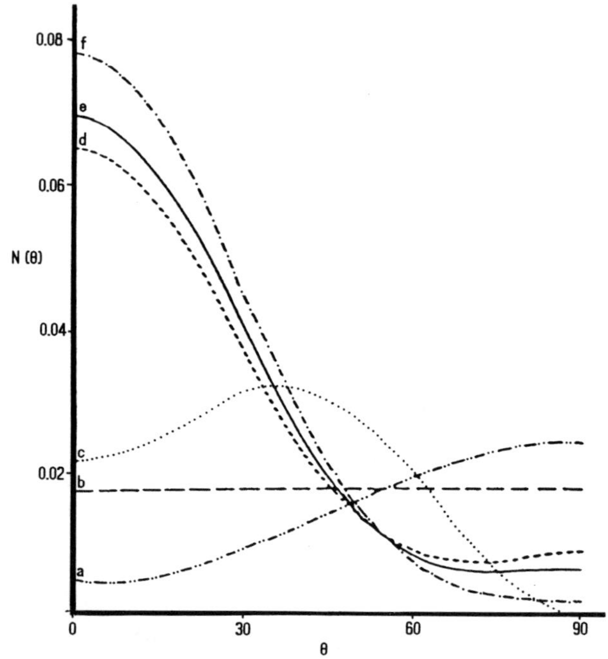

Fig. 7.11 Crystallite orientation distribution functions for oriented PTFE calculated from spectra like those in Fig. 7.10: (a) $\lambda = 0.8$; (b) $\lambda = 1.0$; (c) $\lambda = 1.4$; (d) $\lambda = 1.75$; (e) $\lambda = 2.1$; (f) $\lambda = 2.4$. (Reprinted from [58, 59], copyright John Wiley & Sons, Inc., © 1983.)

7.5 ISOTROPIC CHEMICAL SHIFT

The suppression of the effects of the intrinsic anisotropic character of the chemical shift by magic-angle spinning (MAS), when combined with high-power dipolar decoupling, leads to a 'high-resolution' solid-state spectrum, the sharp bands of which fall at the expected isotropic shifts. In effect, the coherent rotation at a specific angle mimics the effect of random rapid fluctuations of the local chemical-shift axes relative to the magnetic field due to Brownian motion in solution, in that it suppresses the orientation-dependent part of the chemical-shift interaction [62]. For an insoluble material like a thermoset resin, this may be the only way to record its 'high-resolution' NMR spectrum. Spectra of solid polymers almost never approach the degree of resolution seen in solution spectra, but this is due, at least in part, to the fact that the solid state often contains structures that fluids do not. This is particularly true for polymeric materials that have features such as crystalline morphology and amorphous chain conformation distributions.

One obvious application of solid-state NMR is to the characterization of

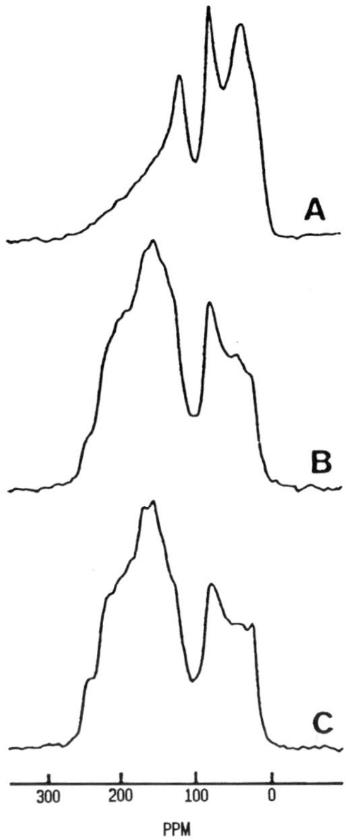

Fig. 7.12 ^{13}C NMR spectra of biaxially oriented poly(ethylene terephthalate) with film oriented (A) perpendicular and (B), (C) parallel to the magnetic field direction. (Reprinted with permission from [61], copyright 1987 American Chemical Society.)

insoluble polymers, in particular, to network polymers or thermosets. The kinetics of isocyanurate ring formation, for example, has been studied by both ^{13}C and ^{15}N [52–54]:

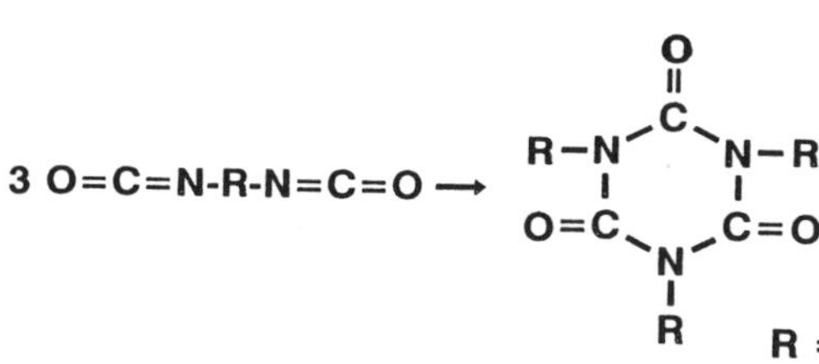

R = $-C_6H_4-CH_2-C_6H_4-$

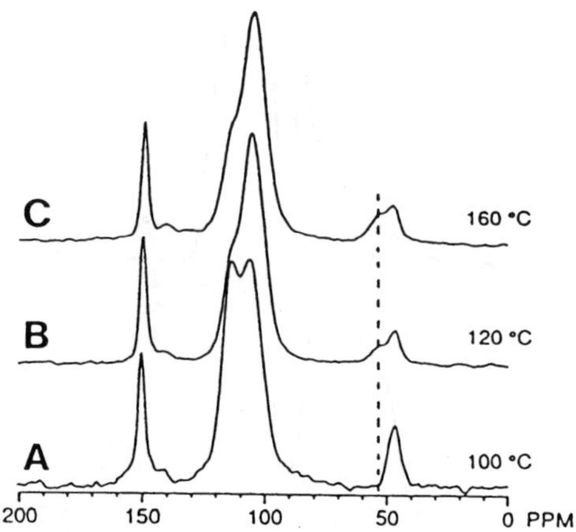

Fig. 7.13 ^{15}N chemical-shift spectra of ^{15}N-enriched polyisocyanurate resins cured at (A) 100° C; (B) 120° C; (C) 160° C. (Reprinted with permission from [54], copyright 1990 American Chemical Society.)

The aromatic regions of the ^{13}C spectra are very complex, but they are consistent with an increase in ring formation and a decrease in monomer concentration with temperature, although it is difficult to draw more definitive conclusions from the ^{13}C data alone. For these systems, the ^{15}N resonances are more descriptive of chemical structure. As discussed in section 7.2, ^{15}N is not a particularly easy nucleus to observe, primarily because of its low natural abundance and low gyromagnetic ratio. The spectra of artificially enriched samples, on the other hand, do show the effects of reaction temperature even more clearly, as illustrated in Fig. 7.13.

Treatment of polymers can lead to chemical structural modifications which are reflected in the ^{13}C spectrum, as in the formation of alkalicellulose [63]. Normal ramie fibers exhibit intramolecular hydrogen bonds between the hydroxyl group on C3 and the ring oxygen of the next monomer unit and intermolecular bonds between the C2 and C6 hydroxyl groups on adjacent chains:

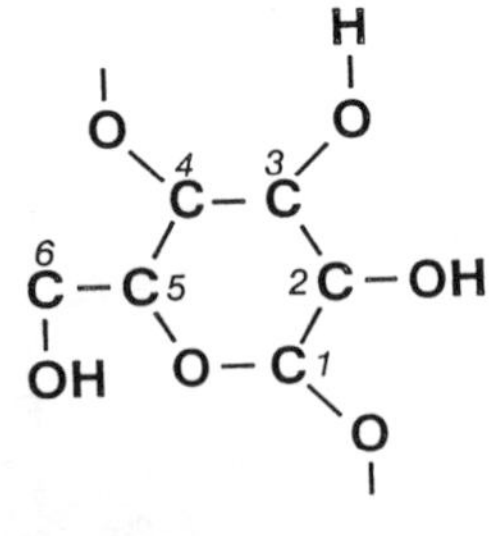

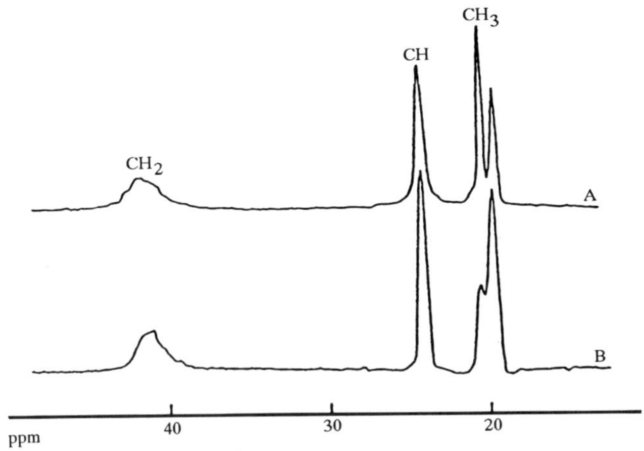

Fig. 7.14 ^{13}C chemical-shift spectra of polypropylene: (A) α-form and (B) γ-form. (Reprinted from [64] by permission of Huetig and Wepf Verlag, 1990.)

If, upon treatment with NaOH solution, these hydrogen bonds are disrupted by co-ordination with Na^+, the resulting increase in shielding is expected to produce a lower chemical shift. The intermolecular hydrogen bond is most easily broken at lower NaOH concentrations, which do not disrupt the intramolecular bonds.

Just as the presence of weak bonds between neighboring atoms affects the chemical shift, the spatial arrangement of neighboring groups in distinct crystalline modifications of polymers also gives rise to distinct chemical shifts. This is clearly demonstrated by a comparison of the ^{13}C spectra of the α- and γ-forms of isotactic polypropylene (PP), shown in Fig. 7.14 [64]. No difference is evident in the >CH– resonance, and only a slight asymmetry is observed for the $-CH_2-$ peak, but the $-CH_3$ region of the spectrum clearly exhibits several peaks. These observations are consistent with the known crystal structures of PP. In both forms, two of the three methyl groups are equivalent, which accounts for the approximately 2:1 ratios seen in peak intensities. In the α-form, parallel PP helices are positioned so that the two degenerate methyls are farther apart than the unique methyls on adjacent chains. This leads to decreased shielding at the degenerate centers and to a corresponding higher chemical shift. For the γ-form, the packing situation is reversed, so that the two degenerate methyls appear at lower chemical shift.

Semi-crystalline polymers such as polyethylene (PE) exhibit resonances from not just distinct crystalline polymorphs, but also from the amorphous regions of the polymer. Figure 7.15 compares the spectra of ultra-high-molecular-weight PE fibers (top) and solution-cast film (bottom) [65]. In both cases, the relatively narrow component at 32 ppm

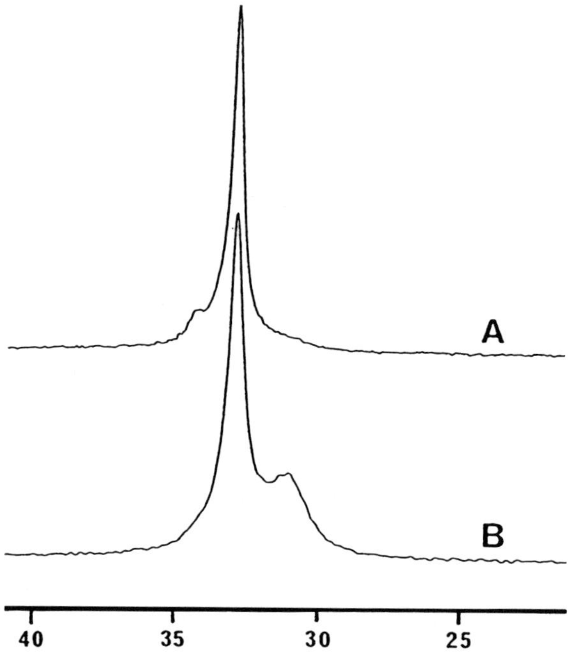

Fig. 7.15 ^{13}C chemical-shift spectra of (A) ultra-high-molecular-weight polyethylene (UHMWPE) fiber; and (B) cast UHMWPE film. (Reprinted from [65] by permission of the Society of Polymer Science, Japan, 1990.)

arises from the orthorhombic crystalline regions of the polymer. Because the chains in this region are arrayed in fairly regular, all-*trans* conformations, they experience very similar environments, leading to a narrow distribution of chemical shifts. The small peak at 34 ppm in the fiber spectrum is due to a small amount of monoclinic PE. The resonance centered at 31 ppm in the spectrum of the cast film sample is due to the amorphous phase of the polymer. On a molecular level, the disorder inherent in this phase arises from a broad distribution of *gauche* and *trans* conformations of the chain. The high degree of order imposed by the drawing process inhibits the adoption of these conformations; this peak is, therefore, not seen in the fiber spectra. These spectra also correlate well with sample crystallinities determined from the sample density: 90% for the fibers and 60% for the cast film.

Most chemical-shift studies of solid polymers have been made for spins without strong homonuclear dipolar coupling (such as ^{13}C, ^{29}Si, or ^{31}P), but there are some examples of abundant-spin spectra. For example, the use of MAS speeds high enough to overcome both homonuclear dipolar coupling and chemical-shift anisotropy allows the ^{19}F spectrum of fluoropolymers to be recorded, as for tetrafluoroethylene-co-hexafluoropropylene [66]. The resolution, which is sufficient to determine co-monomer sequences for the co-polymer, approaches that attainable for solution spectra.

7.6 QUADRUPOLAR COUPLING

All of the nuclei in Table 7.1 whose nuclear spins, I, are greater than 1/2 possess a nuclear quadrupole moment. These nuclei exhibit, in general, broader resonances than spin-1/2 nuclei due to the coupling of this quadrupole moment with the local electric-field gradient at the site of the nucleus. The coupling produces a splitting dependent on the orientation of the electric-field gradient principal axis relative to the magnetic field, in the regime where the quadrupole coupling is a relatively small perturbation. Thus, the resonances of random powders are broad lines, whose shapes depend in a specific manner on molecular dynamics and orientation [67]. Quadrupolar nuclei are not particularly relevant to questions in polymer science since many polymers contain few whose resonances are easily observed. An exception is found in deuterium-labeled organic materials [68].

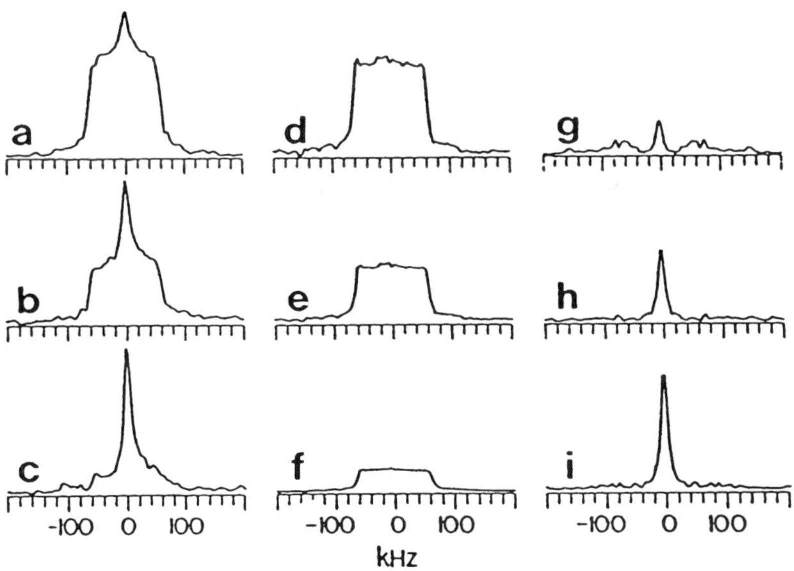

Fig. 7.16 ^2H NMR spectra of polyurethane thermoplastic elastomer: (a) 70 wt% hard segment: (b) 60 wt%; (c) 50 wt%. Hard-segment contributions to the composite line shapes are given in (d)–(f). Spectra of the soft-segment component, calculated by subtracting the 'hard' spectrum from the composite spectrum are shown in (g)–(i). (Reprinted with permission from [69], copyright 1985 American Chemical Society.)

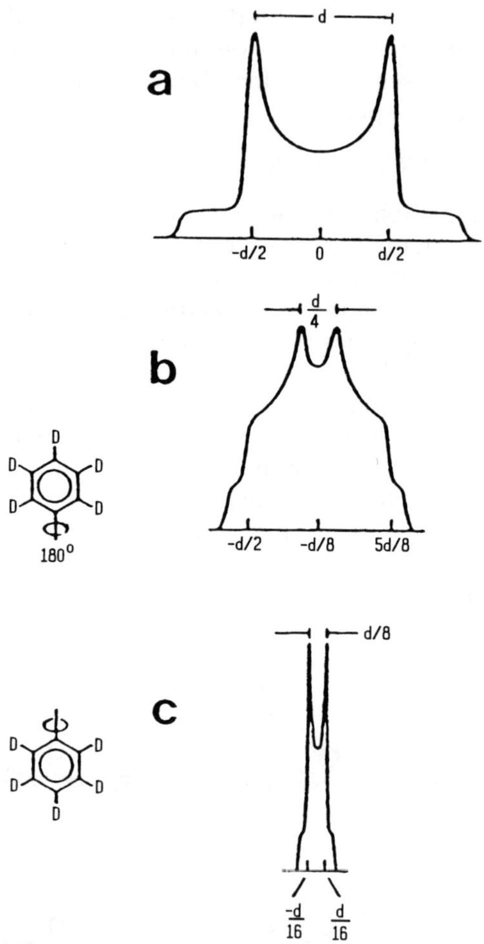

Fig. 7.17 Calculated ^{2}H NMR line shapes for three types of phenyl-ring motion: (a) static; (b) discrete 180° flips; (c) free, rapid rotation about the axis. (Reprinted with permission from [71], copyright 1985 American Chemical Society.)

The labeling approach has the advantage of being able to tag a specific part of the polymer, so that the resonance of nuclei in these regions can be observed individually. For example, segmented polyurethane materials have been labeled in the center of the 'hard' segments, which have the structure:

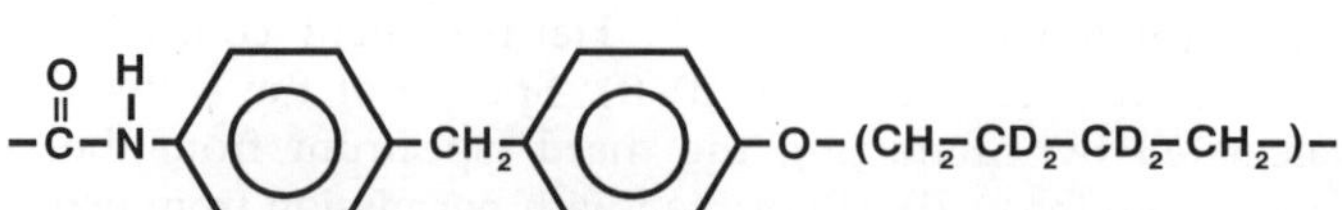

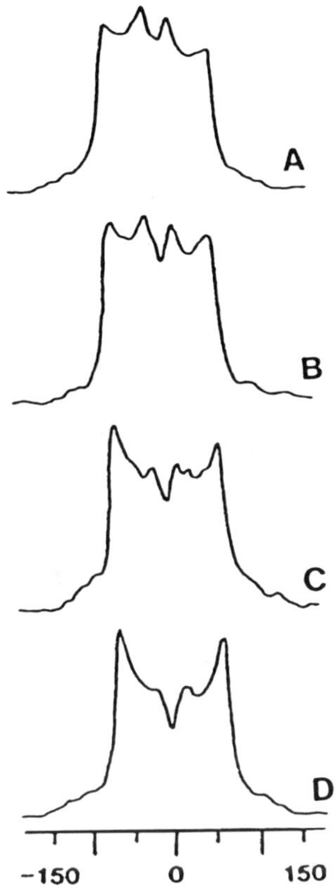

Fig. 7.18 Experimental ^2H spectra for the amorphous deuterons in poly(butylene terephthalate-d_4) at (A) 70° C; (B) 50° C; (C) 30° C; (D) 22° C. (Reprinted with permission from [72], copyright 1984 American Chemical Society.)

The 'soft' segments are blocks of ethylene and propylene oxides, and are unlabeled. In this way, the ^2H resonance observed must arise from the alkoxy portion of the hard segment, and its motion can be studied independently in detail. Relatively slow and/or anisotropic motion gives rise to the very specific line shapes, but relatively rapid, isotropic reorientations produce a relatively narrow, symmetric peak. As shown in Fig. 7.16, the line shapes observed for the deuterated polyurethanes with varying levels of hard segment are all composed of two components – one due to anisotropic and one due to isotropic motion [69]. These results indicate that the phase boundaries in this material are not sharp. Most of the hard segments are in a rigid environment similar to that seen in the hard homo-polymer. However, a significant fraction is in a less restrictive environment, either as interphase regions or as short hard segments

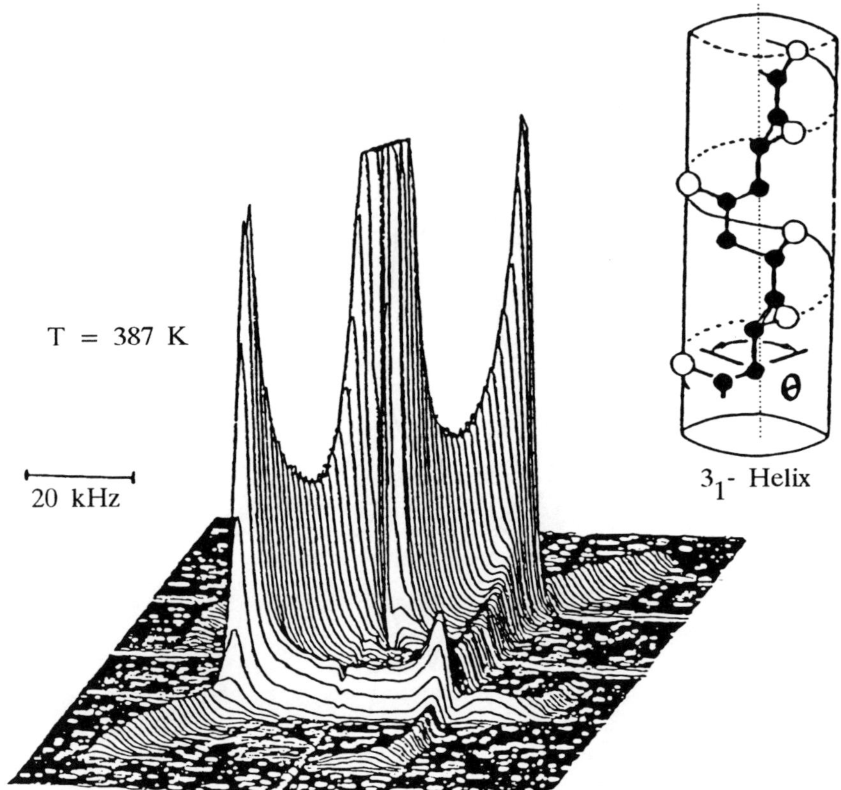

Fig. 7.19 Two-dimensional ^{2}H exchange spectrum of isotactic, deuterated polypropylene. (Reprinted with permission from [73], copyright 1990 American Chemical Society.)

dissolved in the soft phase. Similar effects have recently been observed in other segmented polyurethane materials [70].

The dependence of line shape on the geometry of motion is illustrated in Fig. 7.17 for deuterated phenyl ring motion [71]. These motions can be expressed mathematically, so it is possible, by line-shape simulation, to test models of molecular dynamics. The ^{2}H spectra observed for poly(butylene terephthalate) as a function of temperature are compared to calculated spectra in Fig. 7.18, and the rate of 180° ring flips can be determined from these data [72]. Even more detailed information can be extracted from multidimensional spectra. In the 2D exchange spectrum [73], correlations between two molecular orientations are detected. If a bond remains static during a short time delay, all spectral intensity appears along the diagonal of the 2D spectrum. If, on the other hand, a

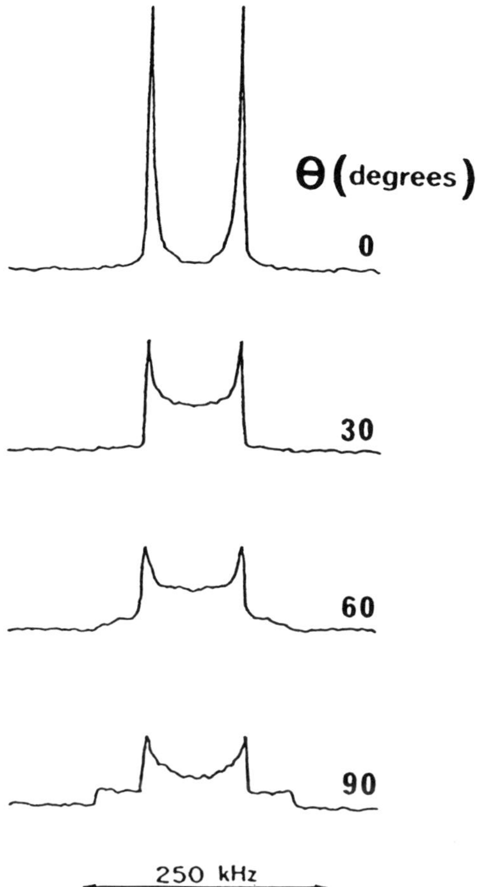

Fig. 7.20 ^{2}H NMR spectra of oriented, deuterated polyvinylidene fluoride ($\lambda = 3$) at various angles between the draw axis and the magnetic field. (Reprinted from [74] by permission of the American Institute of Physics.)

reorientation takes place, a peak will appear at an off-diagonal site that is characteristic of the reorientation angle. Figure 7.19 illustrates the 2D exchange spectrum for isotactic, methyl-deuterated PP. Mathematical simulations of the spectra reveal that this pattern corresponds to discrete 120° jumps about the helical axis of the crystalline polymer chain. Similar studies for atactic PP indicate that, above the glass transition temperature, this material exhibits a broad distribution of reorientational angles, consistent with its amorphous nature.

Yet another application of the quadrupolar interaction is the study of orientation. Since the exact quadrupolar coupling constant depends on the orientation of the principal axis of the electric-field gradient tensor relative to the static magnetic field, in a manner analogous to the chemical-shift tensor, it can be used to study orientation in a similar fashion. In Fig. 7.20, ^{2}H spectra of deuterated poly(vinylidene fluoride) are shown as a function of position relative to the magnetic field [74]. The simulated spectra are based on a 22°-wide Gaussian distribution of chain axes. The orientation in PVF-co-tetrafluorethylene co-polymers and the alignment effects of the electrical polarization of this piezoelectric material have also been studied.

7.7 RELAXATION TIMES

Many of the interactions discussed earlier depend on time because of parameters like the angle between the field and some vector associated with the molecule, typically arising from modulation of the interaction by thermal molecular motion. The result is that the magnetization and other 'constants of motion' (like the net dipolar energy) decay towards equilibrium values with time constants (usually given the cognomen 'relaxation times'). These time constants should not be confused with the relaxation times of dynamic mechanical relaxation measurements. While related, the relaxation of magnetic properties measures molecular mobility indirectly through its influence on the recovery of magnetic parameters to their equilibrium values.

There are primarily four types of relaxation times, which characterize motion in two frequency regimes. The spin-lattice relaxation time, T_1, describes motions which occur at MHz rates, while spin-spin relaxation (T_2), spin-lattice relaxation in the rotating frame ($T_{1\rho}$), and spin-lattice relaxation of dipolar energy (T_{1D}) characterize reorientations occurring at kHz rates. High-frequency motions are, in general, localized over a few bond lengths, and include only minimal contributions from long-range segmental or chain motions. In simple, one-phase systems with one type of relatively isotropic motion, single exponential recovery of the magnetization to equilibrium is observed. For polymer systems, however, more complex behavior is often observed. In some cases, it is fairly clear that relaxation of magnetic properties is best described by the simultaneous or sequential occurrence of two (or more) motional processes. Other systems exhibit more complicated behavior. This has sometimes been explained by multi-exponential behavior, especially if more than two phases have been postulated (crystalline, amorphous, and interphase, for example), and sometimes by an exponential decay of some function of time ($t^{1/2}$, for example). Interpretation of such complicated behavior must be done carefully, since, to some extent, selection of the preferred explanation for multi-exponential relaxation can often be made consistent with preconceived expectations.

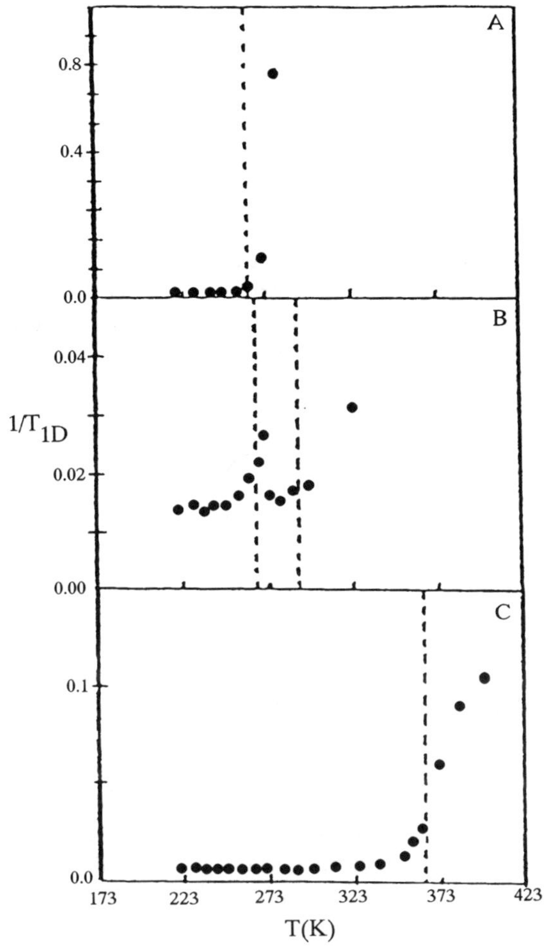

Fig. 7.21 Relaxation rate T_{1D}^{-1}, as a function of temperature for (A) highly amorphous; (B) 20% amorphous; (C) highly crystalline polypropylene. (Reprinted with permission from [75], copyright 1987 American Chemical Society.)

The relation between the relaxation time and the known thermal transitions of PP has been investigated, and is reproduced in Fig. 7.21 [75]. The α transition, which is associated with crystalline regions, is observed at 365 K only for highly isotactic material. Highly amorphous PP exhibits a β transition at 262 K. Semi-crystalline polymer (80% crystalline) shows both a shifted α transition at 289 K and a β transition at essentially the same temperature (265 K) as the amorphous material. These same transitions can be observed by thermal analysis.

Relaxation times are yet other NMR parameters that can provide information about polymer blend miscibility. For PS/PVME blends, there is an increase in ^{1}H T_1 relative to that of the constituent homo-polymers [76]. In solids, this correlation means that the molecular motion has been considerably restricted in the blend. The presence of a second component obviously alters the chain mobility at the molecular level in the MHz region. $T_{1\rho}$, which, as mentioned earlier, characterizes kHz motions, exhibits a clear dependence on temperature, with decreasing $T_{1\rho}$ indicating faster motion at higher temperatures, as expected. For these systems, however, this effect is not due simply to enhanced motions, because this system also exhibits thermally induced phase separation; these NMR data have been used to reconstruct the phase diagram.

Blend miscibility can also be revealed by the exponentiality of relaxation data. $T_{1\rho}$ is one of the time constants that characterize cross polarization, and it can be measured by varying the contact time in a CP experiment and observing the increase in magnetization transfer. Miscible blends of acrylonitrile-co-methyl acrylate-co-butadiene (AMAB) and ethylene-co-maleic acid exhibit single-component relaxation [77]. These times are intermediate between the relaxation times of the pure components and demonstrate a consistent variation with blend composition. On the other hand, partially compatible blends such as AMAB/poly(bisphenol-A carbonate) exhibit multi-exponential behavior. Relaxation times extracted from these data correspond to those measured for the homo-polymers and to intermediate values. This suggests the existence of mixed regions in addition to the two pure components.

Plasticization effects in glassy polymers have also been investigated via NMR relaxation times [78]. One can monitor the molecular-level effects of such additives. Low concentrations of di-*n*-butyl phthalate suppress motions associated with the primary sub-T_g loss peak of poly(bisphenol-A carbonate). In particular, reorientation of the more mobile phenylene groups is restricted.

7.8 MATERIALS IMAGING

While magnetic resonance imaging (MRI) has become a standard tool in medicine, and NMR imaging applications to fluid systems are becoming routine, the application of these techniques to solids has been hampered by the inherent difficulties of solid-state NMR. Progress is being made in macroscopic systems, but the current lower limit on the size of imaged features is on the order of 10 μm. Many interesting features in polymers are much smaller. This size restriction is mostly dependent on the strength of the field gradient imposed over a small sample volume, and this limitation is likely to be difficult to overcome. Given these restrictions, there are relatively few polymer imaging applications, but a few interesting ones have been published.

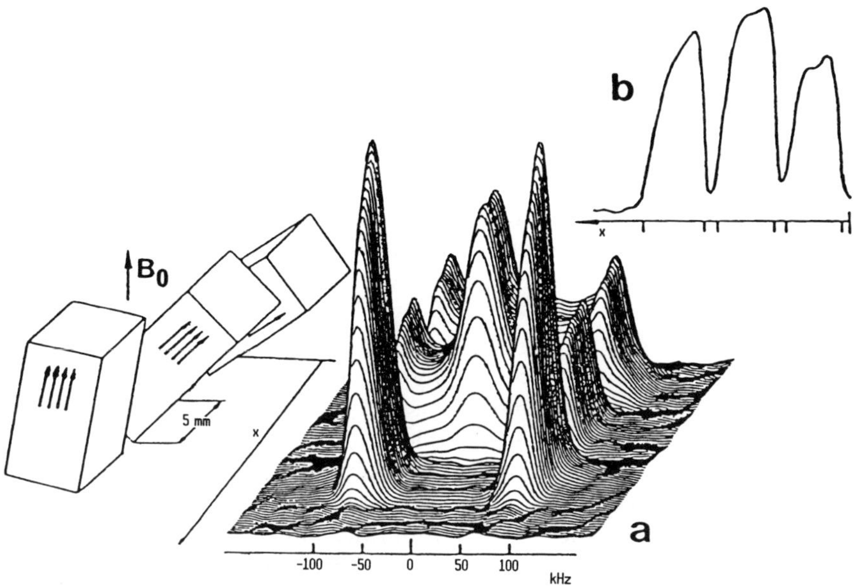

Fig. 7.22 NMR imaging of oriented polyethylene pieces: (a) two-dimensional spectrum; (b) projection on to the *x*-axis. (Reprinted from [87] by permission of Taylor and Francis, Ltd, 1990.)

Elastomers have sufficient molecular motion that they appear, in an NMR experiment, to be more liquid- than solid-like. Accordingly, imaging of elastomers, in which swelling [79–81], vulcanization [82], and interaction with fillers [83–85] have been studied, is the most common application. The increases or decreases in molecular mobility associated with these treatments are reflected in the intensity of the image observed. The gradual disappearance of image intensity with decreased molecular mobility has been exploited to study the polymerization of methyl methacrylate (MMA) [86]. MMA monomer, benzoyl peroxide initiator, and dimethyl aniline accelerator were placed in a glass bottle, and images were recorded as a function of time. The resulting NMR images clearly show that solidification first occurs in the center of the sample, with the ends solidifying later.

The sensitivity of ^2H quadrupolar coupling to orientation and molecular motion in polymers, discussed in detail in section 7.6, has been exploited to produce images of oriented macroscopic samples [87]. An image of a sample comprised of three 1 mm pieces of highly drawn PE oriented at 10°, 35°, and 55° to the magnetic field accurately reflects both the size and relative orientation of the pieces, as shown in Fig. 7.22. While the current limitation on the domain size precludes this approach from being used to study features like crystallite orientation, it does have some potential applications to composite materials. Similarly, samples composed of polymers with inherently different molecular mobilities can also be imaged. Since the room-temperature phenyl-ring motion in polystyrene is much slower than that in polycarbonate, ^2H images of ring-deuterated composites can be used to discriminate between the two components. By choosing the experimental parameters appropriately, either the rigid or mobile component can be highlighted.

7.9 SUMMARY

Perhaps more than any other polymer characterization tool, NMR spectroscopy can provide detailed insight into both the chemical structure and physical properties of solid polymer systems. The studies discussed here do not exhaust the possibilities of NMR, but they do illustrate its breadth of application. Molecular structure can be described by observing the resonances of different nuclei in the polymeric material, additives, or probe molecules. Polymer morphology is reflected in multicomponent NMR line shapes. Spin-diffusion measurements can estimate the size of microscopic features such as crystalline domains. The compatibility of polymer blends and composites can be investigated by various approaches. These approaches depend on the interaction between the individual components' spin populations or on the detection of motional homo- or heterogeneity. In general, the details of molecular dynamics are reflected in several NMR parameters: line width, line shape, coupling constants, and relaxation times. In some cases, known thermal transitions can be correlated with NMR measurements. Polymer-chain orientation is also amenable to NMR analysis, and in some cases, fairly detailed orientation distribution functions can be derived from line-shape analyses.

The relationships between the spin behavior probed during a solid-state NMR experiment and bulk polymer properties are complex. Many of these relationships have been well established; others remain to be defined. In particular, further development of multidimensional representations that illustrate correlations between these interactions and advances in solid-state imaging promise to reveal even more details of polymer structure and dynamics. The ability to manipulate the NMR parameters experimentally – dipolar coupling, chemical-shift anisotropy, isotropic chemical shift, quadrupolar coupling, and relaxation times – ensure that NMR will continue to be applied to polymer systems in new and interesting ways.

ACKNOWLEDGEMENTS

CRD acknowledges the National Science Foundation, Hercules, Inc., and the Department of Energy for support of studies of polymers with NMR spectroscopy.

REFERENCES

1. Gerstein, B.C. and Dybowski, C. (1985) *Transient Techniques in the NMR of Solids: An Introduction to Theory and Practice*, Academic Press, New York.
2. Mehring, M. (1993) *High Resolution NMR in Solids*, Springer–Verlag, Berlin.
3. Haeberlen, U. (1970) *High Resolution NMR in Solids: Selective Averaging*, Academic Press, New York.
4. Munowitz, M. (1990) Approaches to high-resolution NMR in solids. *Trends in Analytical Chemistry*, **9** 253–63.
5. Komoroski, R.A. (ed.) (1986) *High Resolution NMR Spectroscopy of Synthetic Polymers in Bulk*, VCH Publishers, Deerfield Beach, FL.
6. Fyfe, C.A. (1983) *Solid State NMR for Chemists*, CFC Press, Guelph, ON.
7. Koenig, J.L. (1992) *Spectroscopy of Polymers*, American Chemical Society, Washington, DC.
8. Cheng, H.N. (1991) NMR characterization of polymers, in *Modern Methods of Polymer Characterization*, (eds H.G. Barth and J.W. Mays), John Wiley and Sons, New York, pp. 409–93.
9. Yu, T. and Guo, M. (1990) Recent developments in ^{13}C solid-state high-resolution NMR of polymers. *Prog. Polym. Sci.* **15** 825–908.
10. Bluemich, B., Hagemeyer, A., Schaefer, D. *et al.* (1990) Solid-state NMR spectroscopy in polymer science. *Adv. Mat.* **2** 72–81.
11. Laupetre, F. (1990) Applications of high-resolution solid-state carbon-13 NMR to polymers. *Prog. Polym. Sci.* **15** 425–74.
12. Kinsey, R.A. (1990) Solid-state NMR of elastomers. *Rubb. Chem. Tech.* **63** 407–25.
13. Axelson, D.E. and Russell, K.E. (1985) Characterization of polymers by means of ^{13}C NMR spectroscopy. *Prog. Polym. Sci.* **11** 221–82.
14. Jelinski, L.W. and Melchior, M.T. (1987) High-resolution NMR of solids, in *NMR Spectroscopy Techniques*, (eds C. Dybowski and R.L. Lichter). Marcel Dekker, New York, pp. 253–329.
15. Jelinski, L.W. (1982) Solid-state NMR of macromolecules, in *Chain Structure and Conformation of Macromolecules*, (ed. F.A. Bovey), Academic Press, New York, pp. 223–49.
16. Spiess, H.W. (1991) Structure and dynamics of solid polymers from 2D- and 3D-NMR. *Chem. Rev.* **91** 1321–38.
17. Spiess, H.W. (1989) Two-dimensional NMR methods for studying solid polymers. *Makromol. Chem., Macromol. Symp.* **26** 197–207.
18. Harris, R.K. and Mann, B.E. (1978) *NMR and the Periodic Table*, Academic Press, London.
19. Crosby, R.C. and Haw, J.F. (1987) Magic-angle spinning NMR of polyphosphazenes. *Macromolecules*, **20** 2324–8.
20. Powell, D.G. and Mathias, L.J. (1990) Characterization of nylon-6 by ^{15}N solid-state nuclear magnetic resonance. *J. Am. Chem. Soc.* **112** 669–75.
21. Powell, D.G. and Mathias, L.J. (1990) ^{15}N NMR powder spectra of semicrystalline nylon-6. *Polym. Comm.* **31** 58–60.

22. Mathias, L.J. and Johnson, C.G. (1991) Solid-state NMR investigation of nylon-12. *Macromolecules,* **24** 6114–22.
23. Hatfield, G.R., Glans, J.H. and Hammond, W.B. (1990) Characterization of structure and morphology in nylon-6 by solid-state ^{13}C and ^{15}N NMR. *Macromolecules,* **23** 1654–8.
24. Asakura, T., Yoshimizu, H. and Yoshizawa, F. (1988) NMR of silk fibroin. Sequence and conformation analyses of the silk fibroins from *Bombyx mori* and *Philosaia cynthia ricini* by ^{15}N NMR spectroscopy. *Macromolecules,* **21** 2038–41.
25. Beshah, K., Mark, J.E. and Ackerman, J.L. (1986) Topology of poly(dimethylsiloxane) elastomeric networks studied by variable-temperature solid-state nuclear magnetic resonance. *Macromolecules,* **19** 2194–6.
26. Beshah, K., Mark, J.E., Ackerman, J.L. and Himstedt, A. (1986) Characterization of PDMS model junctions and networks by solution and solid-state silicon-29 NMR spectroscopy. *J. Polym. Sci. B: Polym. Phys.* **24** 1207–25.
27. Litvinov, V.M., Whittaker, A.K., Hagemeyer, A. and Spiess, H.W. (1989) Molecular motions in poly(dimethylsiloxane) studied by solid-state ^{29}Si NMR. *Colloid Polym. Sci.* **267** 681–6.
28. Dickinson, L.C., MacKnight, W.J., Connolly, J.M. and Chien, J.C.W. (1987) ^{23}Na NMR of solid-state ionomers: lineshape, shift, and hydration effects. *Polym. Bull.* **17**, 459–64.
29. Dickinson, L.C., Chien, J.C.W. and MacKnight, W.J. (1988) Cross-link point mobility in swollen polymer networks by phosphorus-31 spin-lattice relaxation in the rotating frame. *Macromolecules,* **21** 2959–63.
30. Kambour, R.P., Kelly, J.M., McKinley, B.J. *et al.* (1988) Spectroscopic studies of diluent motion in glassy plasticized blends. *Macromolecules,* **21**, 2937–40.
31. Carduner, K.R., Carter, R.O., Cheung, M.F. *et al.* (1990) Studies on the role of organophosphites in polyester blends: I. ^{31}P nuclear magnetic resonance spectroscopy. *J. Appl. Polym. Sci.* **40** 963–75.
32. Dybowski, C., Bansal, N. and Duncan, T.M. (1991) NMR spectroscopy of xenon in confined spaces: clathrates, intercalates, and zeolites. *Annu. Rev. Phys. Chem.* **42** 433–64.
33. Stengle, T.R. and Williamson, K.L. (1987) Nuclear magnetic resonance of xenon absorbed in solid polymers: a probe of the amorphous state. *Macromolecules,* **20** 1428–30.
34. Kennedy, G.J. (1990) Xenon-129 NMR as a probe of the effect of crosslinking on the amorphous phase structure of polymers. *Polym. Bull.* **23** 605–8.
35. Kentgens, A.P.M., von Boxtel, H.A., Verweel, R.-J. and Veeman, W.S (1991) Line-broadening effect for ^{129}Xe absorbed in the amorphous state of solid polymers. *Macromolecules,* **24** 3713–14.
36. Brownstein, S.K., Roovers, J.E.L. and Worsfold, D.J. (1988) ^{129}Xe line widths in block copolymers. *Magn. Res. Chem.* **26** 392–3.
37. Linder, M., Henrichs, P.M., Hewitt, J.M. and Massa, D.J. (1985) Use of carbon–carbon nuclear spin diffusion for the study of miscibility of polymer blends. *J. Chem. Phys.* **82** 1585–98.
38. Clements, J., Davies, G.R. and Ward, I.M. (1985) A broad line NMR study of oriented poly(vinylidene fluoride). *Polymer,* **26** 208–18.
39. Kauffman, J.S. and Dybowski, C. (1989) Determination of transition temperatures and crystalline content of linear, high-molecular-weight polyethylene by proton NMR spectroscopy. *J. Polym. Sci. B: Polym. Phys.* **27** 2203–9.
40. Goldman, M. and Shen. L. (1966) Spin-spin relaxation in LaF_3. *Phys Rev.* **144** 321–31.
41. Cheung, T.T.P., Gerstein, B.C., Ryan, L.M. *et al.* (1980) *Chem. Phys.* **73** 6059–67.

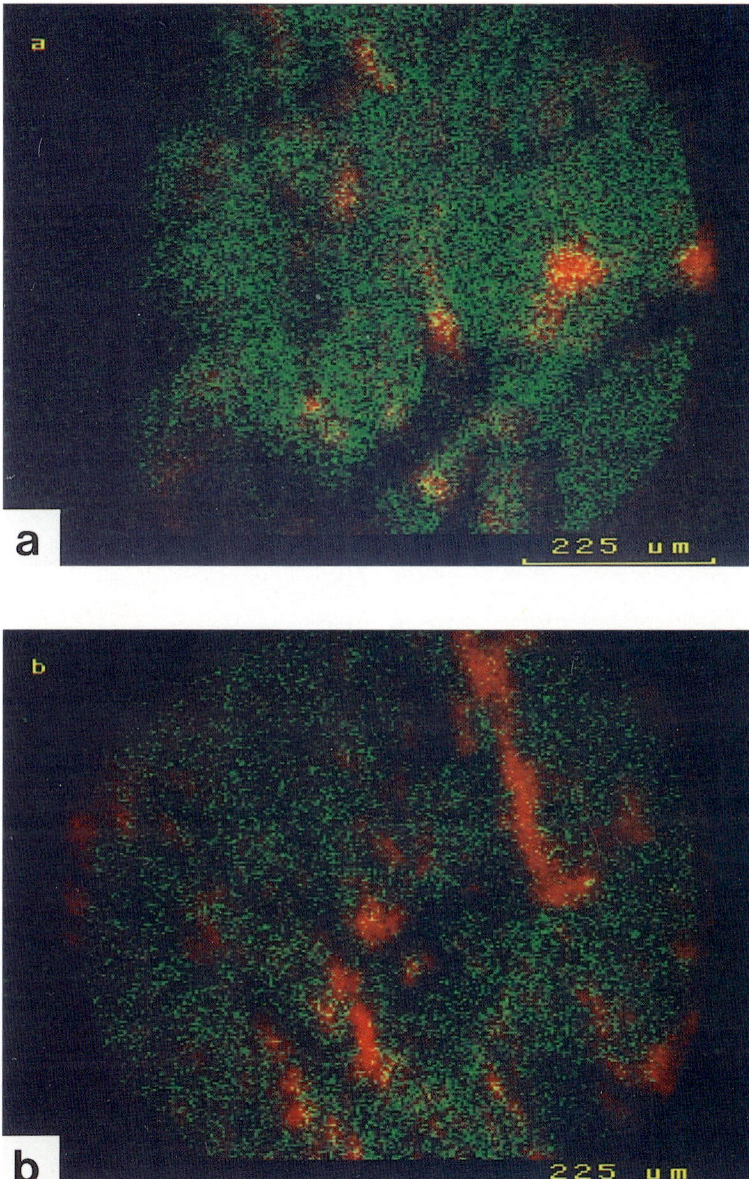

Plate 1 Dry sliding wear studies of a PTFE-filled polycarbonate material. XPS image overlays from unworn samples (a) and (b) and from worn samples (c–f) overleaf; F1s (representing PTFE) is shown in red, O1s (representing polycarbonate) is shown in green. The fluoropolymer spreads over the surface in the experimental sliding direction (right to left for (c) and bottom to top for (d–f)). Black regions in (a,b) arise from X-ray shadowing due to the topography of the roughly cut surface. (Reproduced with permission from [12].)

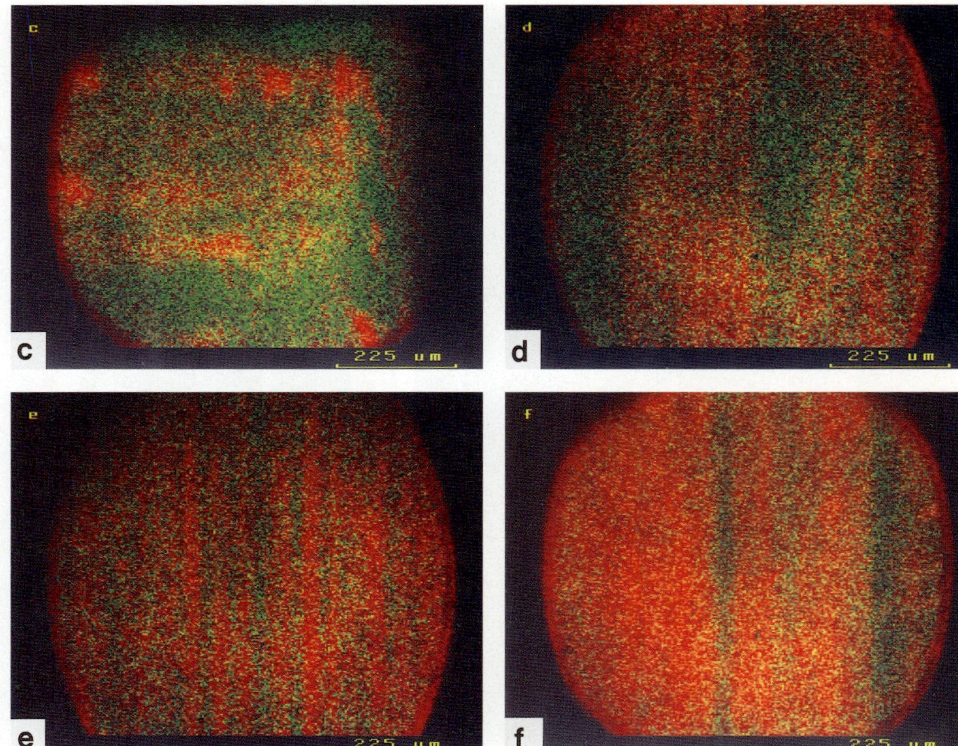

Plate 1 *Contd*

Plate 2 Negative-ion ToF-SIMS images from a three-component laminated film fracture surface. The laminate had the structure 1-3-2-3-1 (representing the component layering) and was failed in a peel test: (a) total negative ions, showing both topography and chemical heterogeneity; and overleaf, (b) mass-resolved images using ions characteristic of layers 1, 2 and 3. Heat scale of intensity used is black (zero), red, orange, yellow, white (highest). (Reproduced with permission from [60].)

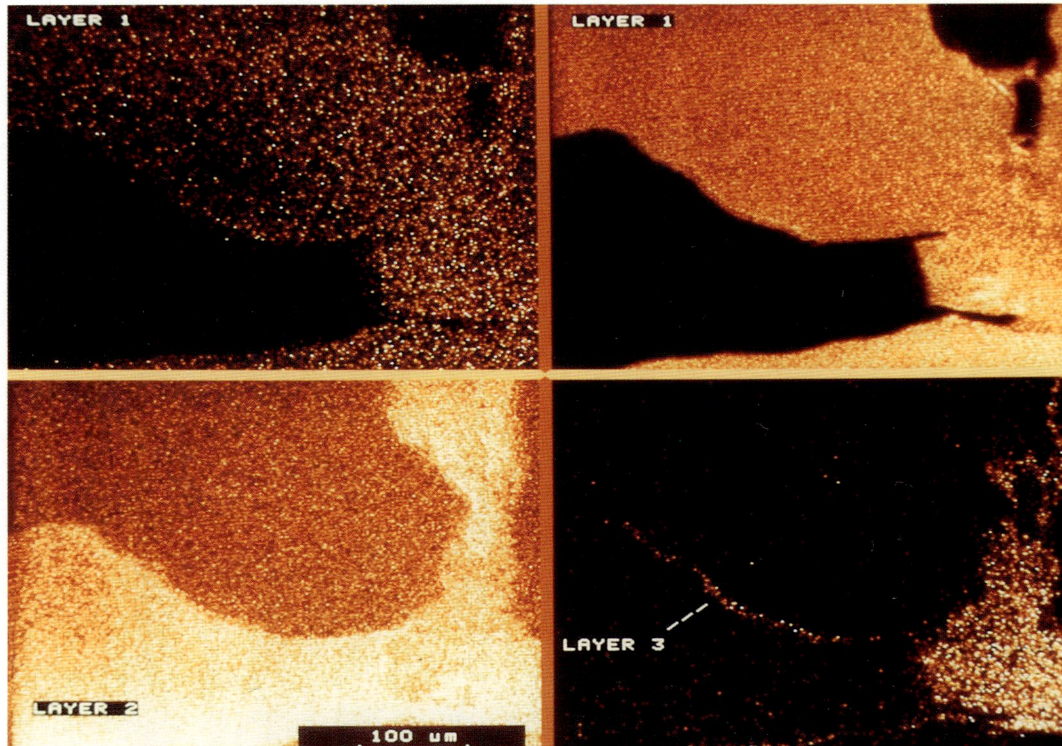

Plate 2 *Contd*

42. van der Hart, D.L. (1990) Proton spin diffusion as a tool for characterizing polymer blends. *Makromol. Chem., Macromol. Symp.* **34** 125–59.
43. Miller, J.B., McGrath, K.J., Roland, C.M. *et al.* (1990) Nuclear magnetic resonance study of polyisoprene/poly(vinylethylene) miscible blends. *Macromolecules*, **23** 4543–7.
44. Caravatti, P., Neuenschwander, P. and Ernst, R.R. (1986) Characterization of polymer blends by selective proton spin-diffusion nuclear magnetic resonance measurements. *Macromolecules*, **29** 1889–95.
45. Caravatti, P., Neuenschwander, P. and Ernst, R.R. (1985) Characterization of heterogeneous polymer blends by two-dimensional proton spin diffusion spectroscopy. *Macromolecules*, **28** 119–22.
46. Jelinski, L.W., Schilling, F.C. and Bovey, F.A. (1981) Carbon-13 nuclear magnetic resonance studies of solid segmented copolymers. 1. Mobile domains of a polyester thermoplastic elastomer. *Macromolecules*, **14** 581–6.
47. Pines, A., Gibby, M.G. and Waugh, J.S. (1973) Proton-enhanced NMR of dilute spins in solids. *J. Chem. Phys.* **59** 569–90.
48. Parmer, J.F., Dickinson, L.C., Chien, J.C.W. and Porter, R.S. (1987) Polymer–polymer miscibility determination via CP-MAS NMR in blends of deuterated and protonated polymers. *Macromolecules*, **20** 2308–10.
49. Gobbi, G.C., Silvestri, R., Russell, T.P. *et al.* (1987) Intermolecular polarization transfer study of polymer blend compatibility. *J. Polym. Sci. C: Polym. Lett.* **25** 61–5.
50. Zhang, X., Natansohn, A. and Eisenberg, A. (1990) Intermolecular cross-polarization studies of the miscibility enhancement of PS/PMMA blends through ionic interactions. *Macromolecules*, **23** 412–16.
51. Zumbulyadis, N. and Reilly, J.M. (1991) Polarization transfer across interfaces. 1. ^{29}Si cross-polarization dynamics at the poly(vinyl alcohol)-silica sol-gel interface. *Macromolecules*, **24**, 5294–8.
52. Duff, D.W. and Maciel, G.E. (1991) Monitoring postcure reaction chemistry of residual isocyanate in 4,4′-methylenebis(phenyl isocyanate) based isocyanurate resins by ^{15}N and ^{13}C CP/MAS NMR. *Macromolecules*, **24** 387–97.
53. Duff, D.W. and Maciel, G.E. (1991) Monitoring the thermal degradation of an isocyanurate-rich MDI-based resin by ^{15}N and ^{13}C CP/MAS NMR. *Macromolecules*, **24**, 651–8.
54. Duff, D.W. and Maciel, G.E. (1990) ^{13}C and ^{15}N CP/MAS NMR characterization of MDI-polyisocyanurate resin systems. *Macromolecules*, **23** 3069–79.
55. Schmidt-Rohr, K., Clauss, J., Bluemich, B. and Spiess, H.W. (1990) Miscibility of polymer blends investigated by ^{1}H spin diffusion and ^{13}C NMR detection. *Magn. Res. Chem.*, **28** S3–9.
56. Hughes, C.D., Sethi, N.K., Baltisberger, J.H. and Grant, D.M. (1989) Polyethylene crystallinity from static, solid-state NMR spectra. *Macromolecules*, **22** 2551–4.
57. Brandolini, A.J., Apple, T.M., Dybowski, C. and Pembleton, R.G. (1982) Solid-state ^{19}F NMR of deformed PTFE. *Polymer*, **23** 39–42.
58. Brandolini, A.J. and Dybowski, C. (1983) Determination of orientation distribution functions using solid-state nuclear magnetic resonance. *J. Polym. Sci.: Lett. Ed.*, **21** 423–8.
59. Brandolini, A.J., Alvey, M.D. and Dybowski, C. (1983) The crystallite orientation distribution functions of deformed polytetrafluoroethylene: multiple-pulse ^{19}F NMR spectroscopy. *J. Polym. Sci. Phys. Ed.*, **21** 2511–34.
60. Brandolini, A.J., Rocco, K.J. and Dybowski, C. (1984) Solid-state ^{19}F NMR investigations of annealed polytetrafluoroethylene. *Macromolecules*, **17** 1455–8.

61. Henrichs, P.M. (1987) Molecular orientation and structure in solid polymers with ^{13}C NMR: a study of biaxial films of poly(ethylene terephthalate). *Macromolecules*, **20**, 2099.
62. Schaefer, J. Stejskal, E.O. and Buchdahl, R. (1977) Magic-angle ^{13}C NMR analysis of motion in solid glassy polymers. *Macromolecules*, **10** 384–405.
63. Takahashi, M., Ookubo, M. and Takenaka, H. (1991) Solid-state ^{13}C NMR spectra analysis of alkalicellulose. *Polymer*, **23**, 1009–14.
64. Brueckner, S., Meille, S.V., Sozzani, P. and Torri, G. (1990) Gamma isotactic polypropylene: an MAS ^{13}C NMR study of a crystalline polymer with non-parallel chains. *Makromol. Chem. Rapid Comm.*, **11** 55–60.
65. Kaji, A., Ohta, Y., Yasuda, H. and Murano, M. (1990) Phase structural analysis of ultra high-molecular weight polyethylene fibers by solid-state high-resolution NMR. *Polym. J.* **22** 455–62.
66. Dec, S.F., Wind, R.A. and Maciel, G.E. (1987) Solid-state fluorine-19 nuclear magnetic resonance study of fluorocarbon polymers. *Macromolecules*, **20** 2754–61.
67. Taylor, P.C., Baugher, J.F. and Kriz, H.M. (1975) *Chem. Rev.* **75** 203.
68. Jelinski, L.W. (1986) Deuterium NMR of solid polymers, in *High-Resolution NMR Spectroscopy of Synthetic Polymers in Bulk*, (ed. R.A. Komoroski), VCH Publishers, Deerfield Beach FL, pp. 335–64.
69. Dumais, J.J., Jelinski, L.W., Leung, L.M. *et al.* (1985) Phase separation in polyurethanes – a deuterium NMR study. *Macromolecules*, **18** 116–9.
70. Kornfeld, J.A., Spiess, H.W., Nefzger, H. *et al.* (1991) Deuteron NMR measurements of order and mobility in the hard segments of a model polyurethane. *Macromolecules*, **24** 4787–95.
71. Dumais, J.J., Cholli, A.L., Jelinski, L.W. *et al.* (1986) Molecular basis of the β-transition in poly(arylene ether sulfones). *Macromolecules*, **29**, 1884–9.
72. Cholli, A.L., Dumais, J.J., Engel, A.K. and Jelinski, L.W. (1984) Aromatic ring flips in a semicrystalline polymer. *Macromolecules*, **17** 2399–404.
73. Schaefer, D., Spiess, H.W., Suter, U.W. and Fleming, W.W. (1990) Two-dimensional solid-state NMR studies of ultraslow chain motion: glass transition in atactic polypropylene versus helical jumps in isotactic polypropylene. *Macromolecules*, **23**, 3431–9.
74. Doverspike, M.A., Conradi, M.S., DeReggi, A.S. and Cais, R.E. (1989) Deuterium magnetic resonance study of orientation and poling in poly-(vinylidene fluoride) and poly(vinylidene fluoride-co-tetrafluoroethylene. *J. Appl. Phys.* **65** 541–7.
75. Ballinger, J.M., Kasuboski, L.A. and Dybowski, C. (1987) Spin-lattice relaxation of dipolar energy of polypropylenes. *Macromolecules*, **20** 1012–4.
76. Chu, C.W., Dickinson, L.C. and Chien, J.C.W. (1990) Study of critical phenomena of polystyrene-polyvinyl methyl ether blends by solid-state NMR. *J. Appl. Polym. Sci.* **41** 2311–25.
77. Percec, S. and Hammond, T. (1990) Phase behavior of blend systems containing nitrile copolymers examined by solid-state nuclear magnetic resonance and differential scanning calorimetry. *Polymer*, **32** 1252–7.
78. Liu, Y., Roy, A.K. and Jones, A.A. (1990) An NMR study of plasticization and antiplasticization of a polymeric glass. *Macromolecules*, **23** 968–77.
79. Webb, A.G. and Hall, L.D. (1991) An experimental overview of the use of nuclear magnetic resonance imaging to follow solvent ingress into polymers. *Polymer*, **32** 2926–38.
80. Weisenberger, L.A. and Koenig, J.L. (1989) NMR imaging of solvent diffusion in polymers. *Appl. Spectrosc.*, **43** 1117–26.

81. Perry, B.C. and Koenig, J.L. (1989) A ^{1}H magnetic resonance imaging study of porosity in polyurethane foams. *J. Polym. Sci. A: Polym. Chem.* **27** 3429–38.
82. Smith, S.R. and Koenig, J.L. (1991) Observation of cross-link density dispersion in sulfur donor-vulcanized polybutadiene using NMR imaging. *Macromolecules,* **24** 3496–504.
83. Bluemler, P. and Bluemlich, B. (1991) Aging and phase separation of elastomers investigated by NMR imaging. *Macromolecules,* **24** 2183–8.
84. Chang, C. and Komoroski, R.A. (1989) NMR imaging of elastomeric networks. *Macromolecules,* **22** 600–7.
85. Garrido, L., Mark, J.E., Sun, C.C. *et al.* (1991) NMR characterization of elastomers reinforced with *in situ* precipitated silica. *Macromolecules,* **24** 4067–72.
86. Jackson, P., Clayden, N.J., Walton, N.J. *et al.* (1991) Magnetic resonance imaging studies of the polymerization of methylmethacrylate. *Polym. Intl.* **24** 139–43.
87. Guenther, E., Bluemich, B. and Spiess, H.W. (1990) Deuteron double-quantum NMR imaging of molecular order and mobility in solid polymers. *Mol. Phys.* **71** 477–89.

8

Polymer surface characterization by XPS and SIMS

D. Briggs

8.1 INTRODUCTION

Despite the vast range of available polymers and the corresponding possibilities for property selection it is a fact that in many applications the surface properties of polymeric structures undermine the benefits derived from bulk properties. Inadequate adhesion is a notable example; consequently, surface pretreatment procedures have been developed to modify surface behaviour. In this example, as in other examples of polymer surface behaviour, the region of structure–property determination generally extends only a few molecular layers (≤ 10 nm) from the surface into the solid.

Clearly, in order to investigate these important aspects of surface behaviour it is necessary to obtain compositional and structural information specifically from this depth regime. In general this can only be achieved by utilizing the ultra-high vacuum (UHV) techniques which have been developed since the late 1960s and which are inherently **surface specific** (as opposed, for instance, to reflection infra-red techniques which in certain favourable cases may have the required **surface sensitivity** but which typically probe 10^3 nm into the solid).

Of these techniques, only two have established themselves for the characterization of polymer surfaces, namely X-ray photoelectron spectroscopy (XPS or ESCA) and secondary ion mass spectrometry operated in the 'static' mode (static SIMS). The term 'characterization' is used here to convey the elucidation of compositional and/or structural information. As will be seen, the techniques are highly complementary and are increasingly used in tandem.

XPS was first applied to polymers in 1971 and therefore has a twenty year history. During the last three years, however, the instrumentation has undergone a step improvement leading to enhanced levels of spectral

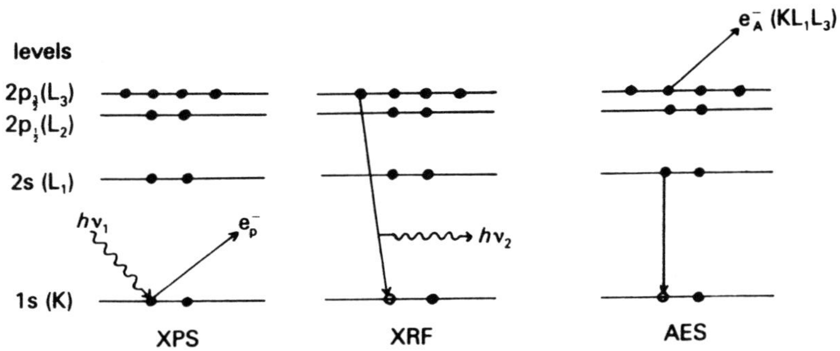

Fig. 8.1 Photoemission and subsequent relaxation processes (X-ray fluorescence and Auger electron emission) in XPS.

information. Since there are numerous earlier reviews (e.g. references 1–4), this chapter will emphasize these recent developments. Static SIMS of polymers is ten years younger and is still very much a developing technique; this will be reflected in the treatment offered here.

8.2 X-RAY PHOTOELECTRON SPECTROSCOPY

8.2.1 Physical basis

In XPS the sample, inside a high-vacuum system (pressure 10^{-5} Pa) is irradiated with soft X-rays, usually Mg Kα (1253.6 eV) or Al Kα (1486.6 eV). The primary event is photoemission of a core electron but relaxation processes lead also to emission of Auger electrons as shown in Fig. 8.1. Electrons are also photoemitted from the molecular orbitals occupying the 'valence band', but with much lower intensity. A typical spectrum from a polymer is shown in Fig. 8.2; this is obtained by passing the emitted electrons into an electrostatic energy analyser and detecting them as a function of their kinetic energy, E_K. The binding energies, E_B, of the photoelectrons are obtained via the Einstein relation:

$$E_B = h\nu - E_k - \phi \tag{8.1}$$

where $h\nu$ is the X-ray photon energy and ϕ the sample work function. The core electron binding energies (BE) are highly characteristic and allow the identification of all elements except hydrogen (there are no peak overlap problems in polymer XPS). The peak intensities are proportional to the number of atoms sampled, and with the aid of appropriate sensitivity factors, atomic compositions can be calculated, with detection limits of ~ 0.2 atom %. The Auger peaks, whilst also element specific, are not a useful feature in polymer characterization and are not considered here.

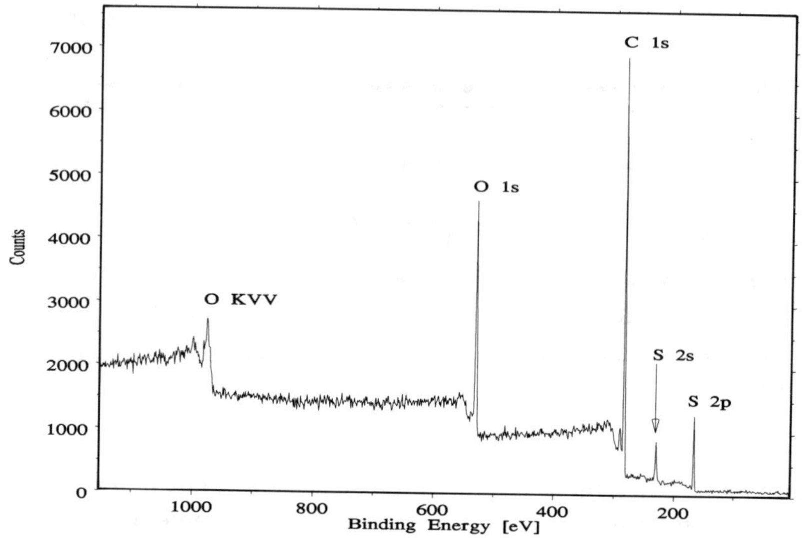

Fig. 8.2 Survey-scan spectrum of poly(ether sulphone) showing C 1s, O 1s, S 2s, 2p core levels.

Valence band spectra have been little utilized to date, but this situation will change (see later).

The information from XPS is surface specific because the electrons that give rise to the peaks in the spectrum have emerged from the polymer elastically (without energy loss). The fraction of electrons coming from various depths below the surface is conveniently summarized as 63, 87 and 95% from λ, 2λ and 3λ respectively where λ is the electron attenuation length (AL) in the sample. The AL is dependent on E_K and the relationship for polymers is still somewhat uncertain; however $\lambda \propto E^{0.5}$ is frequently used [5]. The actual sampling depth, d, also depends on the sample geometry:

$$d = 3\lambda \sin\theta \text{ (95\% of all electrons)} \quad (8.2)$$

where d and θ are defined in Fig. 8.3. Thus the depth sampled using Mg Kα for C 1s ($E_K = 963$ eV, $\lambda = 1.2$ nm) [6] would be 0.6 nm ($\theta = 10°$) and 3.6 nm ($\theta = 90°$).

8.2.2 Instrumentation

Full details of vacuum system design, X-ray sources and analyser electron optics can be found elsewhere [7]. Only the most germane points are discussed here. The photon sources commonly used are Mg Kα and Al Kα. These are actually unresolved doublets with line widths (full width of

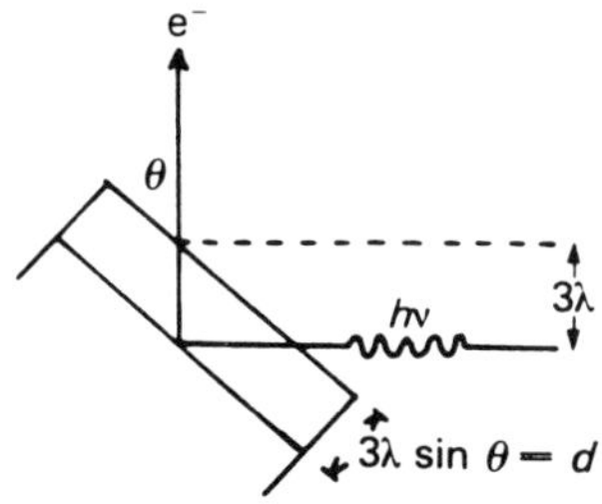

Fig. 8.3 Variation of sampling depth (d) with take-off angle (θ).

line at half maximum intensity, FWHM) of 0.7 and 0.85 eV respectively. Since the width of the X-ray line is the major instrumental contributor to the photoelectron line width (when operating the analyser under conditions of maximum energy resolution) there is an advantage in using Mg Kα. However, Al Kα sources can be run at higher power, giving better signal:noise spectra in a given time. Al Kα can also be monochromated by diffraction from quartz crystals to produce significantly lower line widths (~0.3 eV). Monochromation of the X-rays results in a large loss of intensity and, until recently, this has been a major limitation in polymer characterization where the photoemission cross-sections for core levels of interest (e.g. C 1s, O 1s and N 1s) are low. Instrumental improvements have been accelerated by the introduction of a high-sensitivity instrument employing a high-power, rotating-anode X-ray source (the Scienta ESCA 300) [8, 9]. Even after monochromation the X-ray flux is typically ten times higher than from conventional unmonochromated sources. A schematic of the design is given in Fig. 8.4. Most of the data used in this chapter are from this instrument. The effect of X-ray monochromation is illustrated by Fig. 8.5.

All insulating samples tend to charge up during XPS. With unmonochromated sources this is not a problem, since the X-ray source is also a copious source of low-energy electrons (coming from the thin aluminium window separating the X-ray source from the sample) which stabilize the surface potential at typically (+) 5–10 eV. With monochromated sources this window is now remote and charging can be severe (>100 eV). An external source of charge-compensating electrons is essential, usually from a low-energy (0–5 eV) electron 'flood gun'. This usually leads to a stable surface potential of (−) 2–5 eV.

There are no particular problems in handling polymeric materials in the vacuum conditions demanded by the technique. Outgassing of volatiles is usually slow, although high additive level rubber formulations are an exception. Radiation damage from the X-ray beam is rarely observed during typical spectrum accumulation times, although as high energy or spatial resolution studies become more commonplace (requiring longer

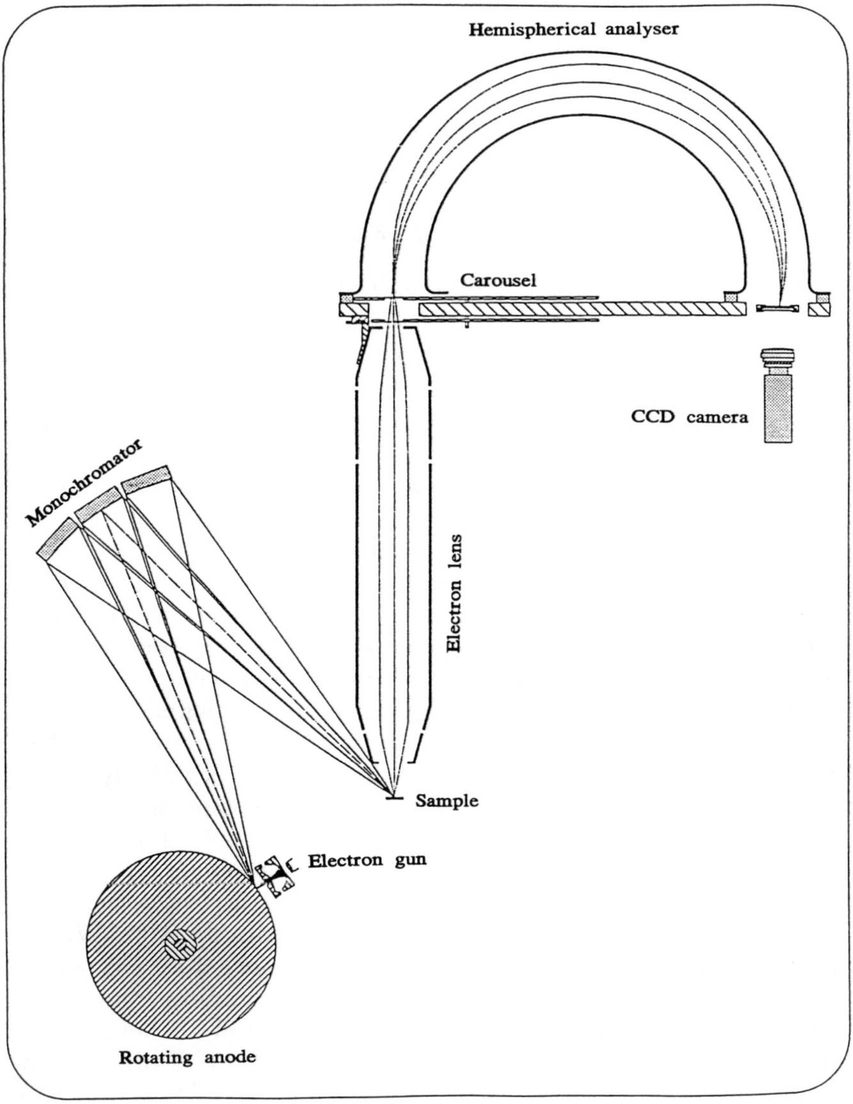

Fig. 8.4 Schematic of the Scienta Instruments AB ESCA300 instrument.

exposure times) damage effects will need to be considered. The higher energy bremmstrahlung radiation present in unmonochromated X-ray sources leads to higher damage rates over monochromated sources. Relative damage rates for over 100 homo-polymers have recently been published [10]. Thermal effects may also be a problem with unmonochromated sources due to the close proximity of the hot X-ray source.

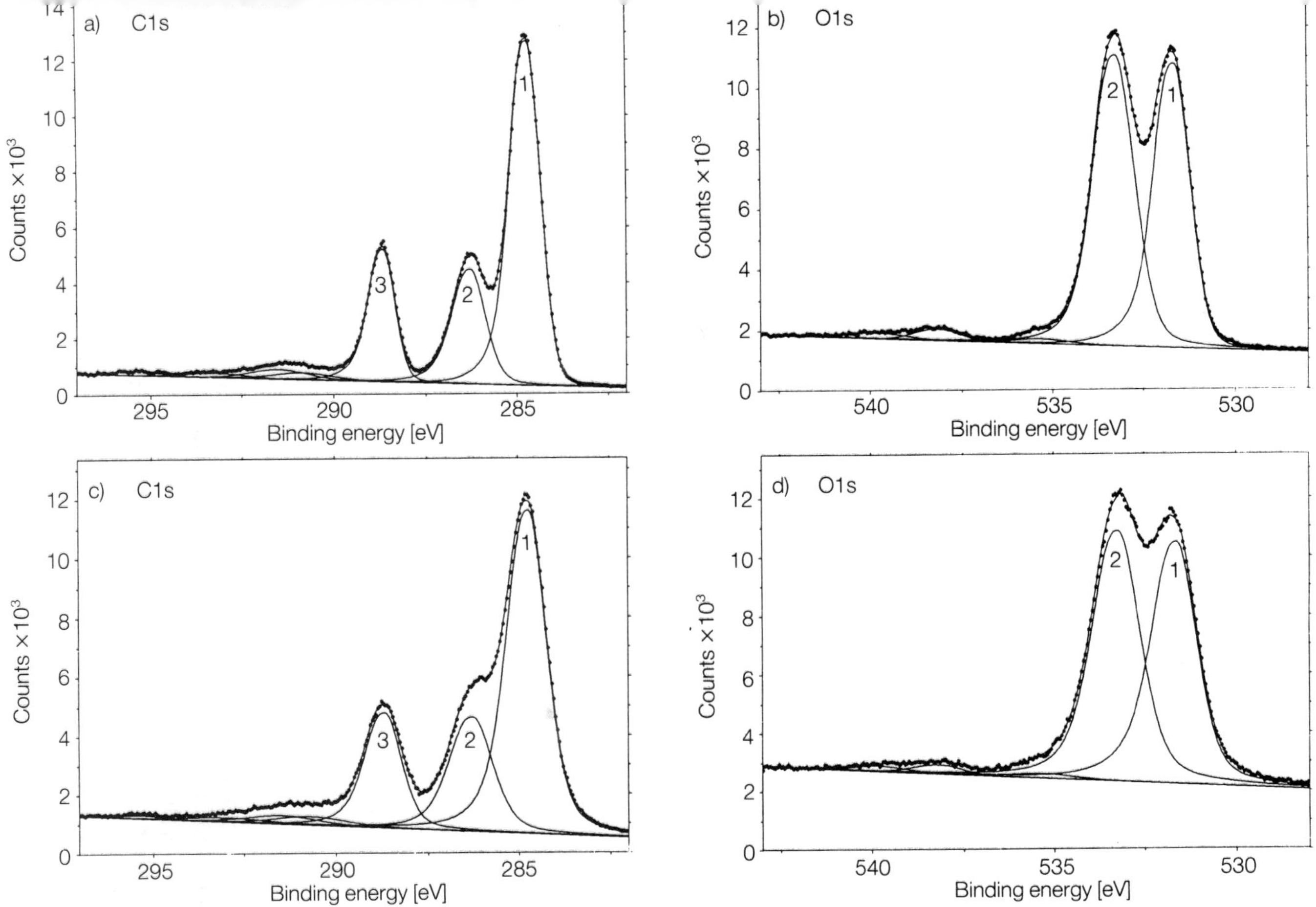

Fig. 8.5 Comparison of C 1s and O 1s spectra from poly(ethylene terephthalate) using monochromated (upper) and unmonochromated Al Kα radiation with otherwise identical instrumental conditions.

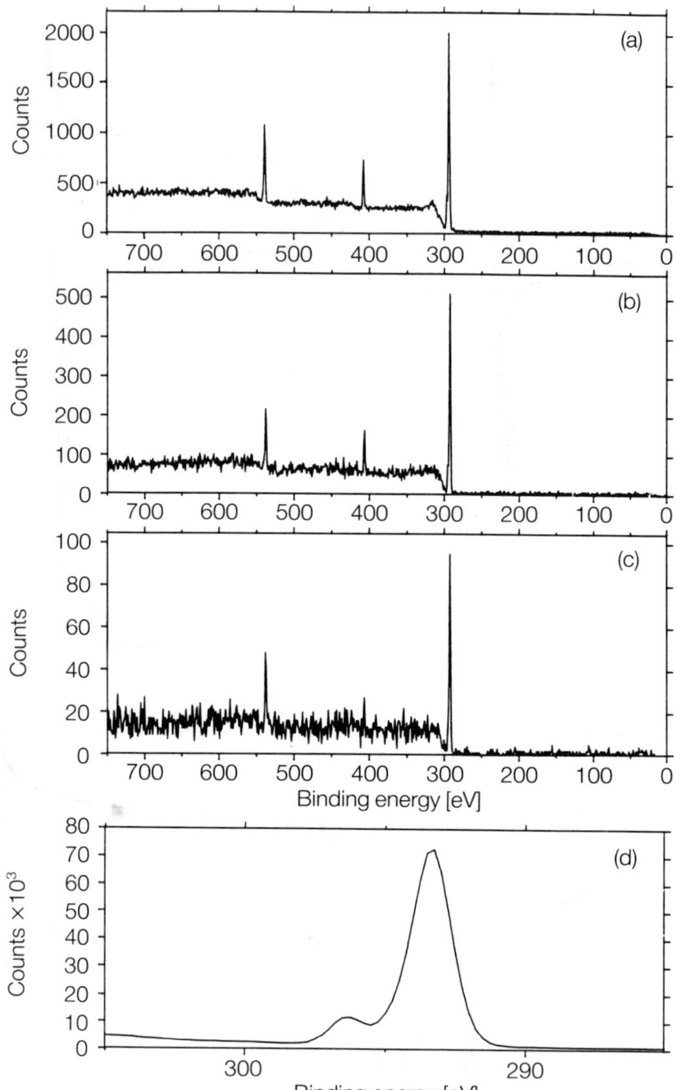

Fig. 8.6 XP spectra of a single nylon 6,6 fibre (50 μm diameter) obtained using the ESCA 300 instrument. Length of fibre analysed is (a) 3 mm, (b) 0.4 mm, (c) 0.04 mm (40 μm) using monochromated Al Kα from a rotating anode operating at 4.8 kW power and a data acquisition time of 9 min in each case. The energy resolution under the conditions used (500 eV pass energy) is illustrated by the C 1s spectrum in (d). (Reproduced with permission from [4].)

A particularly valuable instrumental facility is the ability to alter the electron 'take-off angle' (θ in Fig. 8.3) so that changes in composition within the available sampling depth can be studied – non-destructive depth profiling by angular variation experiments. Although this is only appropriate for samples with reasonably flat surfaces, such situations often pertain in polymer surface characterization.

Problems in polymer technology often involve sample heterogeneity, and hence spatial resolution is required. For some time instruments have offered a degree of spatial selection, typically a circular region with a diameter of 100–300 μm, either by the X-ray focusing action of a single quartz crystal monochromator or by the incorporation of defining apertures within a magnifying electron optical lens. More recently, the use of line focusing via X-ray monochromation combined with a 2D electron detector has allowed minimum spatial dimensions of $\sim 10\,\mu$m to be achieved together with line scanning, i.e. the generation of an intensity distribution along a chosen line on the sample (having some defined width) for any chosen photoelectron. This facility makes use of the imaging properties of the hemispherical electrostatic energy analyser. Alternatively, the full spectrum of the region covered by the line, or part of it, can be obtained [8, 9]. An example of this micro-analytical feature is shown in Fig. 8.6. Furthermore, a true imaging XPS instrument has also been recently introduced which gives a spatial resolution of 5–10 μm within a circular field-of-view of 0.7–1.1 mm. This relies entirely on novel electron optical lens designs [11]. High spatial resolution combined with moderate energy resolution during imaging leads to low detected signal levels, especially for low photoemission cross-section materials such as organic polymers. Nevertheless, useful studies can be performed as illustrated in Plate 1.

8.2.3 Spectral information

(a) Core levels

Apart from Li and Be which have very low photoemission cross-sections, the most intense core levels excited by Mg Kα and Al Kα have a relative sensitivity range of roughly one order of magnitude. Relative sensitivity factors are a function of instrument type and operational parameters, and consequently they are best derived *in situ* by studying homogeneous standards. Less accurately, they can be taken from published tables [13] or calculated [14]. Peak areas can then be converted into surface atomic composition data.

Any given core level also shows relatively small variations in BE depending on the chemical environment, referred to as chemical shifts. This is illustrated by the C 1s and O 1s spectra from poly(ethylene terephthalate) (PET) shown in Fig. 8.4. Much chemical shift data collected mainly during the 1970s has been collated and summarized [4], but this

Table 8.1 Primary C 1s chemical shifts (in eV)

Functional group	*Min.*	*Max.*	*Mean*	*Number of polymers*
C*–O–C	1.26	1.68	1.44	10
C*–OH	1.47	1.73	1.56	4
C*–O–C(=O)	1.48	1.98	1.70	16
C—C (epoxide, bridged by O)	–	–	2.02	1
C=O	2.81	2.97	2.90	3
O–C*–O	2.83	3.06	2.93	5
C–O–C*(=O)	3.84	4.23	4.05	18
HO–C(=O)	4.18	4.33	4.26	2
O–C(–O)–O	4.30	4.34	4.32	2
–C*(=O)–O–C(=O)	4.36	4.46	4.41	3
O–C(=O)–O	–	–	5.35	1

has recently been superseded by a database from over a hundred homo-polymers all run under identical (very high resolution) conditions. Extensive chemical shifts for C 1s, O 1s, N 1s, F 1s, Cl 2p, S 2p and Si 2p peaks are given [10]. By way of illustration, Table 8.1 reproduces the summary of **primary** chemical shifts for C 1s induced by oxygen functionality.

The range of chemical shifts for C 1s and N 1s in polymers is of the order of 10 eV, while for O 1s it is only about 4 eV. Therefore the dynamic range (chemical shift range relative to peak FWHM) is rather poor (cf. NMR spectroscopy) and depends critically on instrumental resolution. The

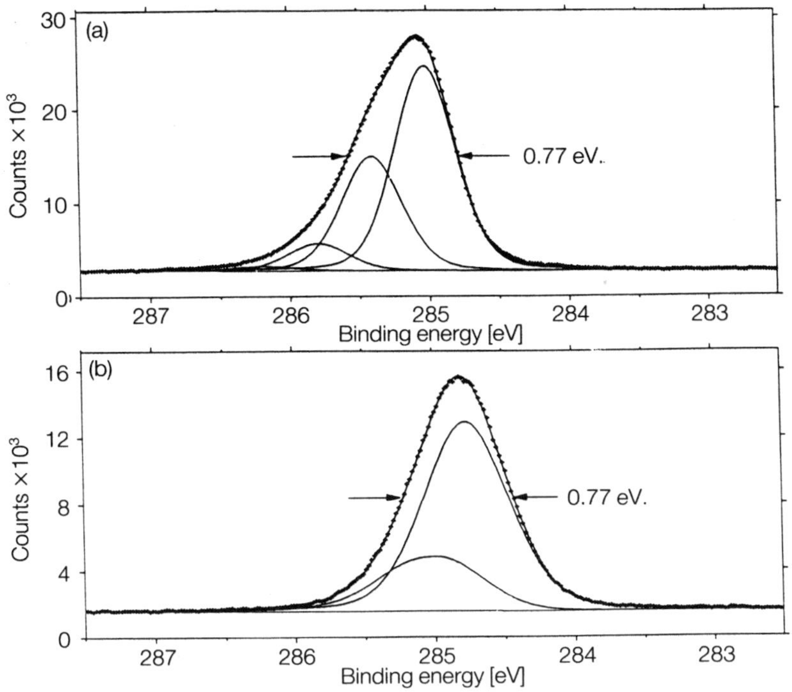

Fig. 8.7 (a) C 1s spectrum from hexatriacontane ($C_{36}H_{74}$) showing fitted vibrational components; (b) C 1s spectrum from polystyrene fitted with components due to aliphatic and aromatic (lower BE) carbon atoms. (Reproduced with permission from [15].)

dynamic range problem leads to complex core level envelopes, as in Fig. 8.4, which need to be 'deconvoluted' in order to accurately assign BE values to the individual components. This is actually done via iterative curve fitting alogorithms which require peak shapes to be defined. Prior to the recent high-resolution studies, component peaks were assumed to be symmetric and either pure Gaussian or a constant Gaussian/ Lorentzian mix. A further common assumption was that all components were of equal FWHM. These assumptions have now been invalidated.

Figure 8.7(a) shows the high-resolution C 1s peak for the linear hydrocarbon hexatriacontane, C_{36} H_{74}. The marked asymmetry on the high BE side has been shown to be due to vibrational fine structure, i.e. to the excitation of mainly C–H stretch vibrations during photoionization [15]. The fitted components have essentially the same relative intensities, at a fixed separation of 0.39 eV (C–H stretch vibration energy), as seen for gas-phase methane in which four components are well resolved [16]. The C 1s spectrum from polystyrene Fig. 8.7(b) is much more symmetrical even though the FWHM is the same as that from hexatriacontane [15]. Two components are fitted to the envelope with a fixed ratio of 6:2 and separated by 0.22 eV; the lower BE component representing the phenyl group is symmetric whilst the higher BE component representing the saturated hydrocarbon backbone is asymmetric as expected. The symmetric nature of the aromatic C 1s component is discussed in detail elsewhere [15]. Although the BE of C 1s in CH_X has been assumed to be independent of hybridization these results show this to be incorrect. *Ab initio* calculation indicates a similar separation of the polystyrene C 1s components to that observed [15], and for a large number of aromatic polymers the ring carbons typically emerged from curve fitting iterations with BEs ~ 0.3 eV lower than for saturated backbone carbons [10]. Furthermore, the C 1s peak shapes for unsaturated aliphatic polymers such as polyisoprene clearly indicate a shift between sp^2 and sp^3 carbon [10].

Vibrational fine structure effects, dominated by aliphatic C–H bonds, lead to significant variations in C 1s component peak widths (when an asymmetric single peak is used to model the vibrational envelope). For instance carboxyl groups and quaternary carbons, having no C–H bonds, give significantly narrower/more symmetric peaks. This has been discussed in detail in a study of PMMA [17]. Figure 8.8 illustrates the effect very well for two of the isomeric butylmethacrylate polymers [18]. These spectra also show two other features. Firstly, it is necessary to fit a component representing $\underline{C} - COOC_4H_9$; this is a secondary or β-substituent effect of the oxygens in the carboxyl group. The reality of such a chemical shift has been hotly debated in the past on the basis of evidence from much lower resolution spectra [17]. A recent study of 24 polymers containing structures which could potentially produce β-oxygen shifts due to C=O or O–C=O groups clearly demonstrated the effect [18]. Moreover, a clear difference was observed between situations where hydrogen bound to the β-shifted carbon atom is replaced by CH_3 (e.g. between acrylate and methacrylate polymers where the β-shift is ~ 0.4 eV and ~ 0.7 eV respectively). A similar methyl substituent effect is observed in Fig. 8.8. where the shift of $\underline{C}$–O is significantly greater for the *t*-butyl isomer than for the *n*-butyl isomer. Under typical unmonochromated XPS conditions these two spectra are virtually undistinguishable [1].

(b) Shake-up satellites

Photoemission of a core electron appears to the valence electrons associated with the atom like an increase in the nuclear charge. The resulting reorganization of the valence electrons (referred to as 'relaxation') may cause one of them to be excited to a higher unfilled level – the so-called 'shake-up' process. The probability of such a process, given

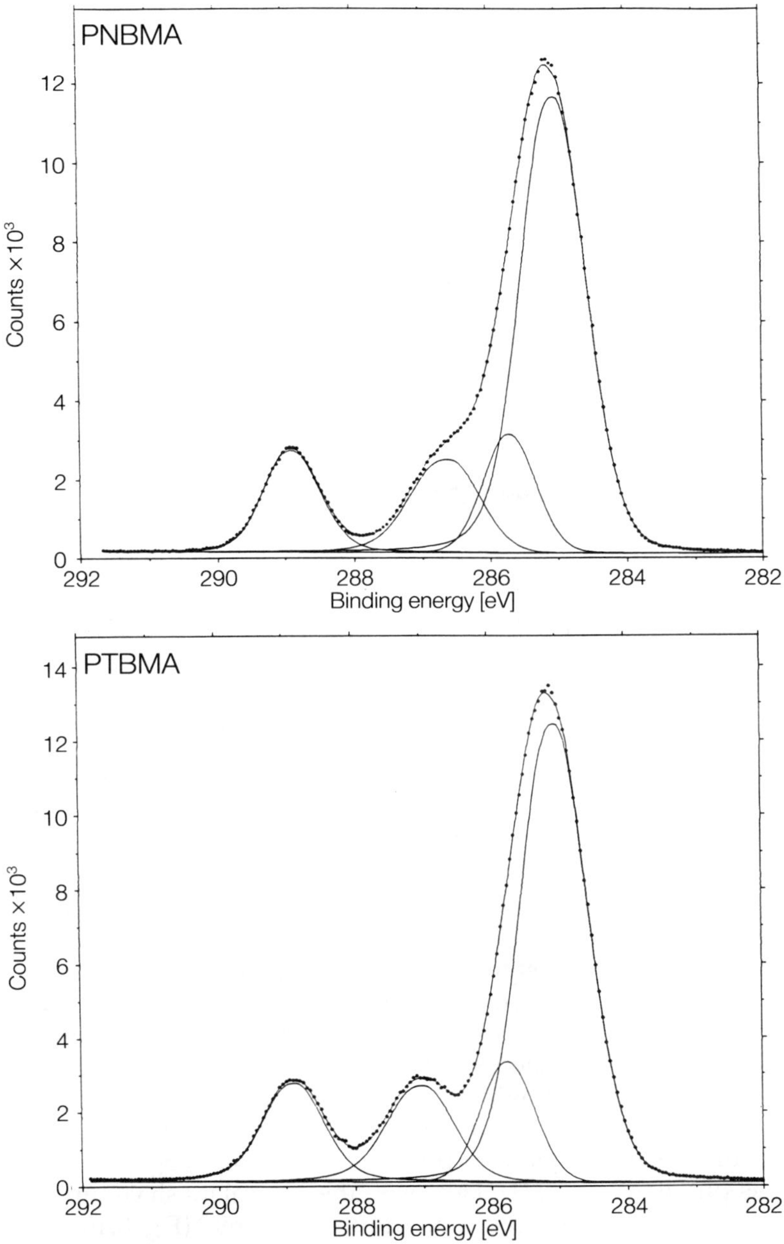

Fig. 8.8 C 1s spectra from poly(butyl methacrylates): *n*-butyl isomer (upper) and *t*-butyl isomer (lower).

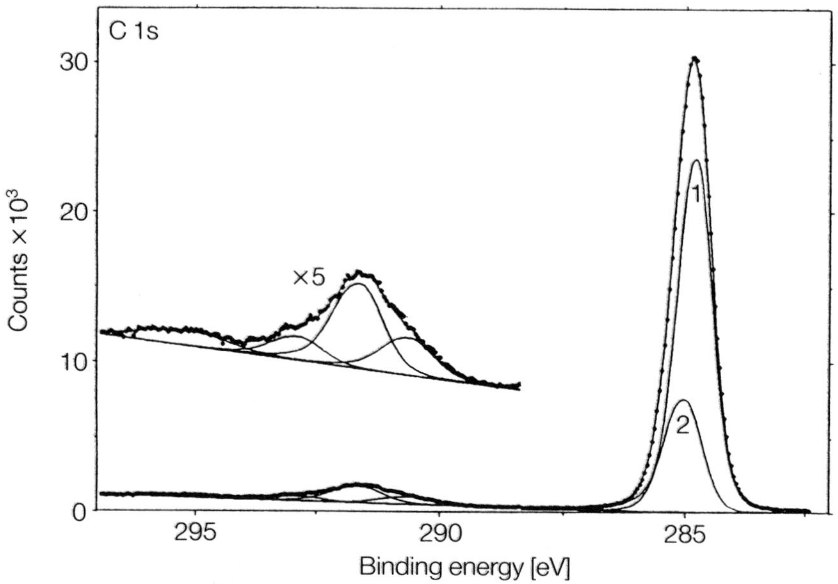

Fig. 8.9 Shake-up structure in the C 1s spectrum of polystyrene.

certain selection rules, increases as the energy gap between filled and unfilled levels decreases. Hence, shake-up is significant for conjugated and, especially, aromatic polymers where the transitions are of the $\pi \rightarrow \pi^*$ variety. The transition energy required is not available to the primary photoelectron and so the two-electron process leads to discrete structures on the low kinetic energy (KE) side of the photoelectron peak (shake-up satellite(s)). In simple aromatic systems such as polystyrene the shake-up intensity is 5–10% of the C 1s peak (Fig. 8.9). High-resolution spectra have shown that the polystyrene shake-up structure is exactly analogous to that of gas-phase benzene [19], showing that the pendant group is unperturbed by the attached hydrocarbon backbone. The four principal components in the benzene spectrum have been assigned by *ab initio* calculations and their origin in polystyrene and related substituted polymers is therefore well understood [20]. Similarly, the shake-up spectra for C 1s and N 1s from poly(4-vinyl pyridine) agree with those from gas-phase pyridine [21]. However, the equivalent shake-up spectra from isomeric poly(2-vinyl pyridine) are quite different (Fig. 8.10) and this illustrates the structure sensitivity of shake-up satellites [22].

Conjugated aromatic polymers can display intense shake-up structure which actually overlaps the photoelectron envelope. Polyimides fall into this class [10]. It is readily apparent that core level components involved in shake-up processes of such high probability have lost significant

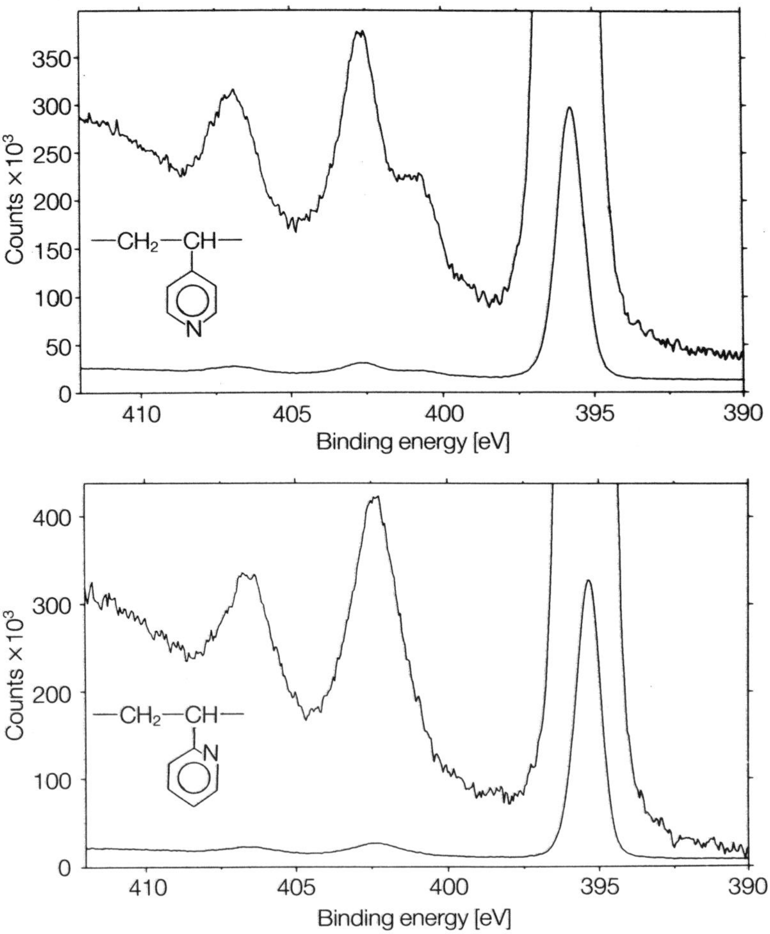

Fig. 8.10 N 1s shake-up spectra from poly(4-vinyl pyridine) (upper) and poly (2-vinyl pyridine) (lower).

intensity and will not, therefore, reflect the polymer stoichiometry. It has not, however, been appreciated that $\pi \rightarrow \pi^*$ shake-up in C=O units of aliphatic polymers can lead to similar effects on C 1s and O 1s component relative intensities. The component intensities derived from the complex C 1s envelope of poly(acetylacetoxyhydroxyethyl methacrylate) (Fig. 8.11) are given in Table 8.2. All the components representing C=O or O–C=O groups are significantly depleted in intensity whereas the other components have the expected stoichiometry [18]. This is an extreme case, but the typical C 1s loss from a study of many polymers was 13% [18]. The effect on the O 1s signal is similar (as seen in carboxylates

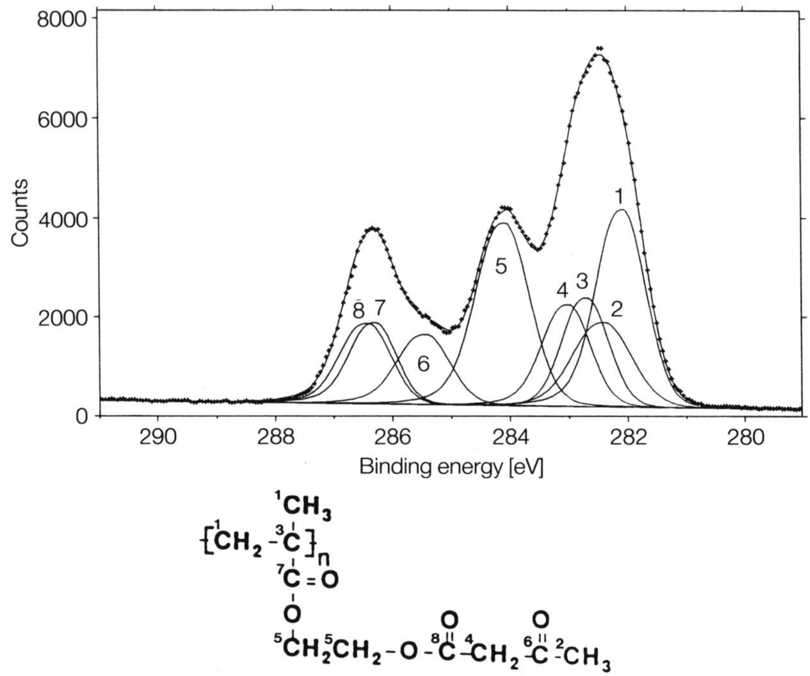

Fig. 8.11 C 1s spectrum from poly(acetylacetoxy hydroxyethyl methacrylate) components are correlated with the numbering system used on the molecular formula.

where only one oxygen is π-bonded) so that the **overall** C:O atomic ratios for these polymers remain close to stoichiometry. [17]. The discrete shake-up satellites from C=O containing polymers are difficult to observe

Table 8.2 Curve fitting parameters for poly(acetylacetoxy ethyl methacrylate)

	Position (eV)	*FWHM (eV)*	*Mix[a]*	*Asym[b]*	*Area (%)*	*Area ratio to ^{3}C*
^{1}C	285.00	0.99	0.91	0.10	21.3	1.99
^{2}C	285.33	1.18	0.92	0.08	10.7	1.00
^{3}C	285.63	0.93	0.97	0.07	10.7	1.00
^{4}C	285.94	0.96	0.92	0.10	10.7	1.00
^{5}C	287.01	1.08	0.92	0.10	21.4	2.00
^{6}C	288.36	1.06	0.92	0.12	8.1	0.76
^{7}C	289.22	0.95	0.94	0.12	8.4	0.79
^{8}C	289.35	1.02	1.00	0.11	8.7	0.81

[a] Gaussian/Lorentzian mixing ratio (1 = pure Gaussian).
[b] Asymmetry para meter, A, used in defining peak shape (see ref. [18]).

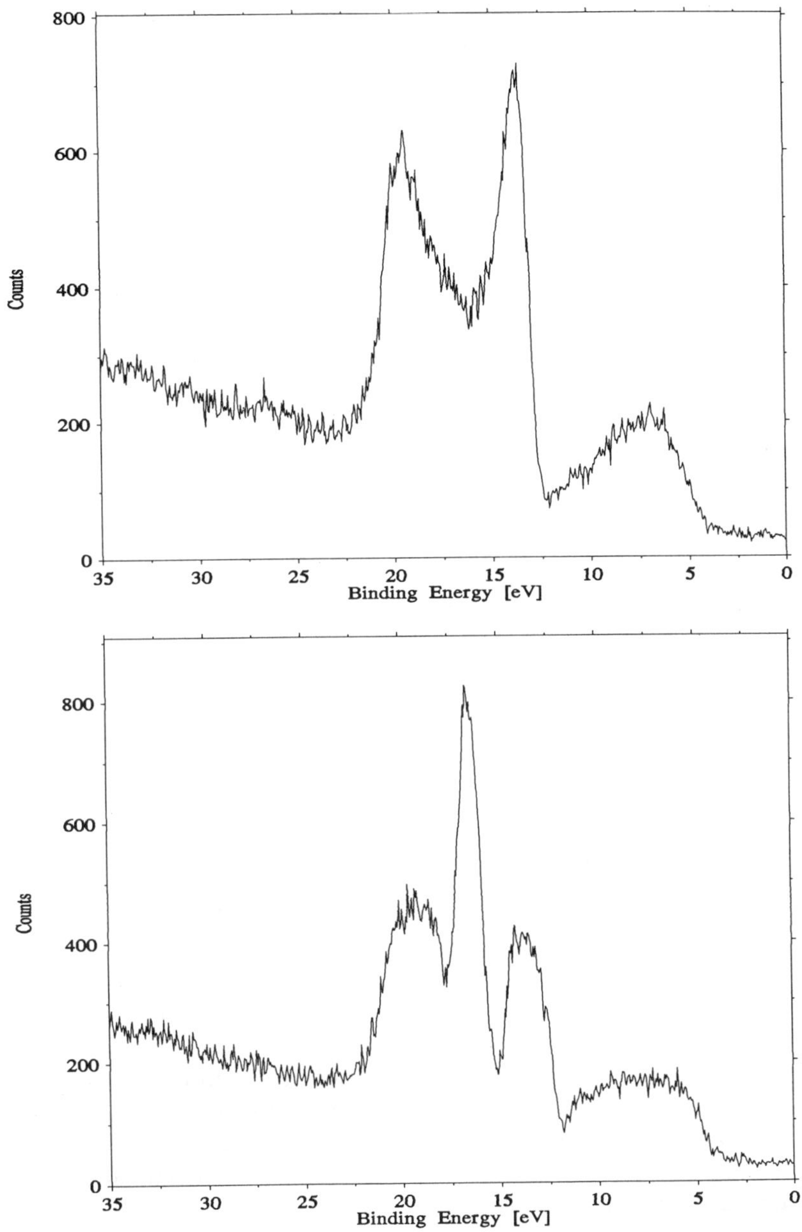

Fig. 8.12 High-resolution valence bands of high-density polyethylene (upper) and isotactic polypropylene (lower).

because they coincide with broad energy loss features, but in favourable cases they can be discerned [19].

(c) Valence band spectra

Photoelectron emission from the molecular orbitals of a polymer gives rise to the valence band (VB) spectrum, a complex envelope of overlapping peaks in the 0–30 eV BE region. Some of the features of the valance band can be quite sharp (e.g. individual peaks < 1 eV FWHM), and this rich structure leads to fingerprinting potential. The VB spectrum is a more direct reflection of polymer structure than the core level spectra and this is evidenced by early high-resolution studies by Pireaux *et al.* [23]. They demonstrated the sensitivity of VB spectra to various types of isomerization (structural-, linkage- and stereo-) as well as to tacticity and geometrical conformation. However, polymer VB spectra are typically 20–100 times less intense than the corresponding C 1s spectrum.

Figure 8.12 compares the VB spectra of polyethylene and polypropylene, whose C 1s spectra, even at very high resolution are essentially identical. Figure 8.13 compares the VB spectra of *cis*- and *trans*-polyisoprene, whose C 1s spectra are also identical. For a C 1s FWHM of ~ 1.1 eV all these hydrocarbon polymers are indistinguishable by their C 1s peaks, i.e. very high-energy resolution is required to observe different C 1s peak shapes [10].

(d) Functional group labelling (derivatization)

Because of the poor dynamic range of core level chemical shifts, many functional groups give C 1s and O 1s BEs which are very similar. Multifunctional surfaces produced by oxidizing polymer surface treatments present severe characterization problems because of the strong overlap of individual components in the core level envelope and the additional peak broadening from secondary shifts. There have been many attempts to overcome these problems through the use of specific derivatization agents which react with one functional group and label it with a distinctive element. Early studies used liquid-phase reactions which carry the risk of surface reorganization through solvent-induced transport effects and material loss. Gas-phase reagents are more difficult to find, but reduce these risks [24]. An example is the labelling of C–OH groups using trifluoroacetic anhydride vapour:

$$-OH + (CF_3CO)_2O \rightarrow O-\underset{\underset{\displaystyle O}{\|}}{C}-CF_3 + CF_3COOH$$

This reaction is reasonably specific and goes to completion within the XPS sampling depth [24a]. Quantification of the $-CF_3$ concentration either

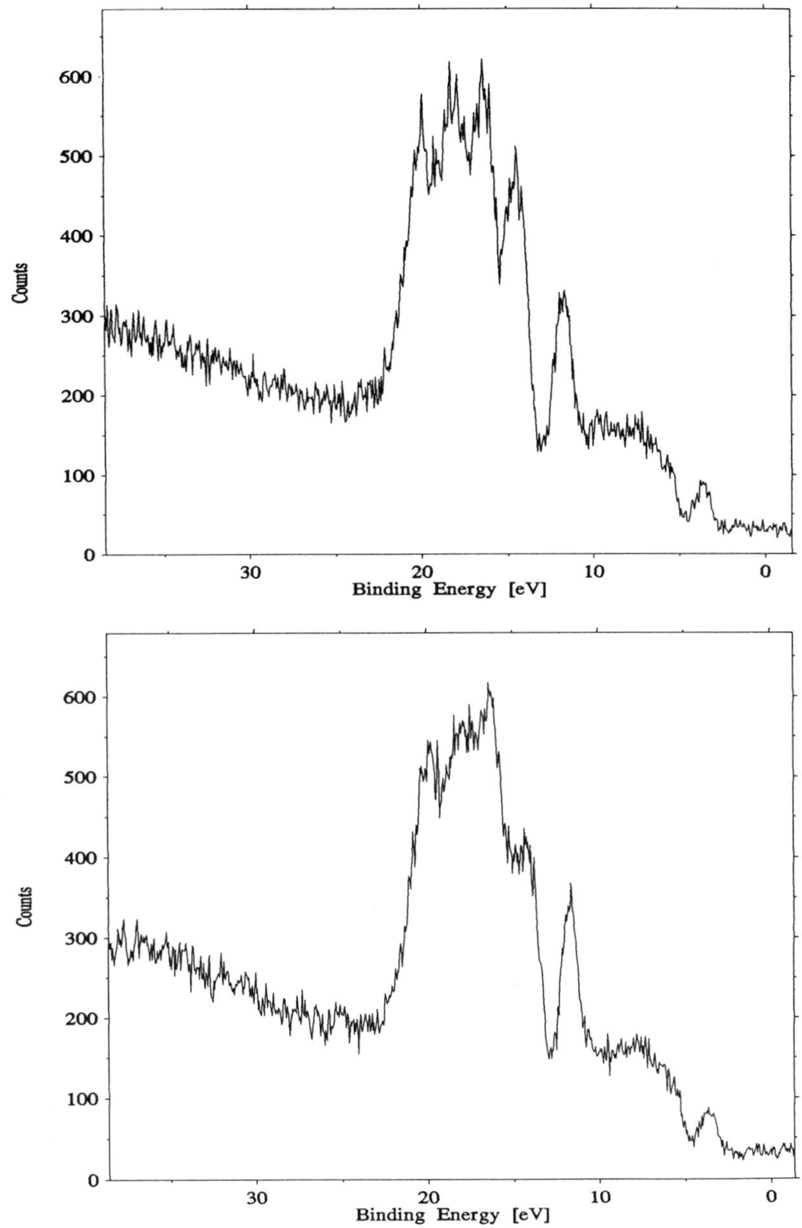

Fig. 8.13 High-resolution valence bands of *cis*-(upper) and *trans*-(lower) poly-isoprenes.

directly from the C 1s (CF_3) component or from the F 1s peak allows the original – OH concentration to be calculated. Use of labels such as F with high photoemission cross-sections leads to increased sensitivity towards the derivatized functional group. A comprehensive listing of reagents and original references is given in reference 4.

8.3 STATIC SECONDARY ION MASS SPECTROMETRY

8.3.1 Physical basis

In SIMS the sample, inside an ultra-high vacuum (UHV) chamber (pressure $<10^{-7}$ Pa) is bombarded by a primary ion beam. The material sputtered from the surface consists of positively and negatively charged atoms (elemental ions) or molecular fragments (cluster ions) and equivalent neutral species. Charged species make up only *c.* 5% of the material leaving the surface but these can be extracted into a mass analyser to yield, separately, positive and negative ion spectra. In the case of polymer surface characterization it is essential that the experiment be carried out in the static mode; i.e. the accumulated ion dose during spectra acquisition is sufficiently low for the surface to be essentially unperturbed during the measurement [25]. This threshold dose is a function of polymer structure and bombardment conditions, but 10^{13} ions cm^{-2} is widely quoted as the 'static SIMS limit' [26]. In the following discussion 'SIMS' is used as shorthand for static SIMS.

Bombardment energies vary from 2–30 keV with both noble gases (Ar, Xe) and liquid metals (Ga) being used to generate primary ions. Under these conditions primary ion penetration is in the nm to tens nm range. Energy deposition within this depth leads to direct 'knock-on' sputtering around the primary ion track, resulting in much bond breaking and the ejection of small fragments. The linear cascade of recoil atoms can lead to the relatively gentle desorption of large molecules or molecular fragments some distance from the primary ion impact site. Desorption of large fragments may also be the result of electronic interactions. At present an exact description of the processes leading to the emission of large secondary ion clusters is unavailable. However, the sputtered fragments can also rearrange and further fragment before detection, leading to a set of stable organic ions which form the spectrum, as in conventional organic mass spectrometry, with very high structural information content.

Increasing either the mass or energy of the primary ion leads to an increase in the total secondary ion yield and to a relative increase in the yield of larger cluster ions; however, the nature of the primary ion has a negligible effect on the type of secondary ions produced (see Table 8.3) [27]. At the same time the rate of sample damage also increases [26]

Table 8.3 Positive secondary ion relative intensities for P^{t}BuMA (after [9])

Ion	*Energy (keV)*	*Cluster mass (m/z)/relative intensity*										$\sum I$ *(Counts* s^{-1}*)**
		15	*29*	*41*	*57*	*69*	*81*	*95*	*109*	*121*	*141*	
He^+	4	7.0	40	75	100	1.8	0.2	–	–	–	–	1.3×10^5
Ne^+	4	6.5	47	66	100	6.5	2.2	2.2	3.4	2.3	–	5.4×10^5
Ar^+	4	7.7	46	63	100	7.7	1.5	1.2	2.4	1.8	0.8	4.6×10^5
Xe^+	4	4.9	41	63	100	22	3.3	4.1	7.3	4.9	3.3	9.4×10^5
Xe^+	2	5.0	31	56	100	8.7	2.5	2.5	4.4	2.9	1.9	5.2×10^5
Xe^+	1	3.8	31	46	100	6.9	2.1	2.3	3.8	2.7	1.5	1.3×10^5
Ga^+	10	5.5	28	50	100	14	5.5	7.8	10	8.3	2.8	1.1×10^6

* Summation of absolute intensities of tabulated peaks using a constant primary ion beam current density of 6.7×10^{-10} A cm^{-2}

and compromises need to be made that also take into account overall detection sensitivity. In particular, these considerations are important for quadrupole SIMS experiments, for which 4 keV Xe^+ at ≤ 1 nA cm^{-2} current density, has become fairly standard practice.

Under charged particle bombardment, insulating polymers charge-up. The positive primary ion current together with secondary electron emission leads to positive charging. Since most of the secondary ions detected are of very low energy (< 30 eV) any significant change in surface potential has serious consequences for spectral intensity, particularly of the high-mass cluster ions. Surface potential is controlled and, therefore charging is avoided, by the use of electron flood sources.

Estimates of the SIMS sampling depth have been made for polymers by two completely different approaches, both of which yield values of the order of 1 nm (see later) [28, 29].

8.3.2 Instrumentation

Vacuum conditions for SIMS are more stringent than for XPS because of the greater need to avoid surface contamination *in situ*. This is due to the greater surface sensitivity of SIMS and the greater confusion of the SIMS spectrum by fragment ions from contamination. True UHV operating conditions are therefore required.

Currently, there are two distinct types of static SIMS instrumentation. All the early developments of polymer SIMS were carried out on instruments employing quadrupole mass spectrometers (QMS). These are relatively compact, have a mass range of $m/z \sim 300$–1000 with unit mass resolution over the whole range, and are particularly convenient for use with multitechnique instruments (e.g. XPS + SIMS). However, the

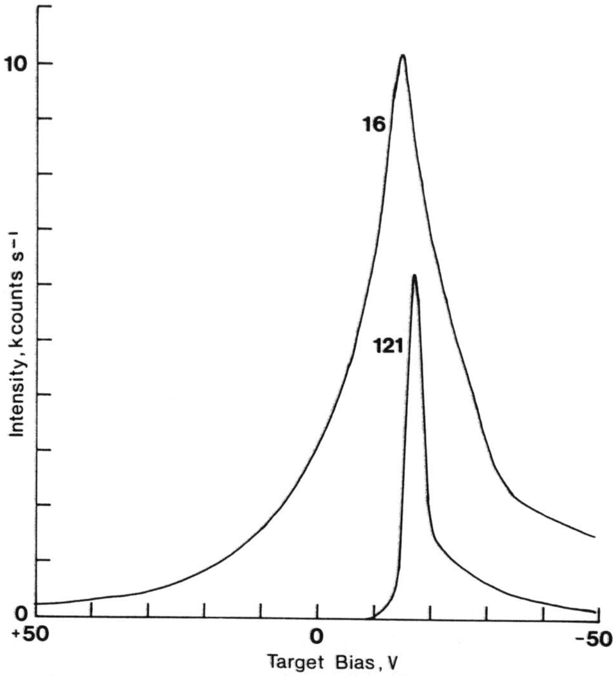

Fig. 8.14 Crude energy distributions of the ions O^-(*m/z* 16) and $C_6H_5COO^-$ (*m/z* 121) appearing in the negative-ion spectrum of poly(ethylene terephthalate).

collection efficiency is relatively low (~ 0.1% of emitted secondary ions detected) and spectrometer transmission falls with increasing mass ($T \propto M^{-1}$ to M^{-2}).

A QMS is a scanning device, i.e. only one mass at a time is detected as the mass range is scanned, and it can only accept ions in a narrow (low) energy band (e.g. 10 ± 2.5 eV). Figure 8.14 illustrates the energy distributions of atomic and cluster ions from a polymer, measured by following the intensity of each ion as the potential applied to the metal sample holder (target bias) is scanned. This convolutes the spectrometer band pass with the true energy distribution. Nevertheless, the narrow energy distribution of the molecular fragment is clear. Optimization of spectral intensity is obviously critical. This is achieved by stabilizing the surface potential using a simultaneous 'low' energy electron flood and fine tuning the potential by adjustment of the target bias. More detailed descriptions of these procedures are given in reference 25.

QMS instruments generally employ noble gas ion sources capable of delivering 1–5 keV primary ions either in a stationary defocused spot or as a focused spot (30–100 μm) raster scanned at TV rate. In either case the irradiated area is many mm² in order to establish a stable low current density (≤ 1 nA cm^{-2}). Some workers have advocated the use of fast atom sources to overcome insulator charging problems [30]. Here the noble gas ion source is coupled to a charge exchange cell such that the ion beam is neutralized by a low pressure of gas, e.g.

$$\overrightarrow{Ar}^{+} + Ar^{\circ} \rightarrow \overrightarrow{Ar}^{\circ} + Ar^{+}$$

Negligible momentum transfer occurs, so the diffuse atom beam produced has the same energy as the original ion beam. Although charging is reduced in positive-ion detection, atom bombardment still requires electron neutralization for negative-ion detection. Either excitation leads to the same spectra, although there is good evidence to suggest that beam damage effects are reduced under atom bombardment [31].

During the last five years time-of-flight (ToF) mass spectrometers have become very important. ToFMS offers unlimited mass range, transmission of 10% or better, parallel mass detection and much better mass resolution than QMS. The ToFMS requires a pulsed ion source. A primary ion pulse of ≥ 1 ns generates a pulse of secondary ions which are mass analysed via their flight time to a detector. Over a period of typically 100 μs the whole spectrum is obtained. During this period the sample can be flooded with low-energy electrons to prevent charging. Originally the high sensitivity of ToFMS was used to provide a high spatial resolution imaging capability in the static mode, by coupling the device to a pulsed fine-focus ($\sim$ 50 nm) liquid metal ion source. More recently ToFMS has been developed to provide high mass resolution (typically $M/\Delta M = 7000$ at m/z 28 and $> 10{,}000$ at $m/z > 100$) such that ions of nominally the same mass are resolved, as in Fig. 8.15. This is particularly important for polymers where many fragments containing O and N have the same nominal mass as common hydrocarbon fragments. Full details of the design and construction of SIMS hardware can be found in reference 32.

Micro-analysis and imaging of heterogeneous polymer surfaces using QMS–SIMS is very difficult [33]. The low detection sensitivity means that, generally, the dose limit of 10^{13} ions cm^{-2} is reached during the acquisition of one spectrum or one image (i.e. the intensity distribution for one value of m/z) from an area of 500 μm square. With highly focused sources, charging becomes a major problem during imaging because of the very high current density at any dwell point of the beam. With QMS, of course, each image requires a separate experiment and increasing damage to the imaged area. These problems are overcome with ToF-SIMS. Parallel mass detection ensures the simultaneous collection of spectra and several images (which may be of single species or summations of species) during a single experiment. Good quality spectra can be obtained from a primary ion dose of only 10^{8} ions. Areas small enough to fully realise the inherent resolution of the technique ($\sim$ 0.2 μm) can be analysed or imaged [34]. An example of polymer imaging is given in Plate 2.

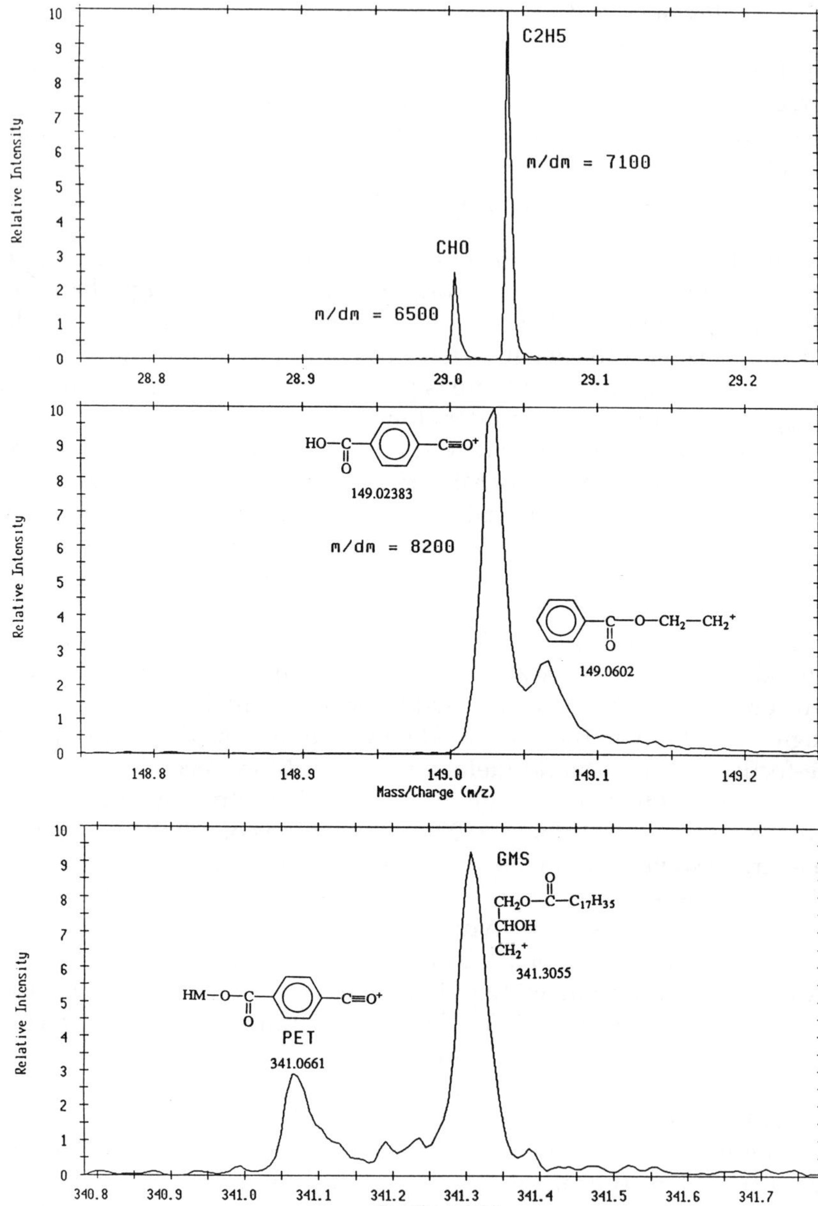

Fig. 8.15 High mass resolution ToF-SIMS spectra from a poly(ethylene terephthalate) (PET) film coated with a thin discontinuous layer of glyceryl monostearate (GMS), in the regions around *m/z* 29, 149 and 341. Note the resolution of two different fragments from PET at *m/z* 149 and of fragments from PET and GMS at *m/z* 341 (in the latter spectrum annotation M represents a repeat unit of the PET molecule). The assignments are possible because of a mass accuracy of 10–20 ppm. Data courtesy of Perkin Elmer, Physical Electronics Division. (See also [63])

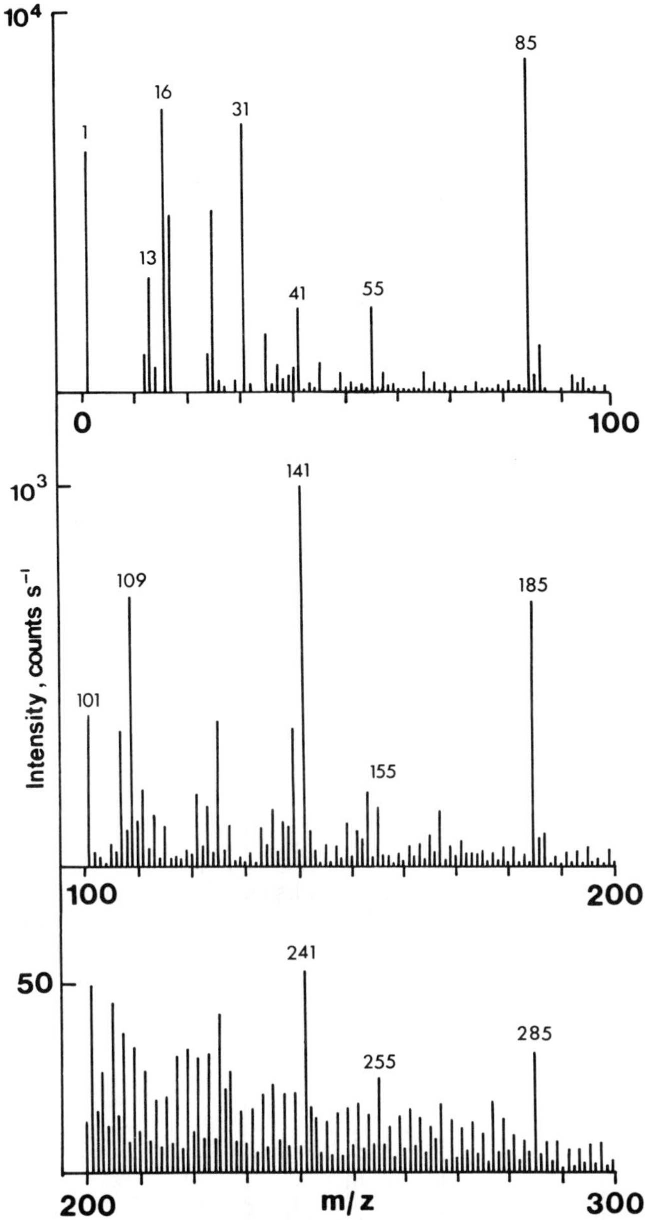

Fig. 8.16 Negative-ion spectrum of poly(methyl methacrylate) obtained using 4 keV Xe^+ at 1 nA cm^{-2} (*m/z* 0–200) and 3 nA cm^{-2} (*m/z* 200–300). (Reproduced with permission from [35].)

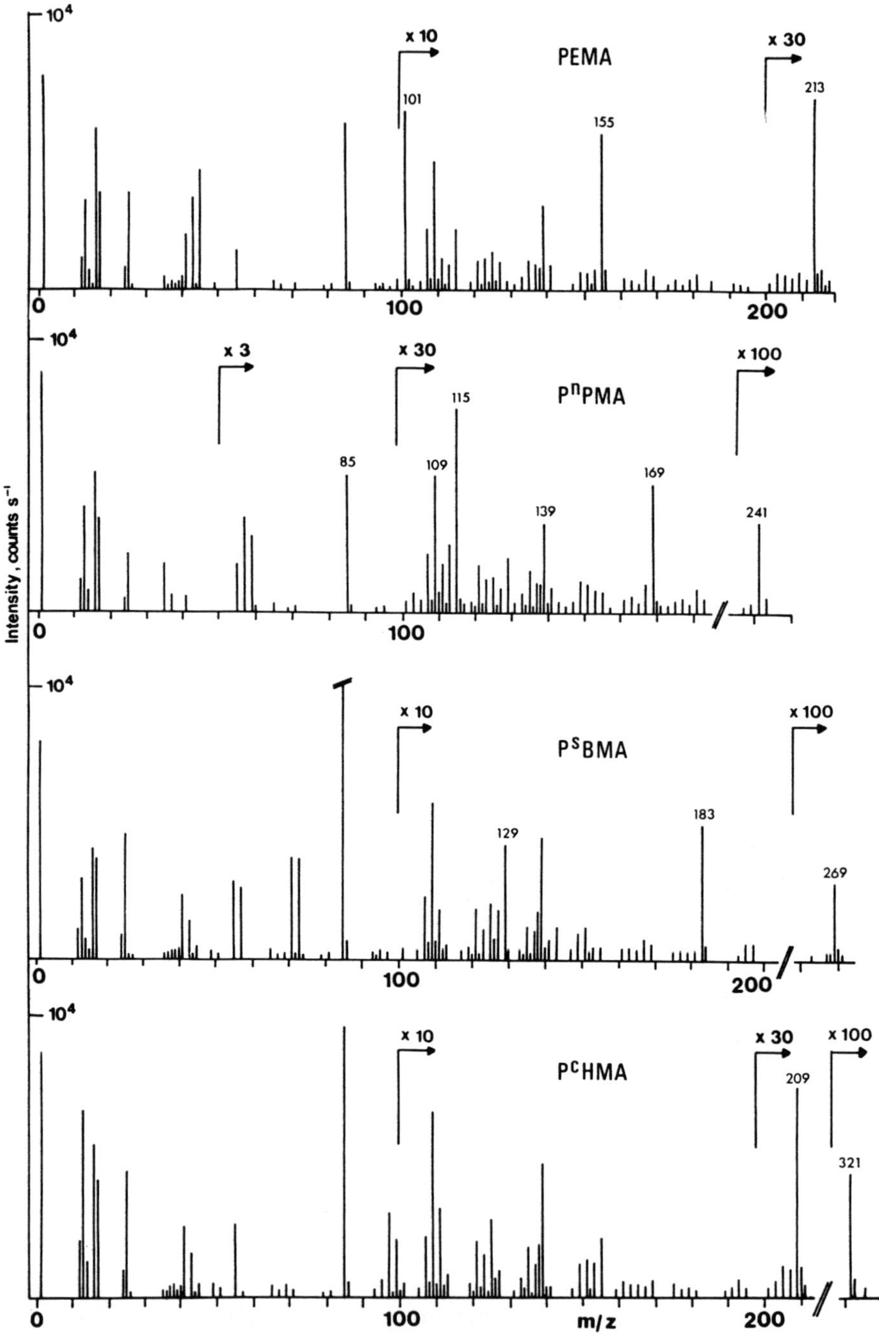

Fig. 8.17 Negative-ion spectra of poly(alkyl methacrylates) where from top to bottom R = C_2H_5, *n*-C_3H_7, *s*-C_4H_9 and cyclo-C_6H_{11}. (Reproduced with permission from [35].)

8.3.3 Spectral interpretation

(a) Fingerprint spectra of homo-polymers

The basis for the development of any new spectroscopic analysis technique is the creation of a spectral database from pure materials. SIMS of polymers is still in this stage of development, although significant progress has been made. The systematic variations in structure which arise naturally in families of polymers, through differing lengths of alkyl side-chains or of hydrocarbon backbone between in-chain functional groups, etc. have been exploited to allow systematic interpretation of fragmentation patterns for different classes of polymers. From these analyses it is clear that fragmentation leads to loss of intact side-chains and backbone cleavage at the point of monomer linkage such that distinctive peaks representing structurally informative fragments appear in the spectra. For example, Fig. 8.16 shows the negative-ion spectrum of PMMA which contains sets of peaks which repeat at intervals of *m/z* 100, which is the mass of MMA monomer [35]. Spectra from some of the higher members of the polymethacrylates are shown in Fig. 8.17. From consistent patterns in the spectra the structures illustrated in Fig. 8.18 have been assigned, with masses given for PMMA (R=CH_3).

For the polymers in Fig. 8.18 R=C_2H_5 (PEMA), *n*–C_3H_7 (P^nPMA), *s*–C_4H_9 (P^sBMA) and cyclo–C_6H_{11} (P^cHMA). (2*M*–15) ions are seen at *m/z* 213, 241, 269 and 321 respectively, whilst (*M*+55) ions appear at *m/z* 155, 169, 183 and 209 respectively. Furthermore, common ions throughout the series appear at *m/z* 85 and 139 which means that these ions do not involve the R group (Fig. 8.19)

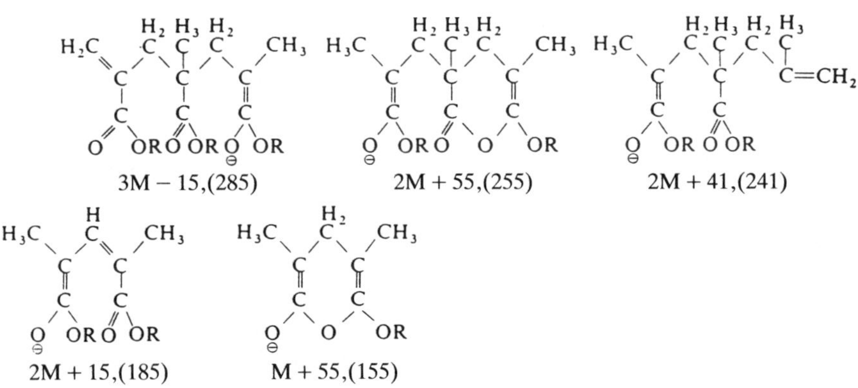

(masses are for R = CH_3)

Fig. 8.18

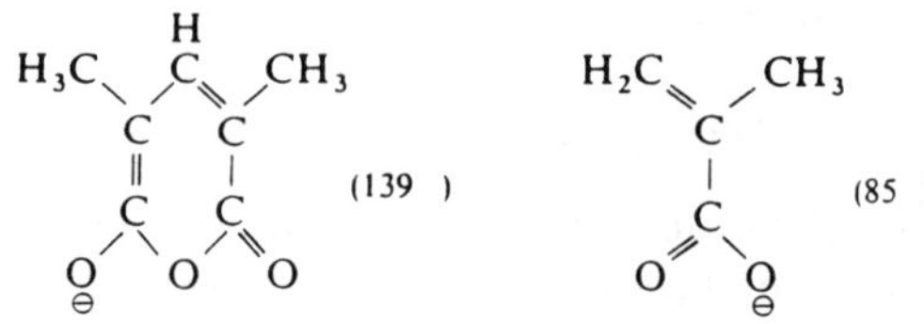

Fig. 8.19

Poly(methacrylic anhydride) contains the proposed ring structure in its repeat unit and the negative-ion spectrum is dominated by *m/z* 139 [35].

The positive-ion spectra of these methacrylate polymers are dominated by hydrocarbon fragments, $C_xH_y^+$, and are quite similar above *m/z* 80. However, for *m/z* < 80 the ions arise mainly from fragmentation of the ester side-chain (R) and the pattern is diagnostic. For instance, the isomers of the four butylmethacrylate polymers are distinguishable [36]. By contrast the positive-ion spectra of polyamides (nylons) are much more informative than the negative-ion spectra [37]. Figure 8.20 compares the spectra of nylon 6 and nylon 66 which have the isomeric structures:

$$[CH_2(CH_2)_3CH_2CONH]_n \text{ and } [NH(CH_2)_6NHCO(CH_2)_4CO]_n$$

nylon 6 nylon 66

Note the prominence of even mass peaks representing N-containing ions. The spectral interpretation is presented elsewhere [37]. This example is used to illustrate the molecular specificity of SIMS; the XPS of these polymers are identical in the core levels and only just distinguishable in the valence band (at the highest energy resolution) [10].

Using the systematic (homologues) approach many types of polymer spectra have been interpreted. These include poly(alkyl acrylates) [35], poly(alkyl-methacrylates) [35], polyamides [37], polyglycols [38], segmented polyurethanes [38], polycarbonates [39], polyesters [40], polyorthoesters [41], polyanhydrides [42], aliphatic hydrocarbon polymers (saturated and unsaturated) [43, 44], polystyrenes [45] and poly(vinyl ethers/ketones/carboxylates) [46]. Two databases of fingerprint spectra plus peak assignments for many common polymers have also been published [25, 47] (although one only contains positive-ion spectra [47]).

In the course of this kind of spectral interpretation most authors have compared and contrasted polymer fragmentation patterns with those of molecular analogues produced by conventional mass spectrometry (particularly electron impact, EIMS). No clear rules emerge, although stable positive ions produced by any ionization technique will obviously feature strongly and suggest a correlation, in terms of ion formation mechanisms, which may be more apparent than real. It has been pointed

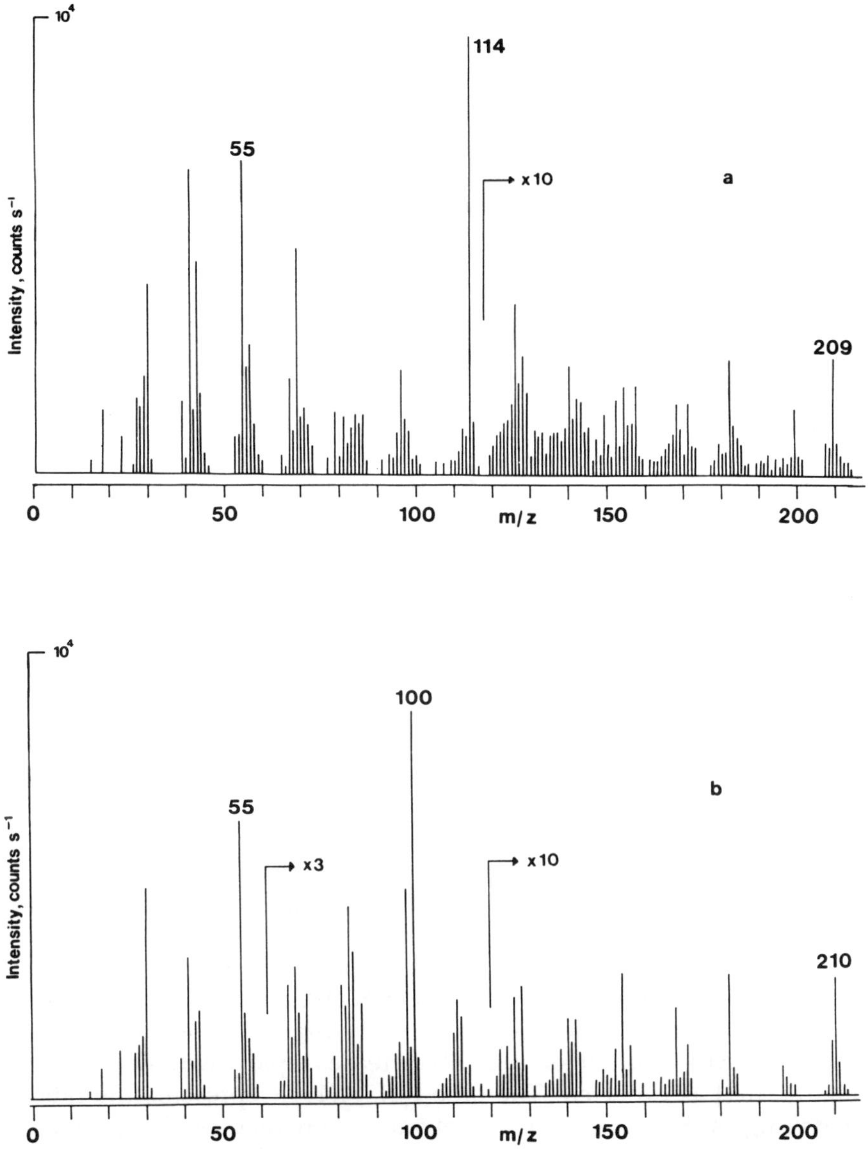

Fig. 8.20 Positive-ion spectra from nylon 6 (upper) and nylon 6,6 (lower). (Reproduced with permission from [37].)

out, for instance, that whilst odd-electron ions dominate over even-electron ions in EIMS, the reverse is true in polymer SIMS [48]. Recently the use of tandem mass spectrometry techniques (MS–MS) to study the collisionally activated decomposition (CAD) of ions generated in the polymer SIMS process has led to several insights into ion formation mechanisms and put previous assumptions about fragmentation pathways on to a firmer footing [49].

ToF-SIMS spectra have the advantage over QMS–SIMS spectra of much higher signal:noise ratios (especially towards the high mass end of the QMS range, say $m/z > 300$), much greater mass range and (depending on design) higher mass resolution. Although these confer significantly greater analytical capabilities, **polymer** spectra tend to be dominated by fragments within the detection capabilities of good QMS-based instruments and for identification via fingerprinting methods QMS instruments are generally perfectly adequate.

(b) *Quantitative aspects*

Although quantitative interpretation of polymer SIMS data is not expected to be straightforward, a number of studies show that the situation is far from hopeless. Relative intensities of atomic or quasi-atomic ions, e.g. O^-:CH^-, F^-:CH^- show good correlations with quantitative XPS atomic concentrations when trends in composition are restricted to similar materials [28, 36, 50]. Matrix effects (extremely important in inorganic SIMS) are observed: it has recently been shown that for poly(vinyl ethers/ketones/carboxylates) the correlations between O^-:CH^- and bulk O:C are good within each class, but clearly different for each functionality [46]. In hydrocarbon polymers the ratio C^-:CH_2^- increases with degree of unsaturation [44].

The relative intensities of cluster ions have been shown to correlate with composition for random methacrylate polymers [51, 52]. Figure 8.21 shows data from a study of random hydroxyethylmethacrylate/ethylmethacrylate (HEMA/EMA) co-polymers [51]. Fragments at m/z 127 and 155 are representative of HEMA and EMA monomers, respectively, and relative intensity plots of the form A/A+B give smooth trends [Fig. 8.21(a)]. The intensities of two fragments believed to represent HEMA–EMA linked monomers are plotted in Fig. 8.21(b); in confirmation of these assignments, both curves maximize in mid-range. The fragment representing EMA–EMA links (m/z 213) increases rapidly with EMA content [Fig. 8.21(c)] as expected. The ratio $I(213)/I(155)$ represents the ratio of diads/singles. For a random AB polymer where diad (AA) signals come from sequences $B-(A)_n-B$ where $n > 2$ a statistical analysis predicts that the ratio of diads to singles (i.e. isolated A units) will be proportional to the mole fraction of A, as seen in Fig. 8.21(c). Clearly the surface of this co-polymer is representative of the bulk composition. Other studies have been done on nylon 6/66 [37] and styrene/hydroxystyrene co-polymers [50]. Similar relative cluster ion intensity plots have been used to follow surface segregation in block co-polymers and blends and of oligomer in poly(ethylene terephthalate) [53].

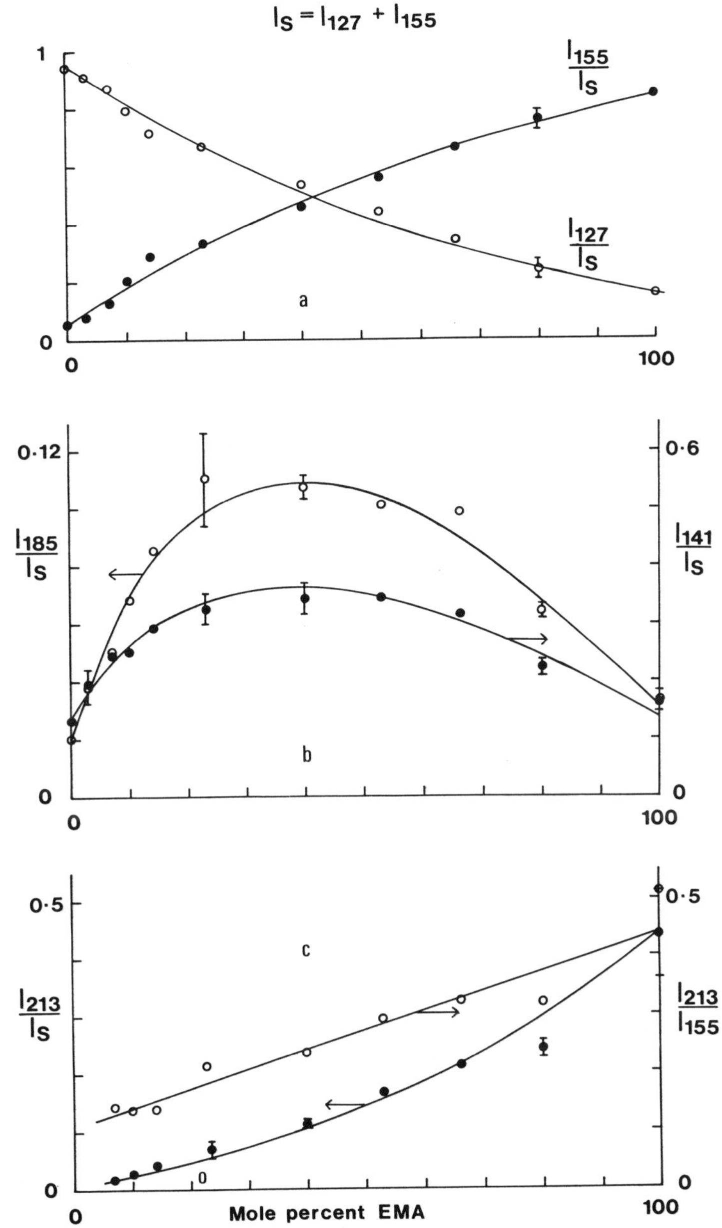

Fig. 8.21 Plots of relative ion intensities from random co-polymers of hydroxyethylmethacrylate (HEMA) and ethylmethacrylate (EMA) as a function of EMA concentration. (Reproduced with permission from [55].)

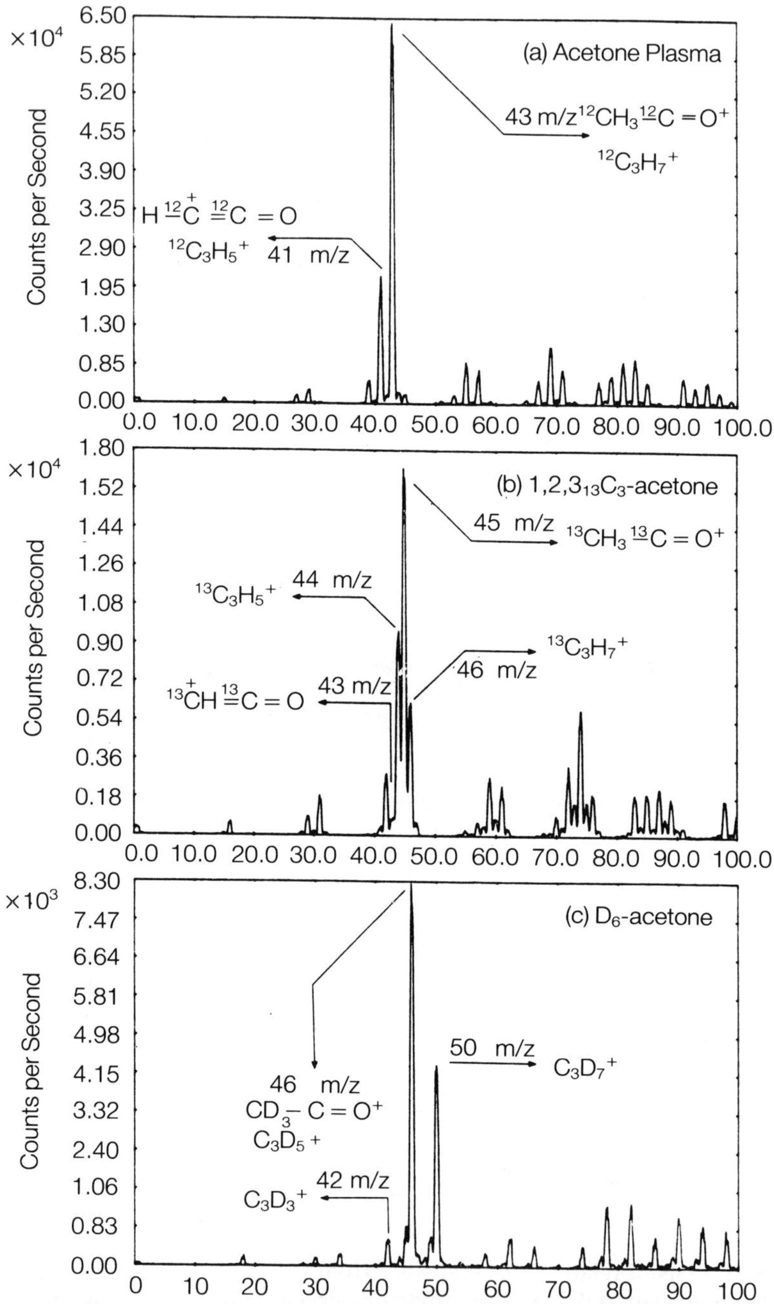

Fig. 8.22 Partial positive-ion spectra (*m/z* 0–100) of plasma-deposited films using acetone as the monomer, either unlabelled, ^{13}C or D labelled as shown. (Reproduced with permission from [57].)

(c) Isotopic labelling

Comparison of spectra of polymers in which some or all of the H atoms have been replaced by D is one way of improving spectral interpretation in the absence of high mass resolution [52, 54]. In aromatic systems, however, facile H–D exchange can add its own complications [45]. The scope for ^{13}C labelling was first outlined in a study of the gas-phase derivatization of -OH groups using acetylchloride labelled at either C atom [55]. Ochiello *et al.* have studied the oxygen-plasma treatment of polypropylene and subsequent ageing by using $^{16}O_2$ and $^{18}O_2$ for treatment and post-treatment exposure [56].

These techniques have been applied to SIMS characterization of plasma-deposited films (PDF) of polymers formed from functional monomers by incorporating stable isotopes into the original monomer(s) [57, 58]. For example, Fig. 8.22 shows partial positive-ion spectra from labelled acetone PDFs which showed that the majority of the m/z 43 peak in the unlabelled spectrum is due to CH_3CO^+ representative of retained monomer functionality.

(d) Surface sensitivity

This is a difficult parameter to measure but many comparisons of XPS and SIMS data from the same surfaces indicate that SIMS is significantly more surface sensitive than XPS (under routine conditions). Only two determinations have been made for polymer systems. The first compared quantitative SIMS data (of the type discussed earlier) from surface segregated segmented polyurethanes labelled with fluorine with quantitative XPS data obtained at different take-off angles. Based on calculated XPS sampling depths, the SIMS sampling depth appeared to be $\sim$ 10Å (i.e. about two monolayers) [28]. The second came from a SIMS study of the reaction of poly(hydroxyethyl methacrylate) (PHEMA) with propionylchloride and led to a value of $\sim$ 9Å [29]. The values were for 4 keV Xe^+ and 12 keV Ar^+ primary ion bombardment, respectively.

(e) Static SIMS and sample damage

In the context of polymer surface characterization, a static SIMS experiment involves the use of a total ion dose which has a negligible effect on the analysed material (i.e. the spectra obtained represent virgin material). Higher doses lead to progressive changes to the spectra via radiation/sputtering damage. Even in those cases where micro-analysis or imaging of small areas are not involved, the experimental parameters are critical for QMS instruments. Thus, Fig. 8.23 shows the effect of ion dose, when analysing an area of $\sim 0.3\,cm^{-2}$, on the negative-ion spectrum of PMMA

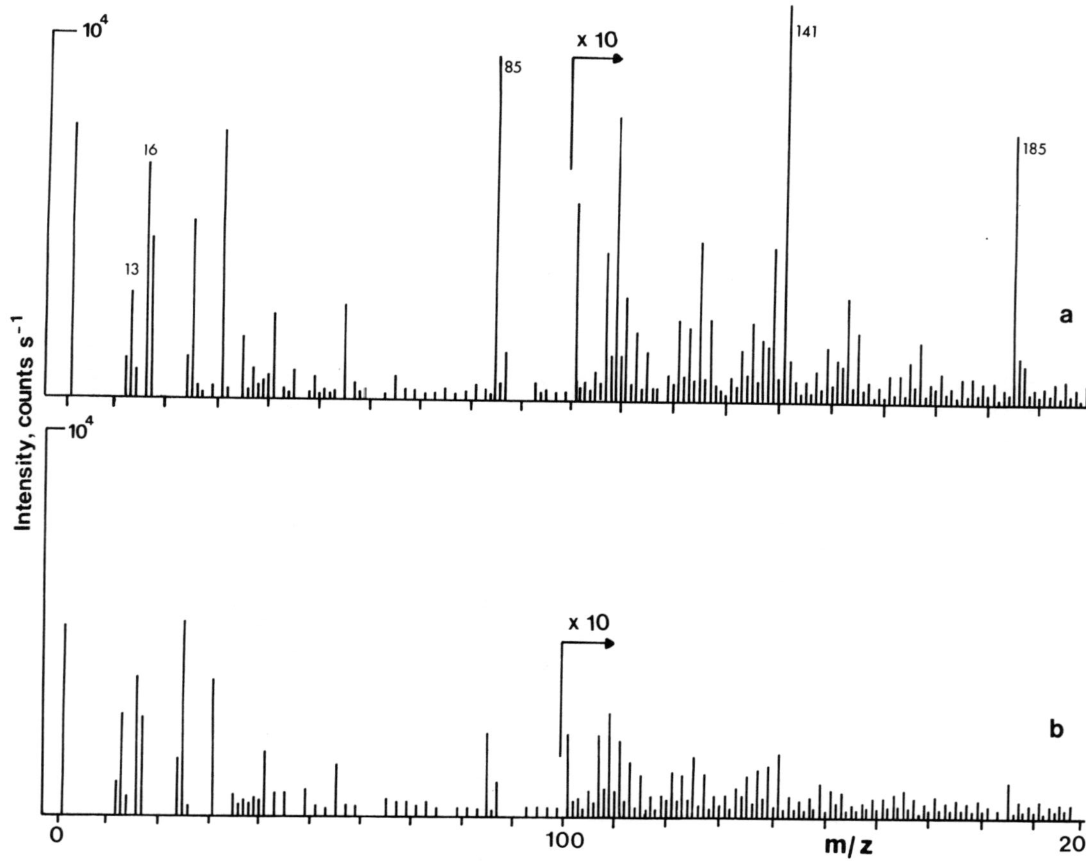

Fig. 8.23 (a) Negative-ion spectra of PMMA using 4 keV Xe^+ (0.5 nA cm^{-2}); total dose during spectral acquisition was 1.5×10^{12} ions cm^{-2}. (b) As (a) but following irradiation with 4 keV Xe^+ at 9 nA cm^{-2}; total dose 5.5×10^{13} ions cm^{-2}. (Reproduced with permission from [26].)

discussed earlier [26]. All the important peaks for structural identification have disappeared after 5.5×10^{13} ions cm^{-2} and the damage threshold is *c.* 10^{13} ions cm^{-2}. This is a fairly general 'limiting' static SIMS dose, although other polymers can be significantly less damage-tolerant (e.g. unsaturated polyolefins [44]). Since spectral quality cannot be degraded significantly, especially for the lower signal:noise peaks in the higher mass region, it is clearly not possible to use a QMS instrument to analyse regions smaller than about 500 μm square. Most polymers selectively lose heteroatoms during ion bombardment and tend towards atomic carbon via a variety of pathways. Figure 8.24 shows the changes to the positive-ion spectrum of PVC; the 'damaged' sample spectrum contains many peaks characteristic of polycyclic aromatic species [26].

With ToF instruments, although the maximum irradiated area is typically much smaller than is the case with QMS instruments ($\leq$ 0.01 cm^2) this is more than compensated for by the sensitivity gains due to higher transmission and parallel mass detection ($\simeq 10^4$ improvement over QMS). Spectra can therefore be acquired for doses of $\sim 10^8$ ions cm^{-2} and micro-analyses carried out within regions 50 μm square [59]. In the imaging mode, therefore, full use can be made of the high spatial resolution ($\sim$ 0.2 μm) available from fine-focus liquid metal ion sources [34].

8.4 APPLICATIONS

8.4.1 General comments

Unfortunately, space does not permit the applications of XPS and SIMS in polymer surface research to be described in any detail. In the case of XPS two reviews with comprehensive bibliographies covering the period 1979–1989 are available [3, 4] and in the case of SIMS a recent review has covered the applications literature up to 1990 [60]. It is worthy of note that the detection of additive molecules and molecular contaminants on polymer surfaces is particularly important in many areas of polymer technology. In this case, as in most applications, the two techniques are highly complementary: XPS detects all elements except H, is quantitative and gives a reasonable level of structural detail; SIMS detects H but not all elements directly, is not quantitative and provides a very high level of structural detail; the two techniques have different, but not mutually exclusive, regimes of surface sensitivity and spatial resolution. These points are best illustrated by referring to two studies of biomedical polymer surfaces, a rapidly growing area of research, in which high-resolution XPS and SIMS were used together.

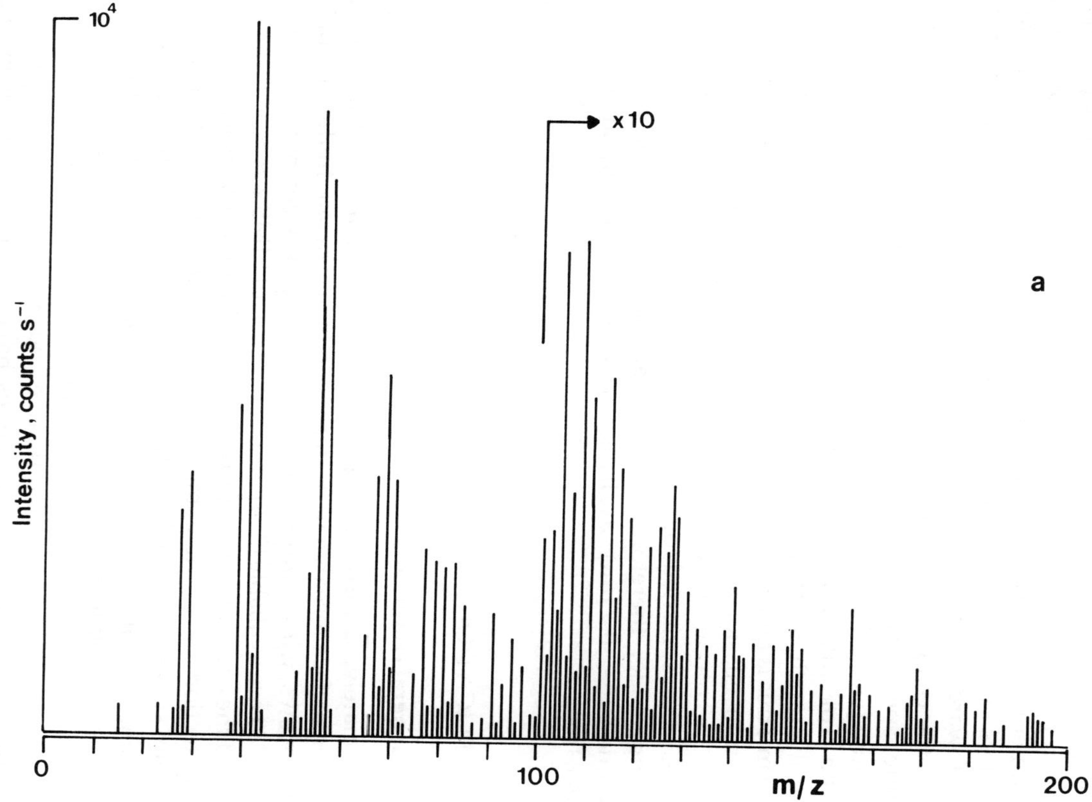
10^4
Intensity, counts s^-1
x10
a
0
100
200
m/z

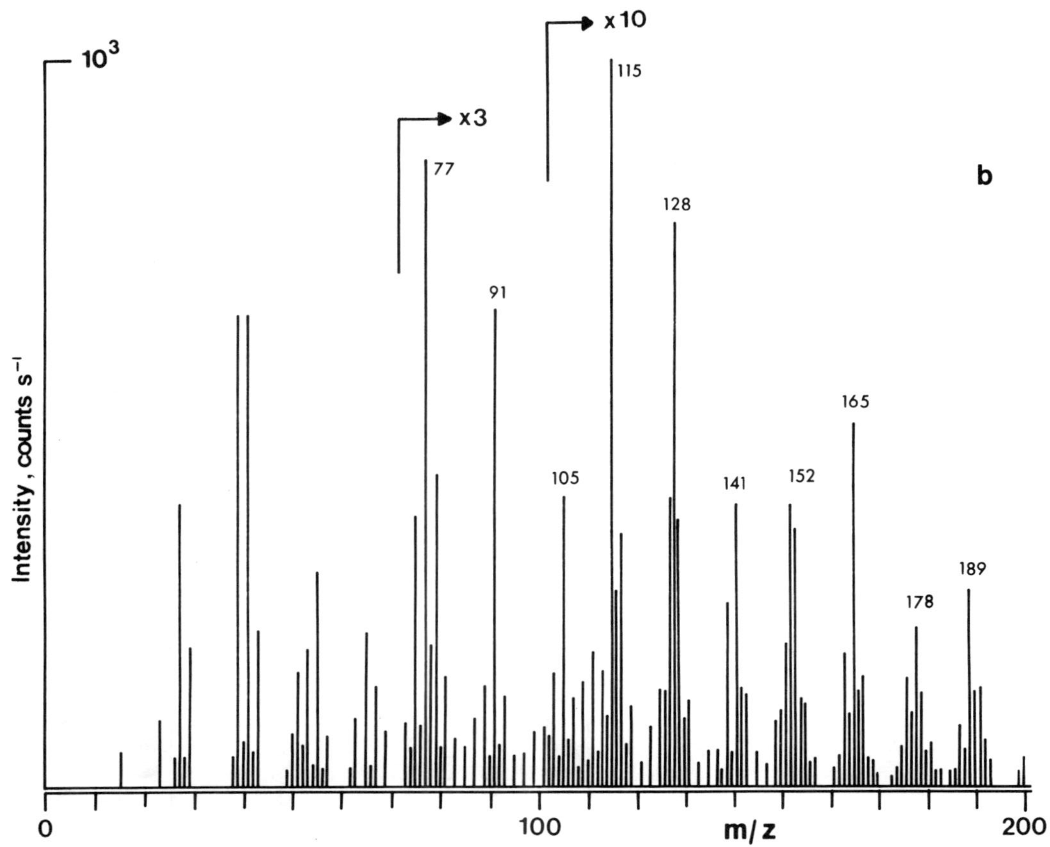

Fig. 8.24 Positive-ion spectra of PVC obtained in the static mode (a) before and (b) following exposure to 7.2×10^{13} ions cm^{-2}; 2 keV Xe^+ used for both radiation damage (10 nA cm^{-2}) and analysis (0.8 nA cm^{-2}). (Reproduced with permission from [26].)

Polyetherurethane Synthesis

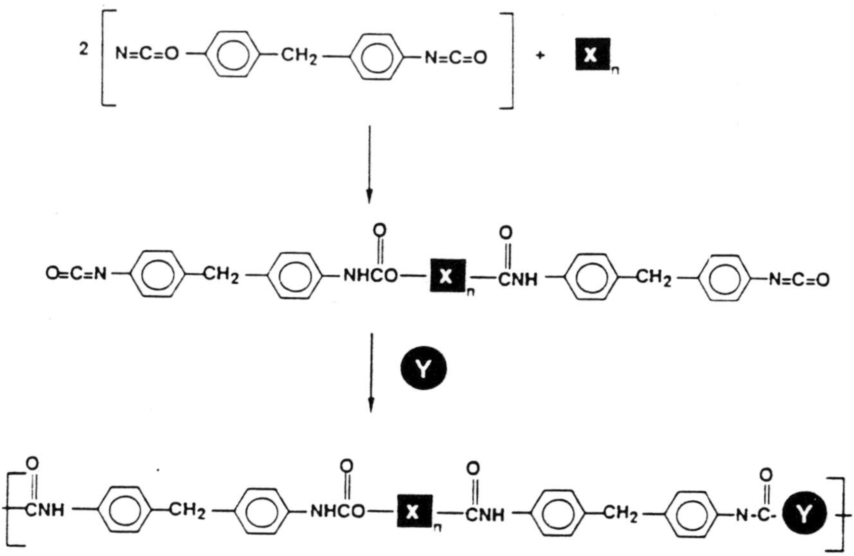

X	Abbreviation	
-(-CH_2CH_2-O-)$_n$	PEG	polyether soft segment
-(-$CH_2CH(CH_3)$-O-)$_n$	PPG	
	DPG TPG	
-(-$CH_2CH_2CH_2CH_2O$-)$_n$	PTMG	
	BD unit	

Y	Abbreviation	
$NH_2CH_2CH_2NH2$	ED	chain extender
$OHCH_2CH_2CH_2CH_2OH$	BD	

Fig. 8.25 (Reproduced with permission from [38].)

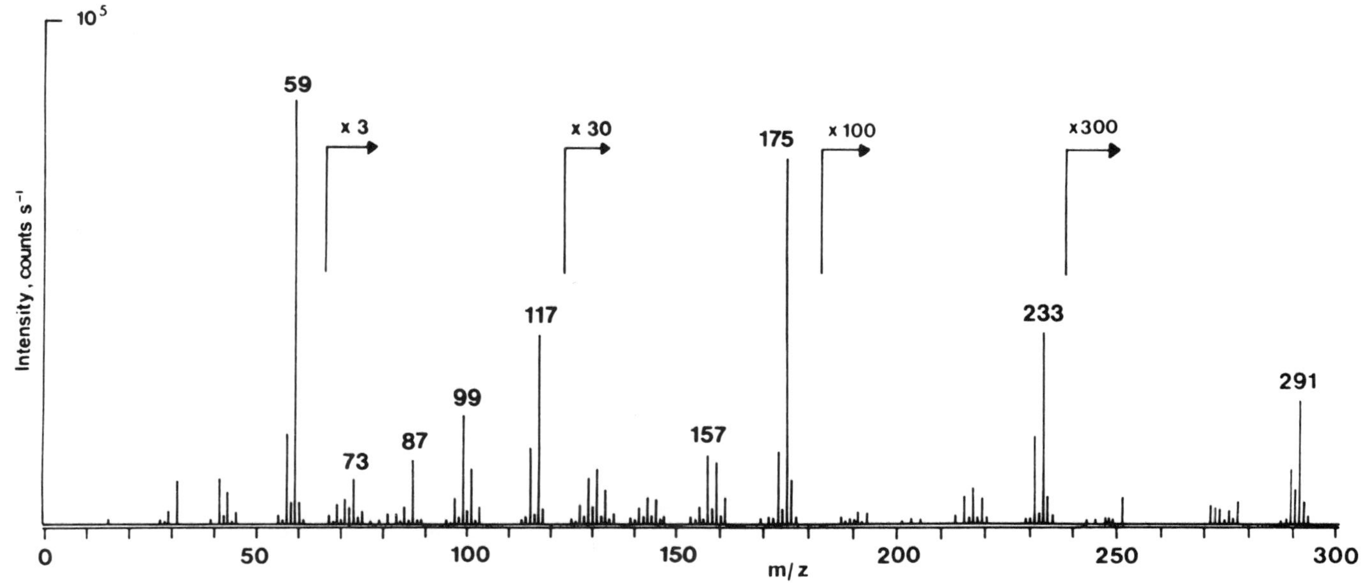

Fig. 8.26 Positive-ion spectrum of PPG molecular weight 425. (Reproduced with permission from [38].)

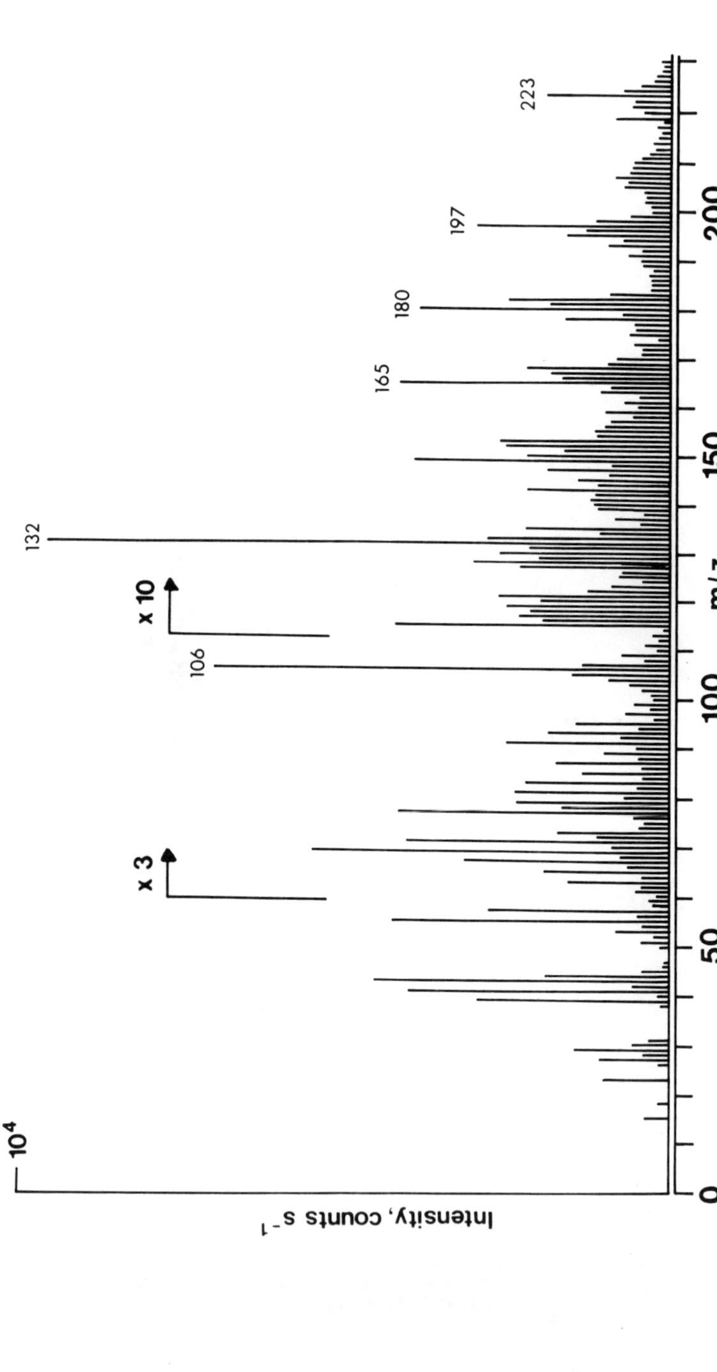

Fig. 8.27 Positive-ion spectrum of butanediol hard-segment model polyurethane. (Reproduced with permission from [38].)

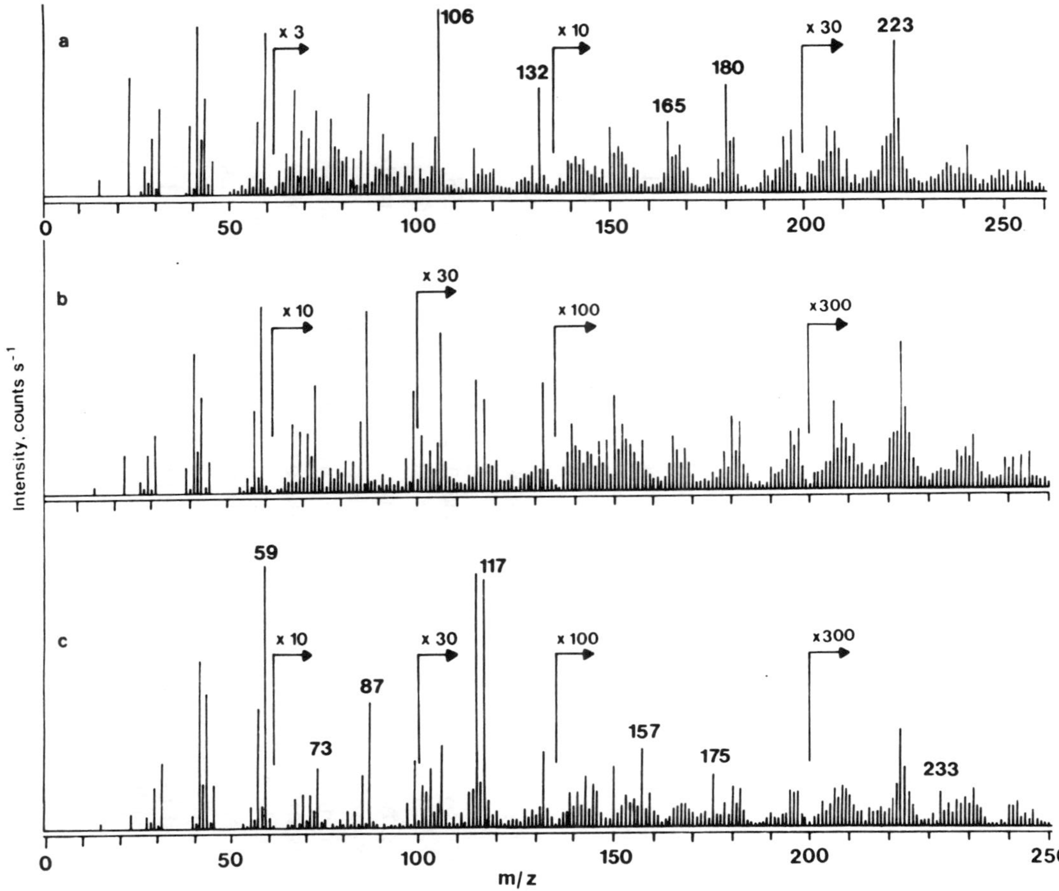

Fig. 8.28 Positive-ion spectra of PPG/MDI/ED polyurethanes prepared with (a) PPG 425 (b) PPG 775 (c) PPG 2000. (Reproduced with permission from [38].)

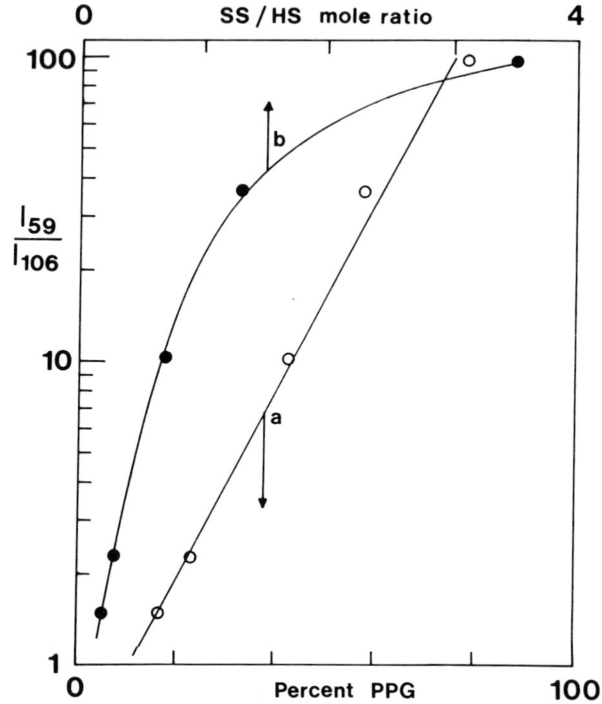

Fig. 8.29 Log (I_{59}/I_{106}) positive-ion peak intensity ratios plotted as a function of bulk concentration expressed in terms of (a) % PPG content and (b) SS/HS ratio. (Reproduced with permission from [38].)

8.4.2 Segmented polyurethane surfaces [38]

The synthesis of these materials is summarized in Fig. 8.25. The polyether unit, typically in the molecular weight range of 400–2000, forms the 'soft segment' of the polyurethane whilst the di-isocyanate plus chain extender (diamine or diol) forms the rigid 'hard segment'. These polymers have a tendency (dependent on the composition and processing history) to form discrete phases, with domain sizes of 10–20 nm, in the bulk. A knowledge of surface structure and of lateral and vertical heterogeneity is of great importance for understanding many aspects of biocompatibility.

Although the pure polyethers can be distinguished by XPS, their identification in the polyurethane is very difficult. However, their molecular fingerprint in SIMS is carried over into the polyurethane. Thus Figs 8.26–28 show SIMS spectra of PPG, a hard-segment model polymer (polyether replaced by butanediol) and a set of PPG/MDI/ED polyurethanes. As the molecular weight of the PPG soft segment increases, so

(a)

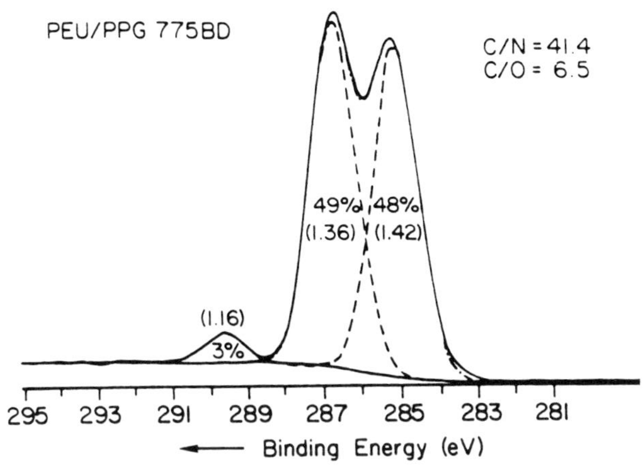

(b)

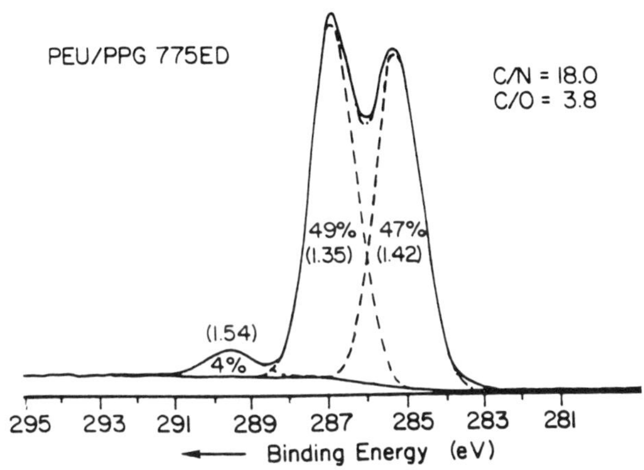

Fig. 8.30 C 1s XP spectra of (a) PPG 775/MDI/BD and (b) PPG 775/MDI/ED polyurethanes. Peak FWHM are indicated in parentheses. (Reproduced with permission from [38].)

the SIMS spectra is dominated by the soft-segment peaks. Partly this is due to the increasing fraction of PPG in the polymer, as shown by curve (a) in Fig. 8.29 where peaks at *m*/*z* 59 and 106 are unique identifiers of the soft and hard segments respectively. However, curve (b) shows that increasing the bulk soft-segment/hard-segment (SS/HS) ratio leads to a dramatic increase in soft-segment peak intensity indicating surface segregation of PPG units. Detailed study of the relative intensities of the PPG fragments in the pure polyether and in the polyurethane surface proves that this is not 'free' PPG and this is confirmed by studies of polyurethanes after solvent extraction or after deliberate doping with free

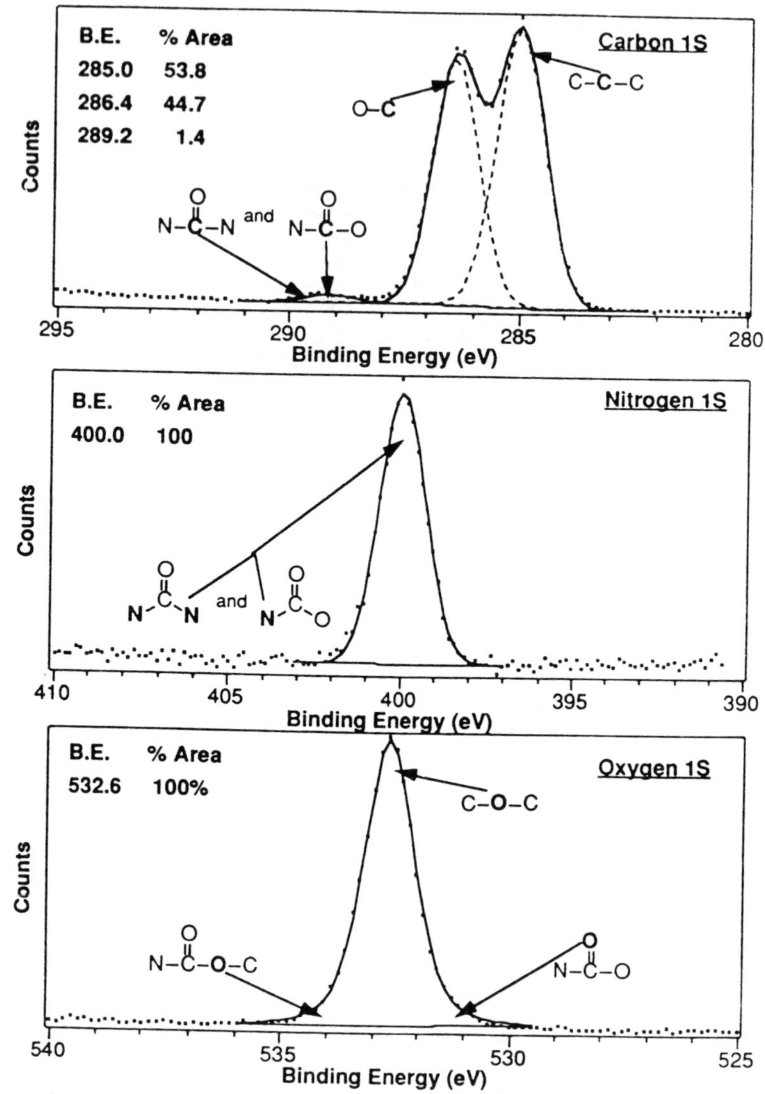

Fig. 8.31 Core-level XP spectra from lot BSP. In the O 1s spectrum, contributions from amide oxygens to high and low BE of the ether oxygen are too small to be accurately fitted. (Reproduced with permission from [61].)

PPG, i.e. the SIMS data is morphologically sensitive. On the other hand, SIMS cannot identify the chain extender unit in the polyurethane but XPS can. In Fig. 8.30 the low-intensity high BE peak in (a) is due to carbamate groups (NH–CO–O) and in (b) is due to carbamate and urea

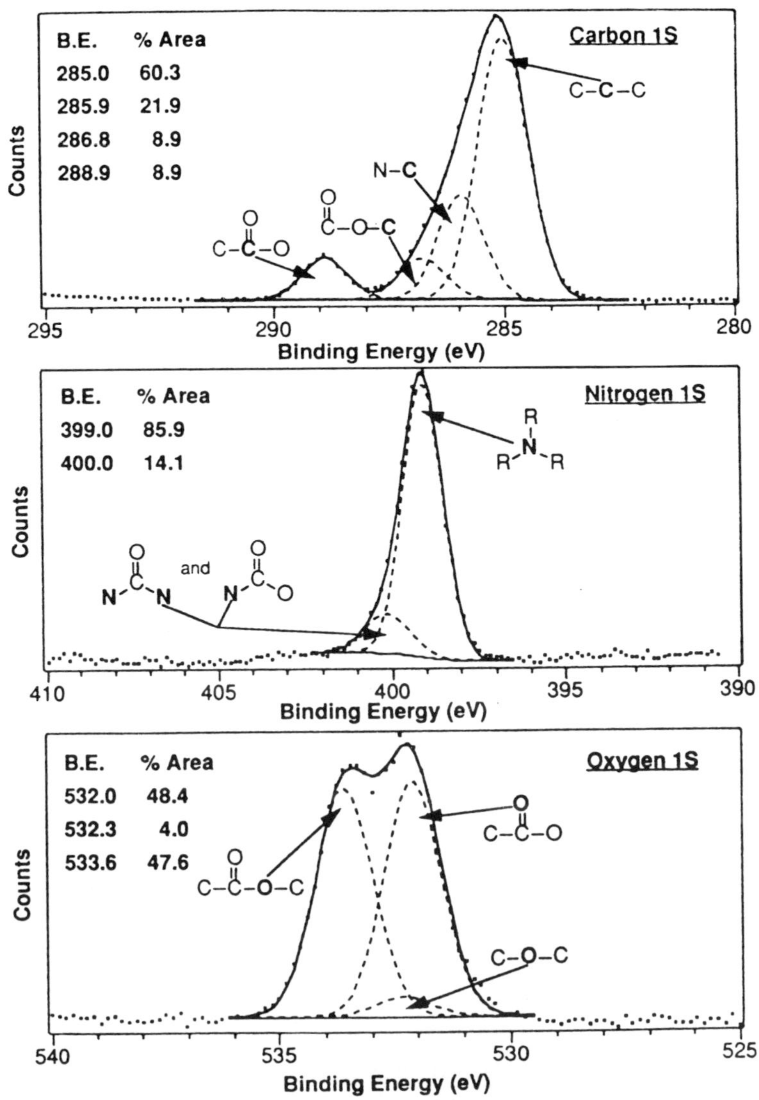

Fig. 8.32 Core-level XP spectra from lot BSUA. (Reproduced with permission from [61].)

(NH–CO–NH). These have slightly different chemical shifts leading to a much broader peak in (b).

The surface segregation within the upper monolayer or so (the SIMS sampling depth) can be quantitatively studied, in principle, by angular-resolved XPS. However, this relies on quantitative N analysis (specific to

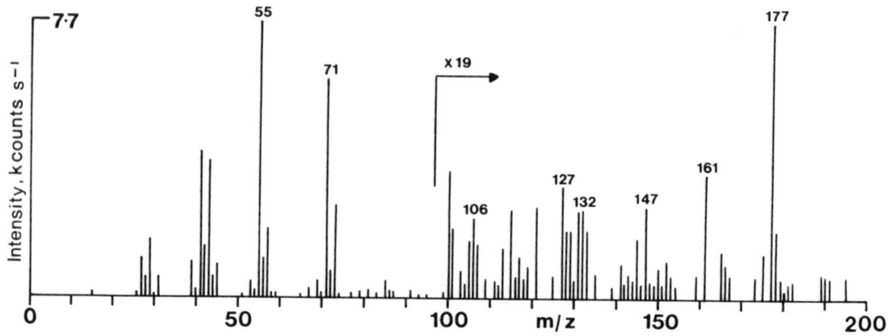

Fig. 8.33 Positive-ion spectrum from lot BSP. Peaks characteristic of PTMG (*m/z* 55, 71, 127), MDI (*m/z* 106, 132) and a hindered phenol antioxidant (*m/z* 147, 161, 177) are all clearly observed. (Reproduced with permission from [61].)

hard segment). For the PPG-PEU with the highest SS/HS ratio, the N content from the maximum XPS sampling depth (50–100Å) is <2 atomic %. Decreasing the sampling depth leads to decreasing N concentration (as expected from the SIMS data) but also to a significant loss of signal:noise with most XPS instruments. The N detection limit is ~ 0.3 atomic %, so low take-off angle data can easily lead to the conclusion that there is no hard segment at the surface. The greater sensitivity of SIMS to hard segment shows this is never actually the case. This has important implications when constructing models of the polyurethane surface microstructure.

8.4.3 Variation in surface chemistry of Biomer [61]

Biomer is a commercial polyurethane of the type described earlier but of undisclosed composition. Variation in the properties of different lots of this medical grade product have been observed and compositional variations detected by pyrolysis mass spectra. Although the bulk composition, measured by FT–IR, of varying lots is the same, the surface composition has recently been shown to vary greatly. Again, this knowledge will be very useful in understanding differences in biological response.

Two lots of Biomer, referred to as BSP and BSUA (details in reference 61) are discussed. The XP-spectra shown in Figs 8.31 and 8.32 are clearly very different. The BSP data are consistent with the expected polyurethane composition based on PTMG [poly(tetramethylene glycol)], MDI and ED (from other analyses) as described in section 8.4.2. The SIMS spectrum (Fig. 8.33) confirms this and also contains other fragments (e.g. *m/z* = 147, 161 and 177) typical of a hindered-phenol antioxidant. The SIMS spectra (Fig. 8.34) for BSUA are very informative. None of the

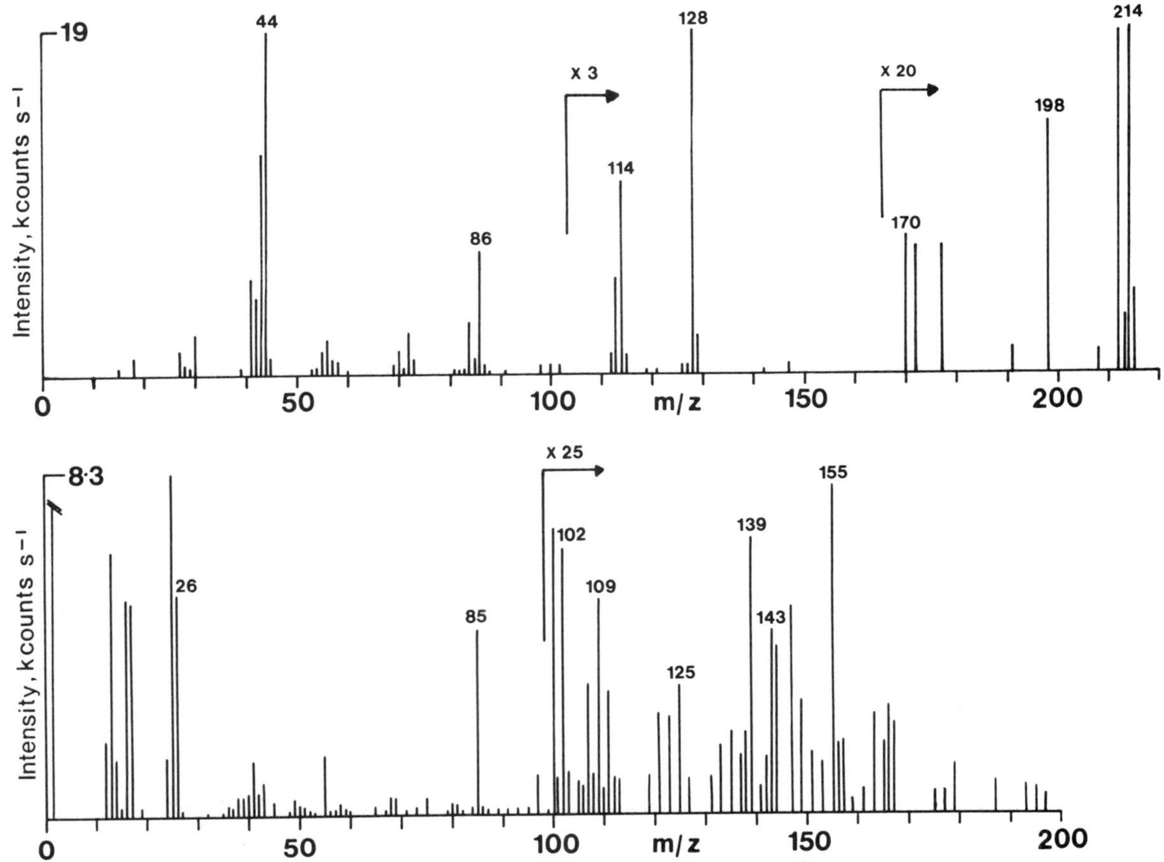

Fig. 8.34 Positive-ion (upper and negative-ion (lower) spectra from lot BSUA. All the peaks are characteristic of DPA-EMA. (Reproduced with permission from [61].)

expected polyurethane peaks are present in the positive-ion spectrum and the negative-ion spectrum (which is uninformative for polyurethanes of this kind) is also rich in information. Both spectra are consistent with the presence of a di-isopropyl amino ethyl methacrylate (DPA-EMA), based on a systematic interpretation of the fragmentation patterns and a knowledge of the spectra from a dimethyl amino ethyl methacrylate co-polymer [62]. Exhaustive extraction experiments failed to convert the BSUA surface into a more BSP-like surface. Hence the controversy in the literature, relating to the biological interactions of Biomer, may be partly explained by these major differences in surface chemistry and the necessity of a full surface characterization of such polymers prior to biological testing is indicated.

ACKNOWLEDGEMENTS

The majority of the work described in this review comes either wholly or partially from my own laboratory. I am indebted to my colleagues at ICI, particularly Dr Graham Beamson and Dr Martin Hearn, for their considerable contributions and to Professor Buddy Ratner and his co-workers at the University of Washington, Seattle, for their stimulating collaboration during the last decade.

REFERENCES

1. Clark, D.T. (1977), in *Handbook of X-ray and Ultraviolet Photoelectron Spectroscopy* (ed. D. Briggs), Heyden, London.
2. Dilks, A. (1988), in *Electron Spectroscopy – Theory, Techniques and Applications*, (eds C.R. Brundle and A.D. Baker), Vol. 4, Academic Press, London.
3. Briggs, D. (1983) in *Practical Surface Analysis*, (eds D. Briggs and M.P. Seah), Wiley, Chichester.
4. Briggs, D. (1990) in *Practical Surface Analysis*, 2nd edn, (eds D. Briggs and M.P. Seah), Vol. 1, Wiley, Chichester.
5. Clark, D.T., Aba-Shbak, M.M. and Brennan, W.J. (1982) *J. Electron Spectrosc. Relat. Phenom.* **28**, 11.
6. Clark, D.T. and Thomas, H.R. (1977) *J. Polymer Sci. Polymer Chemistry Edn*, **15**, 2893; Clark, D.T. and Shuttleworth, D. (1978) ibid. **16**, 1093.
7. Riviere, J.C. (1990) in *Practical Surface Analysis*, 2nd edn, (eds D. Briggs and M.P. Seah), Vol. 1 Wiley, Chichester, p. 19.
8. U. Gelius *et al.* (1990) *J. Electron Spectrosc. Relat. Phenom.* **52**, 747.
9. Beamson, G. *et al.* (1990) *Surf. Interface Anal.* **15**, 541.
10. Beamson, G. and Briggs, D. (1992) *High Energy XPS of Organic Polymers: The Scienta ESCA 300 Database*, Wiley, Chichester.
11. Coxon, P., Krizek, T., Humpherson, M. and Wardell, I.R.M. (1990) *J. Electron Spectronsc. Relat. Phenom.* **52**, 821.
12. Fletcher, I.W., Davies, M. and Briggs, D. (1992) *Surf. Interface Anal.* **18**, 303.
13. Wagner, C.D., Davies, L.E., Zeller, M.V. *et al.* (1981) *Surf. Interface Anal.* **3**, 211.
14. Seah, M.P., Jones, M.E. and Anthony, M.T. (1984) ibid. **6**, 242.

15. Beamson, G., Clark, D.T., Kendrick, J. and Briggs, D. (1991) *J. Electron Spectrosc. Relat. Phenom.* **57**, 79.
16. Gelius, U., Asplund, L., Basilier, E. *et al.* (1984) *Nucl. Instr. Meth.* **229**, 85.
17. Beamson, G., Bunn, A. and Briggs, D. (1991) *Surf. Interface Anal.* **17**, 105.
18. Briggs, D. and Beamson, G. (1992) *Anal. Chem.* **64**, 1729.
19. Beamson, G. and Briggs, D. (1992) *Mol. Phys.* **76**, 919.
20. Nordfors, D., Nilsson, A., Martensen, N. *et al.* (1988) *J. Chem. Phys.* **88**, 2630.
21. Keane, M.P., Naves de Brito, A., Lunell, S. *et al.* (in press).
22. Beamson, G. and Briggs, D. (1992) *High Energy XPS of Organic Polymers: The Scienta ESCA 300 Database*, Wiley, Chichester, pp. 220–223.
23. Pireaux, J.J., Riga, J., Caudano, R. and Verbist, J. (1981) in *Photon, Electron and Ion Probes of Polymer Structure and Properties*, (eds D.W. Dwight, T.J. Fabish and H.R. Thomas), ACS Symposium Series 162.
24. Chilkoti, A., Ratner, B.D. and Briggs, D. (1991) *Chem. Mater.* **3**, 51.
24(a). Chilkoti, A. and Ratner, B.D. (1991) *Surf. Interface Anal.* **17**, 567.
25. Briggs, D., Brown, A. and Vickerman, J.C. (1989) *Handbook of Static SIMS*, Wiley, Chichester.
26. Briggs, D. and Hearn, M.J. (1986) *Vacuum*, **36**, 1005.
27. Briggs, D. and Hearn, M.J. (1985) *Int. J. Mass. Spectrom. Ion Processes*, **67**, 47.
28. Hearn, M.J., Briggs, D., Yoon, S.C. and Ratner, B.D. (1987) *Surf. Interface Anal.* **10**, 384.
29. Lub, J. and van Velzen, P.N.T. (1989) in *Ion Formation from Organic Solids*, IFOSS VI (ed. A. Benninghoven), Wiley, Chichester.
30. Brown, A. and Vickerman, J.C. (1984) *Surf. Interface Anal.* **6**, 1.
31. Leggett, G.J. and Vickerman, J.C. (1992) *Appl. Surf. Sci.* **55**, 105.
32. Jede, R., Ganschow, O. and Kaiser, U. (1992) in *Practical Surface Analysis, Vol. 2: Ion and Neutral Spectroscopy*, 2nd edn, (eds D. Briggs and M.P. Seah), Wiley, Chichester.
33. Briggs, D. (1983) *Surf. Interface Anal.* **5**, 113
34. Briggs, D., Hearn, M.J., Fletcher, I.W. *et al.* (1990) ibid. **15**, 62.
35. Hearn, M.J. and Briggs, D. (1988) ibid. *11*, 198.
36. Briggs, D., Hearn, M.J. and Ratner, B.D. (1984) ibid. **6**, 184.
37. Briggs, D. (1987) *Org. Mass Spec.* **22**, 92.
38. Hearn, M.J., Ratner, B.D. and Briggs, D. (1988) *Macromol.* **21**, 2950.
39. Lub, J., van Vroomhoven, F.C.M.B., van Leyen, D. and Benninghoven, A. (1988) *Polymer*, **29**, 998.
40. Davies, M.C. *et al.* (1989) *Surf. Interface Anal.* **14**, 115.
41. Davies, M.C., Khan, M.A., Lynn, R.A.P. *et al.* (1991) *Biomaterials*, **12**, 305.
42. Davies, M.C., Khan, M.A., Domb, A. *et al.* (1991) *J. Appl. Polym. Sci.* **42**, 1597.
43. Van Ooij, W.J. and Brinkhuis, R.H.G. (1988) *Surf. Interface Anal.* **11**, 430.
44. Briggs, D. (1990) *Surf. Interface Anal.* **15**, 734.
45. Chilkoti, A., Castner, D.G. and Ratner, B.D. (1991) *Appl. Spectrosc.* **45**, 349.
46. Chilkoti, A., Ratner, B.D. and Briggs, D. (1992) *Surf. Interface Anal.* **18**, 604.
47. *Phi Handbook of Polymer SIMS Spectra*, (1991) Perkin Elmer, Physical Electronics Division, Eden Prarie.
48. Short, R.D. and Davies, M.C. (1989) *Int. J. Ion Process. and Format.* **89**, 149.
49. Leggett, G.J., Vickerman, J.C. and Briggs, D. (1992) *J.C.S. Faraday Trans.* **88**, 297.
50. Chilkoti, A., Castner, D.G., Ratner, B.D. and Briggs, D. (1990) *J. Vac. Sci. Technol. A*, **8**, 2274.
51. Briggs, D. and Ratner, B.D. (1988) *Polym. Commun.* **29**, 6.
52. Lub, J., van Vroomhoven, F.C.B.M., van Leyen, D. and Benninghoven, A. (1989) *J. Polym. Sci., Polym. Phys. Ed.* **27**, 2071.

53. Briggs, D. (1986) *Surf. Interface Anal.* **8**, 133.
54. Brinkhuis, R.H.G. and van Ooij, W.J. (1988) ibid. **11**, 214.
55. Briggs, D. and Munro, H.S. (1987) *Polym. Commun.* **28**, 307.
56. Occhiello, E. *et al.* (1989) in *SIMS VII*, (eds A. Benninghoven *et al.*), Wiley, Chichester, p. 789.
57. Chilkoti, A., Ratner, B.D. and Briggs, D. (1991) *Anal. Chem.* **63**, 1612.
58. Chilkoti, A., Ratner, B.D., Briggs, D. and Reich, F. (1992) *J. Polym. Sci., Polym. Chem. Edn*, **30**, 1261.
59. Hearn, M.J. (1992) in *SIMS VIII*, (eds A. Benninghoven *et al.*), Wiley, Chichester.
60. Briggs, D. (1992) in *Practical Surface Analysis*, Vol. 2, 2nd edn, (eds D. Briggs and M.P. Seah) Wiley, Chichester.
61. Tyler, B.J., Ratner, B.D., Castner, D.G. and Briggs, D. (1992) *J. Biomed. Mater. Res.* **26**, 273.
62. Wilding, J.R., Melia, C.D., Short, R.D. *et al.* (1990) *J. Appl. Polym. Sci.* **39**, 1827.
63. Reichlmaier, S. Hammond, J., Hearn, M.J. and Briggs, D. (1993) *Proc. 41st ASMS Conference (San Francisco, USA) and Surf. Interface Anal.* (to be published).

Index